Multiple User Interfaces

Multiple User Interfaces

Cross-Platform Applications and Context-Aware Interfaces

Edited by

Ahmed Seffah and Homa Javahery

Concordia University, Department of Computer Science, Canada

John Wiley & Sons, Ltd

Other Wiley Editorial Offices

John Wiley & Sons Inc., 111 River Street, Hoboken, NJ 07030, USA

Jossey-Bass, 989 Market Street, San Francisco, CA 94103-1741, USA

Wiley-VCH Verlag GmbH, Boschstr. 12, D-69469 Weinheim, Germany

John Wiley & Sons Australia Ltd, 33 Park Road, Milton, Queensland 4064, Australia

John Wiley & Sons (Asia) Pte Ltd, 2 Clementi Loop #02-01, Jin Xing Distripark, Singapore 129809

John Wiley & Sons Canada Ltd, 22 Worcester Road, Etobicoke, Ontario, Canada M9W 1L1

Wiley also publishes its books in a variety of electronic formats. Some content that appears
in print may not be available in electronic books.

Library of Congress Cataloging-in-Publication Data

Multiple user interfaces : cross-platform applications and context-aware interfaces / edited by
    Ahmed Seffah & Homa Javahery.
        p. cm.
    Includes bibliographical references and index.
    ISBN 0-470-85444-8
    1. Computer interfaces. I. Seffah, Ahmed. II. Javahery, Homa.

TK7887.5.M86 2003
004.6 – dc22

2003057602

British Library Cataloguing in Publication Data

A catalogue record for this book is available from the British Library

ISBN 0-470-85444-8

Typeset in 10/12pt Times by Laserwords Private Limited, Chennai, India
Printed and bound in Great Britain by TJ International, Padstow, Cornwall
This book is printed on acid-free paper responsibly manufactured from sustainable forestry
in which at least two trees are planted for each one used for paper production.

Contents

PART IV   MODEL-BASED DEVELOPMENT

Acknowledgements

The help of many people made this book possible, and we are grateful to all of them. We thank our editor Birgit Gruber, at John Wiley & Sons Ltd., who guided us throughout this project.

Daniel Engelberg and Jonathan Benn were indispensable for the editing process, and we thank them for their help in editing various chapters. Daniel Sinnig patiently helped with revising various chapters. Rozita Naghshin, our digital art expert, was a great source of help for advice on image layout and creation. To all the members of the HCSE (Human-Centered Software Engineering) Group, we thank you for participating in the discussion and brainstorming of this project.

We thank FCAR (Le Fonds québécois de la recherche sur la nature et les technologies), NSERC (National Sciences and Engineering Council of Canada), and the Faculty of Engineering, Concordia Research Chair programs, for their financial support.

We are grateful to all the reviewers of this book. We were lucky enough to have a wide spectrum of international reviewers, who patiently reviewed all chapters and gave us crucial feedback. We thank John Grundy from the University of Auckland, who gave us sound advice and feedback for a number of chapters.

Above all, we thank the contributors of this book. Without them, this book would not have been possible. We thank them for patiently modifying chapters, rewriting passages, and putting up with our requests. We acknowledge all of them for their efforts in making this book a success.

Ahmed Seffah   Homa Javahery

About the Editors

Ahmed Seffah is a professor in the department of Computer Science at Concordia University. He is director of the Human-Centered Software Engineering Group and the co-founder of the Concordia Software Usability and Empirical Studies Lab. He holds a PhD in software engineering from the Ecole Centrale de Lyon (France). His research interests are at the crossroads between software engineering and Human-Computer Interaction (HCI), including usability measurement, user interface design, empirical studies on developer experiences with CASE tools, human-centered software engineering, and patterns as a vehicle for integrating HCI knowledge in software engineering practices. Dr. Seffah is the vice-chair of the IFIP working group on user-centered design methodologies. During the last 10 years, he has been involved in different projects in North America and Europe.

Homa Javahery is a researcher and project manager with the Human-Centered Software Engineering Group, including the Usability and Empirical Studies Lab, in the department of Computer Science at Concordia University. She holds a Master's degree in Computer Science from Concordia University, and a Bachelor of Science degree from McGill University. She is combining different design approaches from human sciences and engineering disciplines to develop a pattern-oriented framework for designing a large variety of interfaces. She has been involved in different collaborative projects at the INRIA Research Institute in Nancy, France and the Daimler-Chrysler Research Institute in Ulm, Germany.

Contributors

Gregory D. Abowd
College of Computing
Georgia Institute of Technology
Atlanta, Georgia 30332-0280
USA
abowd@cc.gatech.edu

Marc Abrams
Harmonia, Inc.
PO Box 11282
Blacksburg, VA 24062
USA
marc@harmonia.com

Harry Brignull
University of Sussex
Room 5A3, Interact Lab
School of Cognitive
    and Computing Sciences
Falmer, Brighton BN1 9QH
UK
harrybr@cogs.susx.ac.uk
+44 (0) 1273 877221

Gaëlle Calvary
IIHM Group, CLIPS-IMAG Lab
BP 53, 385 rue de la Bibliotheque
38041 Grenoble Cedex 9
France
Joelle.Coutaz@imag.fr
+33 4 76 51 48 54

Didier Chincholle
Ericsson Research ERA/TVU/U
Torshamnsgatan 23, 164 80 Kista
Sweden
didier.chincholle@era.ericsson.se
+46 8 585 303 76

Joëlle Coutaz
IIHM Group, CLIPS-IMAG Lab
BP 53, 385 rue de la Bibliotheque
38041 Grenoble Cedex 9
France
Joelle.Coutaz@imag.fr
+33 4 76 51 48 54

Charles Denis
INTUILAB
Prologue 1, La Pyrénéenne,
BP 27/01, 31312 Labège Cedex
France
denis@intuilab.com

Anind K. Dey
Senior Researcher, Intel Research
2150 Shattuck Ave, Suite 1300
Berkeley, CA 94704
USA
anind@intel-research.net
+1-510-495-3012

Anke Dittmar
University of Rostock
Department of Computer Science
Albert-Einstein-Str. 21
D-18051 Rostock
Germany
ad@informatik.uni-rostock.de

Min Du
Liverpool John Moores University
School of Computing
    and Mathematical Sciences
Byrom St, Liverpool
L3 3AF UK
edcmdu@livjm.ac.uk
+44 (0) 151 231 2271

Jacob Eisenstein
CEO – RedWhale Software
277 Town & Country Village Palo Alto
CA 94303
USA
jacobe@mit.edu
+1 650 321-3425

Daniel Engelberg
CGI Group Inc.
1130 Sherbrooke West, 7th floor
Montreal, Quebec H3A 2M8
Canada
dan.engelberg@sympatico.ca
+1 514-281-7000, local 5820

David England
Liverpool John Moores University
School of Computing
    and Mathematical Sciences
Byrom St, Liverpool
L3 3AF UK
d.england@livjm.ac.uk
+44 (0) 151 231 2271

Mir Farooq Ali
Virginia Technology Institute
Department of Computer Science (0106)

660 McBryde Hall
Blacksburg, VA 24061
USA
mfali@cs.vt.edu
1(540) 231 1927

Peter Forbrig
University of Rostock
Department of Computer Science
Albert-Einstein-Str. 21
D-18051 Rostock
Germany
pforbrig@informatik.uni-rostock.de

Elizabeth Furtado
Universidade de Fortaleza
NATI – Célula EAD
Washington Soares, 1321
Bairo Edson Queiroz
Fortaleza (Ceará), BR-60455770
Brazil
elizabet@unifor.br

João José Vasco Furtado
Universidade de Fortaleza
NATI – Célula EAD
Washington Soares, 1321
Bairo Edson Queiroz
Fortaleza (Ceará), BR-60455770
Brazil
vasco@unifor.br

Mikael Goldstein
Ericsson Research ERA/TVU/U
Torshamnsgatan 23
164 80 Kista
Sweden
mikael.goldstein@era.ericsson.se
+46 8 757 3679

John Grundy
University of Auckland
Department of Computer Science
Private Bag 92019
Auckland

New Zealand
john-g@cs.auckland.ac.nz
+64-9-3737-599 ext. 8761

Homa Javahery
Department of Computer Science
Faculty of Engineering
  and Computer Science
1455 de Maisonneuve Blvd West
Montreal, Quebec H3G 1M8
Canada
h_javahe@cs.concordia.ca
+1 514-848-3024

Laurent Karsenty
INTUILAB
Prologue 1, La Pyrénéenne,
BP 27/01, 31312 Labège Cedex
France
karsenty@intuilab.com

Quentin Limbourg
Université catholique de Louvain (UCL)
Information System Unit (ISYS-BCHI)
Institut d'Administration et de Gestion
  (IAG)
Place des Doyens, 1
B-1348 Louvain-la-Neuve
Belgium
limbourg@isys.ucl.ac.be
+32-10.47.85.25

Simon Lock
Lancaster University
Computing Department
Lancaster LA1 4YR
UK
lock@comp.lancs.ac.uk
+44-1524-592795

Luisa Marucci
ISTI-CNR
Via G. Moruzzi 1
56100 Pisa
Italy

luisa.marucci@guest.cnuce.cnr.it
+39 050 3153066

Joanna McGrenere
University of British Columbia
Department of Computer Science
201-2366 Main Mall
Vancouver, BC V6J 2E2
Canada
joanna@cs.ubc.ca
604-827-5201

Andreas Müller
University of Rostock
Department of Computer Science
Albert-Einstein-Str. 21
D-18051 Rostock
Germany
Xray@informatik.uni-rostock.de

Gustav Öquist
Bollhusgränd 7
113 31 Stockholm
Sweden
gustav@stp.ling.uu.se
+46 8 739 417 783

Alexandros Paramythis
Foundation for Research and
  Technology – Hellas
Institute of Computer Science
Science and Technology Park of Crete
Heraklion, Crete
GR – 71110 Greece
cs@ics.forth.gr
+30-810-391741

Fabio Paternò
ISTI-CNR
Via G. Moruzzi 1
56100 Pisa
Italy
fabio.paterno@cnuce.cnr.it
+39 050 3153066

Manuel Pérez-Quiñones
Virginia Technology Institute
Department of Computer Science (0106)
660 McBryde Hall
Blacksburg, VA 24061
USA
perez@cs.vt.edu
1(540) 231 2646

Angel R. Puerta
CEO – RedWhale Software
277 Town & Country Village Palo Alto
CA 94303
USA
puerta@redwhale.com
+1 650 321-3425

Carmen Santoro
ISTI-CNR
Via G. Moruzzi 1
56100 Pisa
Italy
C.Santoro@cnuce.cnr.it
+39 050 3153066

Anthony Savidis
Foundation for Research
    and Technology – Hellas
Institute of Computer Science
Science and Technology Park of Crete
Heraklion, Crete,
GR – 71110 Greece
cs@ics.forth.gr
+30-810-391741

Ahmed Seffah
Concordia University
Department of Computer Science
Faculty of Engineering
    and Computer Science
1455 de Maisonneuve Blvd West
Montreal, Quebec H3G 1M8
Canada
seffah@cs.concordia.ca
+1 514-848-3024

Wilker Bezerra Silva
Universidade de Fortaleza
NATI – Célula EAD
Washington Soares, 1321
Bairo Edson Queiroz
Fortaleza (Ceará), BR-60455770
Brazil
wilker@unifor.br

Daniel Sinnig
Concordia University
Department of Computer Science
Faculty of Engineering
    and Computer Science
1455 de Maisonneuve Blvd West
Montreal, Quebec H3G 1M8
Canada
+1 514-848-3024

Constantine Stephanidis
Foundation for Research
    and Technology – Hellas
Institute of Computer Science
Science and Technology Park of
    Crete
Heraklion, Crete, GR – 71110
Greece
cs@ics.forth.gr
+30-810-391741

Leandro da Silva Taddeo
Universidade de Fortaleza
NATI – Célula EAD
Washington Soares, 1321
Bairo Edson Queiroz
Fortaleza (Ceará), BR-60455770
Brazil
taddeo@unifor.br

Daniel William Tavares Rodrigues
Universidade de Fortaleza
NATI – Célula EAD
Washington Soares, 1321
Bairo Edson Queiroz
Fortaleza (Ceará), BR-60455770

Brazil
danielw@unifor.br

David Thevenin
IIHM Group, CLIPS-IMAG Lab
BP 53, 385 rue de la Bibliotheque
38041 Grenoble Cedex 9
France
Joelle.Coutaz@imag.fr
+33 4 76 51 48 54

Jean Vanderdonckt
Université catholique de Louvain (UCL)
Information System Unit (ISYS-BCHI)
Institut d'Administration
    et de Gestion (IAG)
Place des Doyens, 1
B-1348 Louvain-la-Neuve
Belgium

vanderdonckt@isys.ucl.ac.be
+32-10.47.85.25

Vasilios Zarikas
Foundation for Research
    and Technology – Hellas
Institute of Computer Science
Science and Technology Park of Crete
Heraklion, Crete, GR – 71110
Greece
cs@ics.forth.gr
+30-810-391741

Wenjing Zou
University of Auckland
Department of Computer Science
Private Bag 92019
Auckland
New Zealand
wenjingzou@hotmail.com
+64-9-3737-599 ext. 8761

Part I

Basic Terminology, Concepts, and Challenges

Executive Summary and Book Overview

Ahmed Seffah and Homa Javahery

Human-Centered Software Engineering Group, Department of Computer Science,
Concordia University, Canada

1.1. MOTIVATION

In recent years, a wide variety of computer devices including mobile telephones, personal digital assistants (PDAs) and pocket PCs has emerged. Many existing devices are now being introduced as an alternative to traditional computers. Internet-enabled television (WebTV), 3D-interactive platforms with voice capabilities, and electronic whiteboards attached to desktop machines are among the many examples. In addition, we are moving away from the dominance of the WIMP (Windows, Icons, Mouse, and Pointer) system as a main metaphor of human-computer interaction. Novel interaction styles are emerging. These include web applications where users interact with the content, interactive television controlled by hand-held remotes, and PDAs with small screens and styli for gesture-based interaction.

All these variations in devices and interaction styles require changes in design, development and testing frameworks. This book aims to introduce the reader to the current research trends and innovative frameworks being developed to address these changes.

Multiple User Interfaces. Edited by A. Seffah and H. Javahery
© 2004 John Wiley & Sons, Ltd ISBN: 0-470-85444-8

1.2. A FEW DEFINITIONS

This book refers to several context-specific terms including:

- *Multi-device user interfaces*: These allow a user to interact using various kinds of computers including traditional office desktop, laptop, palmtop, PDA with or without keyboards, and mobile telephone.
- *Cross-platform user interfaces*: These can run on several operating systems including Windows, Linux and Solaris, if the user interface (UI) code is portable. For example, Java runs a virtual machine called JVM, and code is compiled into an intermediate format known as Java byte code, which is platform independent. When Java byte code is executed within the JVM, the JVM optimizes the code for the particular platform on which it is running. Microsoft's latest technology,. NET follows the same principles. Code is compiled into Microsoft Intermediate Language (MSIL) and is then executed within the. NET framework as an application domain.
- *Mobile versus stationary/fixed user interfaces*: A mobile platform gives users seamless access to information and services even when they are moving. Mobile computing includes a large variety of mobile phones and PDAs, as well as new devices such as wireless MP3 music players, digital cameras and personal health monitors.
- *Context-aware applications*: These refer to the ability of computing devices to detect, sense, interpret and respond to aspects of a user's local environment and the computing devices themselves.
- *User interface plasticity*: The term plasticity is inspired from the property of materials that expand and contract under natural constraints without breaking, thus preserving continuous use. Applied to HCI, plasticity is the capacity of an interactive system to withstand variations of contexts of use while preserving usability properties.
- *Universal user interfaces*: These can support a broad range of hardware, software and network capabilities with the central premise of accommodating users with a variety of characteristics. These characteristics include diversity in skills, knowledge, age, gender, disabilities, disabling conditions (mobility, sunlight, noise), literacy levels, cultures, income levels, etc. [Hochheiser and Shneiderman 2001].
- *Multiple user interfaces (MUI)*: These provide different views of the same information and coordinate the services available to users from different computing platforms. By computing platform, we refer to a combination of hardware, computing capabilities, operating system and UI toolkit. The hardware includes traditional office desktops, laptops, palmtops, mobile telephones, personal digital assistants (PDAs) and interactive television. In a larger sense, computing platforms include wearable computers and any other real or virtual objects that can interact with the services and information. MUIs can support different types of look-and-feel and offer different interaction styles. These different types of look-and-feel and interaction styles should take into account the constraints of each computing platform while maintaining cross-platform consistency.

1.3. CHALLENGES

Olsen *et al.* [2000], Johnson [1998] and Brewster *et al.* [1998] highlight the design challenges associated with the small screen size of hand-held devices. In comparison to desktop computers, hand-held devices always suffer from a lack of screen real estate. Therefore, new interaction metaphors have to be invented for such devices.

Many assumptions about classical stationary applications no longer apply for hand-held devices due to the wide range of possibilities currently available. This wide range of possibilities is due to hand-held devices having constantly updated capabilities, exploiting additional features of novel generations of networks, and often being enabled for mobile users with varying profiles.

Furthermore, many web-based UIs adapt to client devices, users and user tasks [see Chapters 8 and 10]. This adaptation provides interfaces that run on conventional web browsers using HyperText Markup Language (HTML), as well as on wireless PDAs, mobile phones and pagers using Wireless Markup Language (WML) [see Chapter 5]. In addition, it is important to adapt UIs to different users and user tasks [see Chapters 7 and 8]. For example, it is necessary to hide "Update" and "Delete" buttons if the user is a customer or if the user is a staff member performing only information retrieval tasks. Building such interfaces using current web-based system implementation technologies is difficult and time-consuming, resulting in hard-to-maintain solutions.

Universal design is emerging as an approach where user interfaces of an interactive application have to be designed for the widest population of users in different contexts of use. In particular, the multiplicity of parameters dramatically increases the complexity of the design phase by adding many design options from which to choose. In addition, methods for developing UIs do not mesh well with this variety of parameters as they are not identified and manipulated in a structured way, nor truly considered in the design process [see Chapter 10].

1.4. SPECIFIC OBJECTIVES

Even if the software tools for developing a large variety of interfaces on each computing platform are already available or will be in the near future [Myers 2000], the following are the major development issues that need to be addressed by both academic and industrial researchers:

- Building the ability to dynamically respond to changes in the environment such as network connectivity, user's location, ambient sound and lighting conditions: How can we adapt the UI to the diversity of computing platforms that exists today? How can we maintain or adapt the high level of interactivity of the traditional office desktop in small devices without a keyboard and mouse? How can we make it possible for users to customize a device? When a single device is customized, how can this customization be reflected on all of the other devices available to the user?

- Designing for universal usability: What kinds of design methods are needed for designing for diverse users and a large variety of technologies? Are the design techniques for UI modelling suitable for addressing the problems of diversity, cross-platform consistency and universal accessibility?
- Checking consistency between versions for guaranteeing seamless interaction across multiple devices: Should we strive for uniformity in the services offered, dialogue styles and presentation formats, or should we adapt the interfaces to the constraints and capabilities of each device and/or each context of use? When designing MUIs, what is the best way to take into account the constraints related to each type of device while ensuring maintainability and cross-platform consistency of interfaces?
- Implementing and maintaining versions of the user interface across multiple devices: How can we implement and validate a MUI for **d** devices without writing **p** programs, training an army of developers in **l** languages and UI toolkits, and maintaining **l*p** architectural models for describing the same UI? Are the content markup languages adequate for device-independent authoring?

The book also introduces a variety of development frameworks that have been investigated over the last few years:

- Conceptual and adaptation frameworks for interacting with multiple user interfaces, including visual and awareness metaphors, and specific interaction styles;
- Design frameworks and patterns including widgets, toolkits and tools for multi-device development and in particular for mobile devices;
- Application frameworks that use multi-devices or multiple user interfaces, in particular collaborative work environments, distance education systems and remote software deployment systems;
- Validation frameworks including usability techniques for testing multiple user interfaces, as well as empirical tests and feedback.

1.5. AUDIENCE

This book introduces design and development frameworks for multi-device, context-aware and multiple user interface systems. These frameworks are valuable to researchers and practitioners in usability and software engineering, and generally to anyone interested in the problem of developing and validating multi-devices or cross-platform user interfaces. User interface developers, students and educators can use these frameworks to extend and improve their HCI methodologies, and to learn techniques for developing and evaluating a multiple user interface.

1.6. OVERVIEW

This book is divided into 6 parts:

Part I discusses "Basic Terminology, Concepts, and Challenges". Following the executive summary, in Chapter 2 Ahmed Seffah and Homa Javahery, the co-editors of this book, present a broad overview of multiple user interfaces. They discuss promising development models that can facilitate MUI development while increasing cross-platform usability. This chapter is highly speculative and will provide questions for basic research. This is a selective list of topics, and not exhaustive. The goal is to give researchers and practitioners a glimpse of the most important problems surrounding MUI design and development. The editors' opinions expressed in Chapter 2 do not necessarily reflect all of the contributors' ideas. Complementary and differing opinions are presented by other contributors in their own chapters. After exploring these first two chapters, the reader should have an increased awareness of the diversity of computing platforms and devices, a deeper understanding of the major development challenges and difficulties, and a familiarity with the basic terminology used.

Part II is entitled "Adaptation and Context-Aware User Interfaces", and provides three traditional but comprehensive perspectives on adaptation and context-aware techniques. David Thevenin *et al.* from the CLIPS-IMAG Laboratory in Grenoble, France, introduce the novel concept of user interface plasticity in Chapter 3. This chapter also provides a generous glossary of terms complementing the basic terminology presented in Chapter 2. Chapters 4 and 5 examine two dimensions of multi-interaction and adaptation. David England and Min Du, from Liverpool John Moores University, take a look at the different temporal characteristics of platforms that can affect user performance. They then propose a framework taking into account temporal aspects of interaction in the use of different devices. They describe how the temporal aspects should be incorporated into the interaction design process. In Chapter 5, Constantine Stephanadis *et al.*, from the Institute of Computer Science of the Foundation for Research and Technology – Hellas, Greece, introduce their framework called PALIO (Personalized Access to Local Information and services for tourists), focusing on its extensive support for adaptation. They demonstrate how PALIO has been successfully used in the development of a real-world context-aware information system for tourists using a wide range of devices.

Part III is on "Development Technology and Languages" and consists of three different XML-based development frameworks. In Chapter 6, Mir Farooq Ali *et al.* (from the Virginia Technology Institute and Harmonia Inc.) describe a high level XML-based User Interface Markup Language (UIML) for the development of cross-platform user interfaces. In Chapter 7, Angel Puerta and Jacob Eisenstein, from RedWhale Software, discuss the rationale of XIML, another XML-based language for developing multiple user interfaces by transforming and refining tasks and UI models. These modelling and programming languages distinguish the *concrete aspects* of a user interface such as presentation and dialogue from its *abstract aspects* including the context and the tasks. They are considered by the research community to be a bridge across the gap between the design and development of user interfaces. John Grundy and Wenjing Zou, from the University of Auckland in New Zealand, go a step further by showing how a UI markup language can be interfaced with existing programming languages. They describe how scripts written in their device-independent markup language (AUIT) can be embedded in conventional Java server pages to provide a single adaptable thin-client interface for web-based systems. At run-time, AUIT uses a single interface description to automatically provide an interface for multiple web devices such as desktop HTML and mobile WML-based systems, while highlighting, hiding or disabling the interface elements depending on the current context.

Together, these three chapters show how XML-based markup languages, with the help of model-based techniques, can lead to an advanced framework for the development of multi-platform user interfaces.

Part IV, on "Model-Based Development", includes three chapters describing the state of the art and the needed evolution in model-based development approaches. The basic purpose of model-based approaches is to identify useful abstractions highlighting the main UI aspects that should be considered when designing effective interactive applications. In Chapter 9, Peter Forbrig *et al.* from the University of Rostock, Germany, present two techniques for task modelling and specification. The first technique allows separate descriptions of general temporal relations within a task model versus temporal constraints that are imposed by the target platform. The second technique helps to distinguish between an abstract interaction and specific representations. Using these two techniques, specific features of devices are specified by XML descriptions. In Chapter 10, Vanderdonckt *et al.* (from the Université Catholique de Louvain, Belgium and Universidade de Fortaleza, Brazil) use several models at different levels in their methodological framework for universal design. First, the design process is instantiated at a conceptual level where a domain expert defines an ontology of concepts, relationships and attributes of the domain of discourse, including user modelling. Then at a logical level, a designer specifies multiple models based on the previously defined ontology and its allowed rules. The last step consists of using a physical level to develop multiple user interfaces from the previously specified models, with design alternatives determined by characteristics in the user models. Fabio Paternò, the father of CTT (ConcurTaskTrees) notation, and his colleagues at ISTI-CNR, Italy, explain in Chapter 11 how the user model can be structured for a MUI. In particular, they show how information on user preferences and on the mobile versus stationary environment (such as location and surroundings) can be used to adapt a user interface at run-time and at design time.

Part V is dedicated to "Architectures, Patterns and Development Toolkits". Homa Javahery *et al.* from the Human-Centered Software Engineering Group at Concordia University, discuss in Chapter 12 the role of HCI patterns and software reengineering techniques in migrating traditional GUIs to web and mobile user interfaces. In Chapter 13, Anind Dey, from Intel Research in California, and Gregory D. Abowd, from the Georgia Institute of Technology, present the Context Toolkit, an innovative and integrative infrastructure for the development of context-aware applications. Through the description of a number of built applications, they discuss a low-level widget abstraction that mirrors the use of graphical widgets for building graphical user interfaces and a situation abstraction that supports easier and higher-level application development. In Chapter 14, Simon Lock from Lancaster University and Harry Brignull from Sussex University, UK, describe a run-time infrastructure including a developer-level framework. This infrastructure supports the construction of applications that allow multiple users to interact through a dynamic set of interaction devices.

"Evaluation and Social Impacts" are addressed in Part VI. In Chapter 15, Gustav Öquist *et al.* (from Uppsala University in Sweden and Ericsson Research) discuss "Assessing Usability Across Multiple User Interfaces". They present their practical experiences with stationary versus mobile usability evaluations. In particular, they outline four typical contexts of use that they characterized by monitoring four environmental usability factors. By assessing these factors, it was possible to obtain a profile of how useful a given interface can be in a certain context of use. The usability profiles for several different interfaces for input and output are also presented in this chapter, partly to illustrate how usability can be assessed over multiple user interfaces, and partly as an illustration of how different interfaces have been adapted to mobile environment attributes. In Chapter 16, Joanna McGrenere from the University of British Columbia summarizes her MUI evaluation experiment combining three usability studies. In the first study, McGrenere conducted a broad-based assessment of 53 users of MS-Word 97. Based on the findings from this study, she developed a first multiple-interface prototype for MS-Word 2000 including one personalizable interface. In the last study, personalization was achieved through the Wizard of Oz technique. Even if McGrenere's definition of a multiple user interface is restrictive compared to the ones proposed in this book, her empirical results are very promising. They demonstrate how users were better able to navigate through the menus and toolbars and learn a multiple-interface prototype.

Continuing part VI, Charles Denis and Laurent Karsenty from IntuiLab Inc., argue in chapter 17 that the usability of individual devices is not sufficient: a multi-device system needs to also be *inter-usable*. They define inter-usability as the ease with which users transfer what they have learned from previous uses of a service when they access the service on a new device. Based on theoretical considerations and empirical observations gathered from a study with France Telecom, they propose an analysis grid combining two types of continuity, namely knowledge and task, with ergonomic design principles including consistency, transparency, and dialogue adaptability. The concept of inter-usability is very similar to the concept of horizontal usability introduced in Chapter 2 by the editors. Inter-usability or horizontal usability is a new dimension for studying the usability of MUIs and multi-device user interfaces.

REFERENCES

Brewster, S., Leplâtre, G. and Crease, M. (1998) *Using Non-Speech Sounds in Mobile Computing Devices*. Proceedings of the First Workshop on Human Computer Interaction with Mobile Devices, May 21–23, 1998, Glasgow, UK.

Hochheiser, H. and Shneiderman, B. (2001) Universal Usability Statements: Marking the trail for all users. *ACM Interactions* 8(2), March–April 2001, 16–18.

Johnson, P. (1998) *Usability and Mobility: Interactions on the move*. Proceedings of the First Workshop on Human Computer Interaction With Mobile Devices, May 21–23, 1998, Glasgow, UK.

Myers, B., Hudson, S. and Pausch, R. (2000) Past, Present, and Future of User Interface Software Tools. *ACM Transactions on Computer-Human Interaction*, 7, 3–28.

Olsen, D., Jefferies, S., Nielsen, T. *et al.* (2000) *Cross-modal Interaction using XWeb*. Proceedings of the 13th Annual ACM Symposium on User Interface Software and Technology, UIST'2000, November 5–8, 2000, San Diego, USA.

Multiple User Interfaces: Cross-Platform Applications and Context-Aware Interfaces

Ahmed Seffah and Homa Javahery

Human-Centered Software Engineering Group, Department of Computer Science,
Concordia University, Canada

2.1. MUI: CHARACTERIZATION AND EVOLUTION

We introduced the concept of "Multiple User Interface" (MUI) at the IHM-HCI 2001 workshop [Seffah *et al.* 2001]. Others are also using the term MUI with varying definitions [McGrenere *et al.* 2002; Vanderdonckt and Oger 2001]. For the purposes of this book, a Multiple User Interface is defined as an interactive system that provides:

- access to information and services using different computing platforms;
- multiple views of the same information on these different platforms;
- coordination of the services provided to a single user or a group of users.

Each view should take into account the specific capabilities and constraints of the device while maintaining cross-platform consistency and universal usability. By computing platform, we refer to a combination of computer hardware, an operating system and

Multiple User Interfaces. Edited by A. Seffah and H. Javahery
© 2004 John Wiley & Sons, Ltd   ISBN: 0-470-85444-8

a user interface (UI) toolkit. Different kinds of computing platforms include traditional office desktops, laptops, palmtops, mobile telephones, personal digital assistants (PDAs), and interactive television. A MUI provides multiple views of the same information on these different platforms and coordinates the services provided to a single user or a group of users. Each view should take into account the specific capabilities and constraints of the device while maintaining cross-platform consistency and universal usability. The information and services can reside on a single server or computer, or can be distributed among independent and heterogeneous systems. The desired views are made available on different computing platforms via the traditional client/server protocol or a direct peer-to-peer access. The concept of MUIs is highly promising in a variety of fields, such as cooperative engineering, e-commerce, on-site equipment maintenance, remote software deployment, contingency management and assistance, as well as distance education and telemedicine.

As an example of MUI use, a civil engineer can use a Palm Pilot on PalmOS for gathering data when inspecting a new building. He/She can then use a mobile telephone to add comments, fax, or upload information to the office headquarters. Finally, the same engineer or any other employee can use an office workstation under Windows/Linux to analyze the data and prepare a final report. During this workflow, the engineer interacts with the same information and services using different variations of the UI. These variations can support differences in *look-and-feel*, and to a certain extent, differences in interaction style. The following is a scenario that further clarifies the MUI concept and its use, based on [Ghani 2001]:

*You are riding in a car with your colleague who is driving. Suddenly, your **mobile phone** comes on, asking if you can take a video conference call from your team in Canada to discuss a project on which you are working. You take the call and as you communicate with them via the integrated **car video system**, you find out that they need one of the spreadsheets you have saved on your **laptop**, which is in the trunk of the car. Using your wireless technology, you are able to transfer the file to your **PDA**, and then send it to your team. A few minutes later your team will receive and open your spreadsheet, using an **office desktop** and you can start discussing options with them once again via your **interactive television** from your comfortable home.*

A MUI includes the following characteristics:

- It allows a user to interact with server-side services and information using different interaction/UI styles. For example, the civil engineer in our previous example could use pen and gestures on a PDA, function keys or single characters on a mobile phone, and a mouse for a desktop computer.
- It allows an individual or a group to achieve a sequence of interrelated tasks using different devices. For example, our civil engineer could use a mobile telephone on the road to confirm an appointment, a desktop computer to email information about an interview, and a PDA to gather information about the user needs when interviewing stakeholders. Finally, a laptop/desktop computer can be used to synthesize the information and write the requirements report.

Figure 2.1. An example of a MUI.

- It presents features and information that behave the same across platforms, even though each platform/device has its specific look-and-feel.
- It feels like a variation of a single interface, for different devices with the same capabilities.

Figure 2.1 shows a MUI for an Internet financial management system. The interface consists of three views: desktop, PDA with keyboard, and mobile phone. Ideally, from the user perspective, these three different interfaces should be as similar as possible. However, this is not realistic because of the capabilities and constraints imposed by each platform. The MUI here can be seen as a compromise between customized and platform-dependent UIs. The effort required to keep all interfaces consistent and to maintain them increases linearly with the number of interfaces, while the functionality of the underlying financial system is expanding.

The MUI provides multiple views of the same model and coordinates the user actions gathered from different devices/computers. The model can reside on a single information repository, or can be distributed among independent systems. Each view can be seen as a complete UI for a specific combination of hardware, operating system, and UI toolkit. All these interfaces form a unique and single MUI. The list of features, the interaction style, and the displayed information and feedback can vary from one platform to another.

2.1.1. INTERACTION STYLES

A MUI can support different interaction styles for different computing platforms. The following three styles are commonly used:

- *Graphical User Interface* (GUI) or WIMP interfaces. This style is the most popular and generally used for the office desktop. It employs four fundamental elements: windows, icons, menus, and pointers.
- *Web-based User Interface* (WUI). The Web was originally conceived as a hypertext information space. However, the development of increasingly sophisticated client and server-side technologies has fostered its use as a remote software interface. A WUI is generally a mixture of markup (i.e. HTML, XML syntax), style sheets, scripts (i.e. ECMAScript, VBScript), as well as embedded objects such as Java applets and plug-ins. An application using WUI style information is typically displayed in a single GUI window called a browser, although multiple browser windows can be used by an application to display information. The browser provides basic navigation. Different browsers for small devices and mobile phones are being developed, and such browsers are able to display a customized version of a WUI. The Yahoo Stock Manager is an example of a WUI that has a customized version for mobile phones. The Web clipping architecture and WAP framework provide basic services for dynamically generating customized HTML documents that can be displayed on mobile phones.
- *Handheld User Interface* (HUI), such as on mobile phones and PDAs. There are two major classes of PDAs and mobile phones in use today – those with a true GUI-style appearance and behaviour, and those that use a GUI or WUI subset. Both classes of UIs employ gesture-based interaction using a stylus and/or a touch screen. Even if it is not yet clear what style of UI will dominate handheld computers, the use of GUI and WUI styles should be re-evaluated due to lack of screen space and low bandwidth.

We expect that in the near future, designers will be asked to combine these three different styles. Figure 2.2 shows an example of a MUI that combines the GUI and WUI styles. Atomica Slingshot, formerly called GuruNet, is a pop-up application for retrieving reference information (dictionary, thesaurus and encyclopaedia) and real-time information such

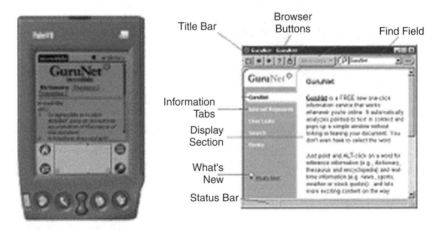

Figure 2.2. GuruNet: An example of a MUI.

as news, weather, and stock quotes across the Internet from inside any application. Atomica Slingshot (www.gurunet.com/us/products/slingshot/index.html) offers a WUI. There also exists a GUI version for office desktops and an optimized HUI version for PDAs.

Dealing with different UI styles complicates both the development and validation of a MUI. Making a trade-off between these different styles when they come into conflict is not an easy task. Due to device constraints, especially bandwidth capability, the UI on a wireless handset has to be fundamentally different from the UI on a PC. It is more restrictive since it cannot replicate the rich graphics, content, and functionality of the UI of a traditional GUI application. The device's characteristics and constraints also need to be considered with respect to their immediate effect on the task execution time, in order to achieve the same usability for mobile UIs as for GUIs. In traditional desktop UIs, the keyboard and the mouse are the primary input devices. Changing from a mouse to a stylus on a touch screen requires different interaction techniques. Display resolution and smaller design space are the mobile computing platform's most limiting constraints. Because of screen resolution differences, an optimal layout for one display may be impossible to render on another device. Reducing the size of a graphic might make it meaningless, while scrolling large graphics both horizontally and vertically is an unwieldy, unusable approach.

For all these reasons, some of the current development techniques that make implicit assumptions about device characteristics need to be rethought. New presentation and interaction techniques for MUIs are needed.

2.1.2. FUNDAMENTAL CHARACTERISTICS

The following are the major intrinsic characteristics of a MUI:

- *Abstraction*: All information and services should be the same across computing platforms supporting the same level of interactivity, even if not all information and services are shown or needed for all platforms. For example, a product listing might include only the best-selling items on a handheld device, with the rest relegated to a secondary "more products" page. For an office desktop, the product list includes all the items for sale.
- *Cross-platform consistency*: A MUI can have a different look-and-feel while maintaining the same behaviour over different platforms. For example, all user preferences must be preserved across the interfaces. If the end-user has specified a particular access mechanism using one interface, it should be used on all interfaces.
- *Uniformity*: A MUI should offer support for the same functionality and feedback even if certain features or variations are eliminated on some platforms [Ramsay and Nielsen 2000]. For example, let us assume that an airline reservation system presents choosing a flight and buying the ticket in two separate steps. This separation should be preserved on all versions instead of unifying the two tasks into a single step on a simplified version of the interface.
- *User awareness of trade-off*: It would be acceptable to have a simplified version of a program that excludes certain less-important features (such as specifying a seating preference) that are present in the more advanced version. Missing these features is a

trade-off that the user would be willing to make in return for the benefits of being able to use the system in mobile contexts.

- *Conformity to default UI standards*: It is not necessary for all features to be made available on all devices. For example, a PDA interface could eliminate images or it might show them in black and white. Similarly, text can be abbreviated on a small display, although it should be possible to retrieve the full text through a standardized command.

These characteristics and constraints are not artefacts of current development technologies, but are intrinsic to the MUI concept. Together, they characterize a MUI and complicate its development.

2.1.3. VERTICAL VERSUS HORIZONTAL USABILITY

MUI usability issues can be considered to have two dimensions: vertical and horizontal. Vertical usability refers to usability requirements specific to each platform while horizontal usability is concerned with cross-platform usability requirements.

Many system manufacturers have issued design guidelines to assist designers in developing usable applications. These guidelines can be categorized according to whether they advocate a design model (i.e. "do this") or whether they discourage a particular implementation (i.e. "don't do this"). For the PalmOS platform (www.palmsource.com), several design guidelines address navigation issues, widget selection, and use of specialized input mechanisms such as handwriting recognition. Microsoft Corporation has also published usability guidelines to assist developers with programming applications targeted at the Pocket PC platform. However, 'give the user immediate and tangible feedback during interaction with an application' is either too general or too simplistic. In many cases, the use of several different guidelines could create inconsistencies. Guidelines can come into conflict more than usual, and making a trade-off can become an unsolvable task for MUI developers.

Sun's guidelines for the Java Swing architecture (http://java.sun.com) describe a look-and-feel interface that can overcome the limitations of platform-dependent guidelines. However, these guidelines do not take into account the distinctiveness of each device, and in particular the platform constraints and capabilities. An application's UI components should not be hard-coded for a particular look-and-feel. The Java PL&F (Pluggable Look and Feel) is the portion of a Swing component that deals with its appearance (its *look*); it is distinguished from its event-handling mechanism (its *feel*). When you run a Swing program, it can set its own default look by simply calling a UIManager method named *setLookAndFeel*.

2.1.4. RELATED WORK

Remarkably, although research on MUIs and multi-device interaction can be traced to the early 1980s, there are relatively few examples of successful implementations [Grudin 1994]. Perhaps the main cause of this poor success rate is the difficulty of integrating the overwhelming number of technological, psychological and sociological factors that affect MUI usability into a single unified design.

In the evolution of user interfaces, a *multi-user* interface has been introduced to support groups of devices and people cooperating through the computer medium [Grudin 1994]. A single user in the context of a MUI corresponds to a group of users for a multi-user interface. The user is asynchronously collaborating with himself/herself. Even if the user is physically the same person, he/she can have different characteristics while working with different devices. For example, a mobile user is continuously in a rush, impatient, and unable to wait [Ramsay and Nielsen 2000]. This user needs immediate, quick, short and concise feedback. In the office, the same user can afford to wait a few seconds more for further details and explanations.

The MUI domain can benefit from the considerable number of studies done in the area of context-aware (or context-sensitive) user interfaces. This is still an active research topic, with many emerging models such as plastic user interfaces [Thevenin and Coutaz 1999] and the moderator model [Vanderdonckt and Oger 2001]. In a recent essay, Winograd [2001] compared different architectures for context of use. As characterized in the previous section, a MUI is a context-sensitive UI. This does not mean that a MUI should adapt itself magically at run-time to the context of use (and in particular to platform capabilities and constraints). The MUI can be either adaptive or adaptable. As we will discuss in the next section, the adaptation can be done during specification, design or development by the developer. The adaptation can also occur before or after deployment, either by the end-user or the developer.

The concept of a compound document is also a useful technology that can support the development and integration of the different views that form a MUI. A compound document framework can act as a container in which a continuous stream of various kinds of data and components can be placed [Orfali *et al.* 1996]. To a certain extent, a compound document is an organized collection of user interfaces that we consider as a specialization of a MUI. Each content form has associated controls that are used to modify the content in place. During the last decade, a number of frameworks have been developed such as Andrew, OLE, Apple OpenDoc, Active X and Sun Java Beans.

Compound document frameworks are important for the development of a MUI for several reasons. They allow the different parts of a MUI to co-exist closely. For example, they keep data active from one part to another, unlike the infamous cut and paste. They also eliminate the need for an application to have a viewer for all kinds of data; it is sufficient to invoke the right functionality and/or editor. Views for small devices do not have to implement redundant functions. For example, there is no need for Microsoft Word to implement a drawing program; views can share a charting program. Compound document frameworks can also support asynchronous collaboration between the different views and computers.

McGrenere *et al.* [2002] illustrate the use of two versions of the same application with two different user interfaces as follows:

One can imagine having multiple interfaces for a new version of an application; for example, MS-Word 2000 could include the MS-Word 97 interface. By allowing users to continue to work in the old interface while also accessing the new interface, they would be able to transition at a self-directed pace. Similarly, multiple interfaces might be used to provide a competitor's interface in the hopes of attracting new customers. For example, MS-Word could offer the full interface of a word processor such as Word Perfect (with single button access to switch between the two), in order to support users gradually transitioning to the Microsoft product.

Our definition of a MUI is different from McGrenere's definition. The common basis is the fact that the user is exposed to two variations of the same interface. McGrenere considers only the variations, which are referred to as versions, for the same computing platform; while in our definition, the two variations can be either for the same computing platform or for different ones.

2.2. FERTILE TOPICS FOR RESEARCH EXPLORATION

We will now discuss promising development models that can facilitate MUI development while increasing their usability. This section of the paper is highly speculative and will raise far more fundamental research questions than it will provide answers. Furthermore, this is a selective list of topics, and not exhaustive. Our goal is to give researchers a glimpse of the most important problems surrounding potential MUI development models.

In the migration of interactive systems to new platforms and architectures, many modifications have to be made to the user interface. As an example, in the process of adapting the traditional desktop GUI to other kinds of user interfaces such as Web or handheld user interfaces, most of the UI code has to be modified. In this scenario, UI model-based techniques can drive the reengineering process. Reverse engineering techniques can be applied, resulting in a high-level model of the UI. This model can then be used to help reengineer the user interface.

2.2.1. CONTEXT-AWARE DEVELOPMENT

Context-aware UI development refers to the ability to tailor and optimize an interface according to the context in which it is used. *Context-aware computing* as mentioned by Dey and Abowd refers to the "ability of computing devices to detect and sense, interpret and respond to, aspects of a user's local environment and the computing devices themselves" [Dey and Abowd 2000]. *Context-aware applications* dynamically adapt their behaviour to the user's current situation, and to changes of context of use that might occur at run-time, without explicit user intervention. Adaptation requires a MUI to sense changes in the context of use, make inferences about the cause of these changes, and then to react appropriately.

Two types of adaptation have to be considered for MUIs:

- *Adapting to technological variety* Technological variety implies supporting a broad range of hardware, software, and network access. The first challenge in adaptation is to deal with the pace of change in technology and the variety of equipment that users employ. The stabilizing forces of standard hardware, operating systems, network protocols, file formats and user interfaces are undermined by the rapid pace of technological change. This variety also results in computing devices (e.g. mobile phones) that exhibit drastically different capabilities. For example, PDAs use a pen-based input mechanism and have average screen sizes around three inches. In contrast, the typical PC uses a full sized keyboard and a mouse and has an average screen size of 17 inches. Coping with such drastic variations implies much more than mere layout changes. Pen-based input mechanisms are slower than traditional keyboards and are therefore inappropriate for applications such as word processing that require intensive user input.

- *Adapting to diversity in context of use* Further complications arise from accommodating users with different skills, knowledge, age, gender, disabilities, disabling conditions (mobility, sunlight, noise), literacy, culture, income, etc. [Stephanidis 2002]. For example, while walking down the street, a user may use a mobile phone's Internet browser to look up a stock quote. However, it is highly unlikely that this same user would review the latest changes made to a document using the same device. Rather, it would seem more logical and definitely more practical to use a full size computer for this task. It would therefore seem that the context of use is determined by a combination of internal and external factors. The internal factors primarily relate to the user's attention while performing a task. In some cases, the user may be entirely focused while at other times, the user may be distracted by other concurrent tasks. An example of this latter point is that when a user is driving a car, he/she cannot use a PDA to reference a telephone number. External factors are determined to a large extent by the device's physical characteristics. It is not possible to make use of a traditional PC as one walks down the street. The same is not true for a mobile telephone. The challenge to the system architect is thus to match the design of a particular device's UI with the set of constraints imposed by the corresponding context of use.

A fundamental question is when should a MUI be tailored as a single and unique interface? The range of strategies for adaptation is delimited by two extremes. Interface adaptation can happen at the factory, that is, developers produce several versions of an application tailored according to different criteria. Tailoring can also be done at the user's side, for instance, by system administrators or experienced users. At the other extreme, individual users might tailor the interfaces themselves, or the interface could adapt on its own by analyzing the context of use. The consensus from our workshop was that the adaptation of a MUI should be investigated at different steps of the deployment lifecycle [Seffah *et al.* 2001]:

- *User customization after deployment* Here, tailoring operations are the entire responsibility of the user. While this laissez-faire approach avoids the need for system support, it lacks a central arbitrator to resolve incompatible and inconsistent preferences between devices. The arbitrator should have the ability to make global changes (cross-platform changes) based on local adaptations. This makes MUIs more difficult to write, and the adaptation fails to repay the development cost of support.
- *Automatic adaptation at run-time* The idea is to write one UI implementation that adapts itself at run-time to any computing platform and context of use. The drawback of this strategy is that there may be situations where adaptation performed by the system is inadequate or even counterproductive.

- *Just-in-time customization during development or deployment* Developers can use a high-level language to implement an abstract and device-independent UI model. Then, using a rendering tool, they can generate the code for a specific platform. The *User Interface Markup Language*, UIML [Abrams and Phanouriou 1999], and the *eXtensible Interface Markup Language*, XIML [Eisenstein *et al.* 2001], aim to support such an approach.
- *Customization during design and specification* This approach requires the development of an appropriate design methodology and multi-platform terminology to properly build a task model of a MUI. This model may be expressed in one or more notations. Tailoring can be done at the stage of abstract interface specification where the dialogue gets modified, for example to shortcut certain steps, to rearrange the order for performing steps, etc.

Efforts have already begun to develop frameworks that support the building of context-aware applications. The *Context Toolkit* [Dey and Abowd 2000] is an infrastructure that supports the rapid development of context-aware services, assuming an explicit description of a context. This framework's architecture enables the applications to obtain the context they require without knowledge about how the context was sensed. The Context Toolkit consists of *context widgets* that implicitly sense context, *aggregators* that collect related context, *interpreters* that convert between context types and interpret the context, *applications* that use context and a *communications infrastructure* that delivers context to these distributed components. The toolkit makes it easy to add the use of context or implicit input to existing applications.

2.2.2. MODEL-BASED DEVELOPMENT

Model-based approaches for UI development [Bomsdorf and Szwillus 1998; Müller *et al.* 2001] exploit the idea of using declarative interface models to drive the interface development process. An interface model represents all the relevant aspects of a UI using a user interface modelling language. Model-based development approaches attempt to automatically produce a concrete UI design (i.e. a concrete presentation and dialogue for a specific platform) from the abstract "generic" representation of the UI (i.e., generic task, domain and dialogue model). This is done by mapping the abstract model onto the concrete user interface or some of its elements [Bomsdorf and Szwillus 1998]. For example, given user task t in domain d, the mapping process will find an appropriate presentation p and dialogue D that allows user u to accomplish t. Therefore, the goal of a model-based system in such a case is to link t, d, and u with an appropriate p and D. Model-based UI development could be characterized as a process of creating mappings between elements in various model components. The process of generating the concrete interface and UI model involves levels as shown in Figure 2.3.

Model-based approaches, in particular the related automatic or semi-automatic UI generation techniques, are of interest to MUI development. UI modelling will be an essential component of any effective long-term approach to developing MUIs. Increased user involvement in the UI development process will produce more usable UI models. Model-based UI systems take an abstract model of the UI and apply design rules and data

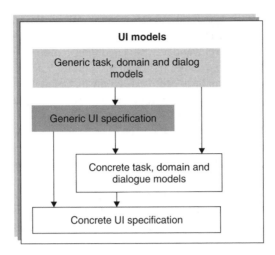

Figure 2.3. Examples of models and mappings in model-based development.

about the application to generate an instance of the UI. Declarative model-based techniques use UI modelling techniques for abstractly describing the UI. A formal, declarative modelling language should express the UI model.

Current model-based techniques, which most frequently use task and domain models, do not generate high-quality interfaces. Furthermore, task analysis is performed to obtain a single UI that is adapted for a single context of use. We need to model tasks that can be supported in multiple contexts of use, considering multiple combinations of the contextual conditions. Knowledge bases for domain, presentation, dialogue, platform and context of use need to be exploited to produce a usable UI that matches the requirements of each context of use.

UI models that support mobility contain not only the visual look-and-feel of the UI, but also semantic information about the interface. The model-based techniques proposed for mobile UIs range from relatively low-level implementation solutions, such as the use of abstract and concrete interactor objects, to high-level task-based optimization of the interface's presentation structure. UI models should factor out different aspects of UI design that are relevant to different contexts of use and should isolate context-independent issues from context-specific ones.

As a starting point for research in the field of model-based development for MUIs, the focus should be on task-based models [Paternò 2001]. Such models can foster the emergence of new development approaches for MUIs, or at least help us to better understand the complexity of MUI development. A task model describes the essential tasks that the user performs while interacting with the UI. A typical task model is a hierarchical tree with sub-trees indicating the tasks that the user can perform. Task models are a very convenient specification of the way problems can be solved.

Early investigations show that in the case of a MUI, we should make a distinction between four kinds of task models [Müller *et al.* 2001]: general task models for the problem domain, general task models for software support, device-dependent task models and environment-dependent task models. The general task model for the problem domain is the result of a very detailed analysis of the problem domain. It describes how a problem can be tackled in general. All relevant activities and their temporal relations are described. Such a model can be considered as the representation of an expert's knowledge. The state of the art for the problem domain is captured within this model.

Certain approaches transform whole applications from one platform to another one without considering the tasks that will be supported. However, sometimes it is wise to look at the tasks first and to decide which tasks a device can support optimally. This information is captured in the device-dependent task model. The environment-dependent task model is the most specific one. It is based on design decisions in previous models and describes computer-supported tasks for a given device. This model describes the behaviour of a system based on the available tools, resources, and the abilities of the user. It can be interpreted statically (environmental influences are defined during design time) or dynamically (environmental influences are evaluated during run-time).

2.2.3. PATTERN-DRIVEN DEVELOPMENT

In the field of UI design, a pattern encapsulates a proven solution for a usability problem that occurs in various contexts of use. As an illustration, the *convenient toolbar* pattern (used on web pages) provides direct access to frequently used pages or services. This pattern, also called Top Level Navigation [Tidwell 1997], can include navigation controls for News, Search, Contact Us, Home Page, Site Map, etc. UI design patterns can be used to create a high-level design model, and can therefore facilitate the development and validation of MUIs. Discussion of patterns for software design started with the software engineering community and now the UI design community has enthusiastically taken up discussion of patterns for UI design. Many groups have devoted themselves to the development of pattern languages for UI design and usability. Among the heterogeneous collections of patterns, those known as *Common Ground, Experience, Brighton,* and *Amsterdam* play a major role in this field and have significant influence [Tidwell 1997; Borchers 2000]. Patterns have the potential to support and drive the whole design process of MUIs by helping developers select proven solutions of the same problem for different platforms.

Pattern-driven development should not be considered as an alternative approach to model-based and context-aware development. In the context of MUI development, patterns can complement a task model by providing best experiences gained through end-user feedback. Furthermore, patterns are suitable for transferring knowledge from usability experts to software engineers who are unfamiliar with MUI design, through the use of software tools. For instance, CASE tools have long been available to assist software developers in the integration of the many aspects of web application prototyping [Javahery and Seffah 2002].

However, the natural language medium generally used to document patterns, coupled with a lack of tool support, compromises these potential uses of patterns, as well as the pattern-oriented design approach. These well-known weaknesses of UI patterns should motivate researchers to investigate a systematic approach to support both pattern writers and users alike by automating the development of pattern-assisted design. We should also provide a framework for automating the development of pattern-oriented design. The motivation of such automation is to help novice designers apply patterns correctly and efficiently when they really need them. One approach to pattern-oriented design automation is being able to understand during the design process when a pattern is applicable, how it can be applied, and how and why it can or cannot be combined with other related patterns.

2.2.4. DEVICE-INDEPENDENT DEVELOPMENT

Currently, different development languages are available (Figure 2.4). Under the umbrella of platform-dependent languages, we classify the wide variety of existing mark-up languages for wireless devices such as the Wireless Markup Language (WML) or the light HTML version. These languages take into account the platform constraints and capabilities posed by each platform. They also suggest specific design patterns for displaying information and interacting with the user in specific ways for each device.

Platform-independent languages are mainly based on UI modelling techniques. Their goal is to allow cross-platform development of UIs while ensuring consistency not only between the interfaces on a variety of platforms, but also in a variety of contexts of use. They provide support for constraints imposed not only by the computing platforms themselves, but also by the type of user and by the physical environment. They should help designers recognize and accommodate each context in which the MUI is being used. Such languages provide basic mechanisms for UI reconfigurations depending on variations of the context of use. They address some of the problems raised by context-aware development.

XML-based languages such as XIML and UIML are promising candidates for MUI development. Some of the reasons are that such XML-based languages:

- Can contain constraint definitions for the XML form itself, and also for the external resources;
- Allow the separation of UI description from content, by providing a way to specify how UI components should interact and a way to spell out the rules that define interaction behaviour;

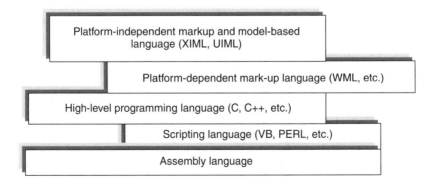

Figure 2.4. Evolution of UI development languages.

- Provide an abstraction level that allows the UI to adapt to a particular device or set of user capabilities;
- Support model-based development.

MUI design pattern implementations should exist in various languages and platforms. Rather than using different programming languages for coding the different implementations, we should use an XML-based notation as a unified and device-independent language for documenting, implementing and customizing MUI design patterns. By using XML-compliant implementations, patterns can be translated into scripts for script-based environments like HTML authoring tools, beans for Java GUI builders like VisualAge, and pluggable objects like Java applets and ActiveX components. Generating a specific implementation from an XML-based description is now possible because of the availability of XML-based scripting languages. Among them, we consider UIML and XIML as potential candidates.

UIML and XIML languages permit a declarative description of a UI in a highly device-independent manner. They allow portability across devices and operating systems, and use a style description to map the interface to various operating systems and devices. UIML separates the UI content from its appearance. UIML does this by using a device-independent UI definition to specify the UI content and a device-dependent style sheet that guides the placement and appearance of the UI elements. UIML descriptions of a UI can be rendered in HTML, Java and WML. Tools that generate the code from design patterns, such as the IBM-Automatic code generator [Budinsky *et al.* 1996], are a starting point for automating the development of pattern-oriented design. Furthermore, using an XML-based language for documenting patterns has already been explored. However, the XML-based descriptions force all pattern writers and users to closely adhere to and master a specific format and terminology for documenting and implementing patterns.

2.3. CONCLUDING REMARKS

Understanding MUIs is essential in our current technological context. A MUI imposes new challenges in UI design and development since it runs on different computing platforms accommodating the capabilities of various devices and different contexts of use. Challenges are also presented because of the universal access requirements for a diversity of users. The existing approaches to designing one user interface for a single user profile for one computing platform do not adequately address the MUI challenges of diversity, cross-platform consistency, universal accessibility and integration. Therefore, there is an urgent need for a new integrative framework for modelling, designing, and evaluating MUIs for the emerging generation of interactive systems.

As outlined in this chapter, effective MUI development should combine different models and approaches. MUI architectures that neglect these models and approaches cannot effectively meet the requirements of the different users. Unfortunately, adoption of a MUI application is contingent upon the acceptance of all of the stakeholders. Researchers should focus on ways to assist developers in creating effective MUI designs for a large variety of computing platforms. Existing methods work well for regular software development and have thus been adapted for MUIs. However, these methods usually result in tools that do not capture the full complexity of the task. Pattern hierarchies seem to be an exception to this finding. Whereas an individual pattern provides a solution to a specific problem, hierarchically organized patterns guide the developer through the entire architectural design. In this way, they enforce consistency among the various views and break down complex decisions into smaller, more comprehensible steps.

ACKNOWLEDGEMENTS

We thank Dr. Peter Forbrig for his contribution to the MUI effort.

REFERENCES

Abrams, M. and Phanouriou, C. (1999) *UIML: An XML Language for Building Device-Independent User Interfaces*. Proceedings of XML 99, December 1999, Philadelphia.

Bomsdorf, B. and Szwillus, G. (1998) From Task to Dialogue: Task-Based User Interface Design. *SIGCHI Bulletin*, 30(4).

Borchers, J.O. (2000) *A Pattern Approach to Interaction Design*. Proceedings of the DIS 2000 International Conference on Designing Interactive Systems, August 16–19, 2000, 369–78. New York, ACM Press.

Budinsky, F., Finnie, F.J., Vlissides, J.M. and Yu, P.S. (1996) Automatic Code Generation from Design Patterns. *Object Technology*, 35(2).

Dey, A.K. and Abowd, G.D. (2000). *Towards a Better Understanding of Context and Context-Awareness*. Proceedings of the CHI'2000 Workshop on Context Awareness. April 1–6, 2000, The Hague, Netherlands.

Eisenstein, J., Vanderdonckt, J. and Puerta, A. (2001) *Applying Model-Based Techniques to the Development of UIs for Mobile Computers*. Proceedings of the ACM Conference on Intelligent User Interfaces, IUI'2001, January 11–13, 2001, 69–76. New York, ACM Press.

Ghani, R. (2001) 3G: 2B or not 2B? The potential for 3G and whether it will be used to its full advantage. *IBM Developer Works: Wireless Articles*, August 2001.

Grudin, J. (1994) Groupware and Social Dynamics: Eight Challenges for Developers. *Communications of the ACM*, 37(1), 92–105.

Javahery, H. and Seffah, A. (2002) *A Model for Usability Pattern-Oriented Design*. Proceedings of the Conference on Task Models and Diagrams for User Interface Design, Tamodia'2002, July 18–19 2002, Bucharest, Romania.

McGrenere, J., Baecker, R. and Booth, K. (2002) *An Evaluation of a Multiple Interface Design Solution for Bloated Software*. Proceedings of ACM CHI, 2002, April 20–24, 2002, Minneapolis, USA.

Müller, A., Forbrig, P. and Cap, C. (2001) *Model-Based User Interface Design Using Markup Concepts*. Proceedings of DSVIS 2001, June 2001, Glasgow, UK.

Ramsay, M. and Nielsen, J. (2000) *WAP Usability Déjà Vu: 1994 All Over Again. Report from a Field Study in London*. Nielsen Norman Group, Fremont, USA.

Orfali, R., Harkey, D. and Edwards, J. (1996) *The Essential Distributed Objects Survival Guide*. John Wiley & Sons Ltd., New York.

Paternò, F. (2001) Task Models in Interactive Software Systems in *Handbook of Software Engineering & Knowledge Engineering* (ed. S.K. Chang). World Scientific Publishing Company.

Seffah, A., Radhakrishan T. and Canals, G. (2001) Multiple User Interfaces over the Internet: Engineering and Applications Trends. Workshop at the IHM-HCI: French/British Conference on Human Computer Interaction, September 10–14, 2001, Lille, France.

Stephanidis, C. (ed) (2002) User Interfaces for all: *Concepts, Methods, and Tools*. Lawrence Erlbaum Associates Inc., Mahwah, USA.

Thevenin, D. and Coutaz, J. (1999) *Plasticity of User Interfaces: Framework and Research Agenda*. *Proceedings of IFIP TC 13 International Conference on Human-Computer Interaction, Interact'99*, 110–117, August 1999 (eds A. Sasse and C. Johnson), Edinburgh, UK. IOS Press, London.

Tidwell, J. (1997) Common Ground: A Pattern Language for Human-Computer Interface Design. http://www.time-tripper.com/uipatterns.

Vanderdonckt, J. and Oger, F. (2001) *Synchronized Model-Based Design of Multiple User Interfaces*. Workshop on Multiple User Interfaces over the Internet: Engineering and Applications Trends. IHM-HCI: French/British Conference on Human Computer Interaction, September 10–14, 2001, Lille, France.

Winograd, T. (2001) Architectures for Context. *Human-Computer Interaction*, 16, 2–3.

Part II

Adaptation and Context-Aware User Interfaces

A Reference Framework for the Development of Plastic User Interfaces

David Thevenin, Joëlle Coutaz, and Gaëlle Calvary

CLIPS-IMAG Laboratory, France

3.1. INTRODUCTION

The increasing proliferation of fixed and mobile devices addresses the need for ubiquitous access to information processing, offering new challenges to the HCI software community. These include:

- constructing and maintaining versions of the user interface across multiple devices;
- checking consistency between versions to ensure a seamless interaction across multiple devices;
- designing the ability to dynamically respond to changes in the environment such as network connectivity, user's location, ambient sound and lighting conditions.

These requirements create extra costs in development and maintenance. In [Thevenin and Coutaz 1999], we presented a first attempt at cost-justifying the development process

Multiple User Interfaces. Edited by A. Seffah and H. Javahery
© 2004 John Wiley & Sons, Ltd ISBN: 0-470-85444-8

of user interfaces using the notion of plasticity as a fundamental property for user interfaces. The term *plasticity* is inspired from materials that expand and contract under natural constraints without breaking, thus preserving continuous usage. Applied to HCI, plasticity is the "capacity of an interactive system to *withstand variations of contexts of use while preserving usability*" [Thevenin and Coutaz 1999].

Adaptation of user interfaces is a challenging problem. Although it has been addressed for many years [Thevenin 2001], these efforts have met with limited success. An important reason for this situation is the lack of a proper definition of the problem. In this chapter, we propose a reference framework that clarifies the nature of adaptation for plastic user interfaces from the software development perspective. It includes two complementary components:

- A taxonomic space that defines the fundamental concepts and their relations for reasoning about the characteristics and requirements of plastic user interfaces;
- A process framework that structures the software development of plastic user interfaces.

Our taxonomic space, called the "plastic UI snowflake" is presented in Section 3.3, followed in Section 3.4 by the description of the process framework. This framework is then illustrated in Section 3.5 with ARTStudio, a tool that supports the development of plastic user interfaces. In Section 3.2, we introduce the terminology used in this chapter. In particular, we explain the subtle distinction between plastic user interfaces and multi-target user interfaces in relation to context of use.

3.2. TERMINOLOGY: CONTEXT OF USE, PLASTIC UI AND MULTI-TARGET UI

Context is an all-encompassing term. Therefore, to be useful in practice, context must be defined in relation to a purpose. The purpose of this work is the adaptation of user interfaces to different elements that, combined, define a context of use. Multi-targeting focuses on the technical aspects of user interface adaptation to different contexts of use. Plasticity provides a way to characterize system usability as adaptation occurs. These concepts are discussed next.

3.2.1. CONTEXT OF USE AND TARGET

The context of use denotes the run-time situation that describes the current conditions of use of the system. A target denotes a situation of use as intended by the designers during the development process of the system.

The *context of use* of an interactive system includes:

- the people who use the system;
- the platform used to interact with the system;
- the physical environment where the interaction takes place.

A *target* is defined by:

- the class of user intended to use the system;
- the class of platforms that can be used to interact with the system;
- The class of physical environments where the interaction is supposed to take place.

In other words, if at run-time the context of use is not one of the targets envisioned during the design phase, then the system is not able to adapt to the current situation (person, platform, physical environment).

A *platform* is modelled in terms of resources, which in turn determine the way information is computed, transmitted, rendered, and manipulated by users. Examples of resources include memory size, network bandwidth and input and output interactive devices. Resources motivate the choice of a set of input and output modalities and, for each modality, the amount of information made available. Typically, screen size is a determining factor for designing web pages. For DynaWall [Streitz *et al.* 1999], the platform includes three identical wall-sized tactile screens mounted side by side. Rekimoto's augmented surfaces are built from a heterogeneous set of screens whose topology may vary: whereas the table and the electronic whiteboard are static surfaces, laptops may be moved around on top of the table [Rekimoto and Saitoh 1999]. These examples show that the platform is not limited to a single personal computer. Instead, it covers all of the computational and interactive resources available at a given time for accomplishing a set of correlated tasks.

An *environment* is 'a set of objects, persons and events that are peripheral to the current activity but that may have an impact on the system and/or users behaviour, either now or in the future' [Coutaz and Rey 2002]. According to this definition, an environment may encompass the entire world. In practice, the boundary is defined by domain analysts. The analyst's role includes observation of users' practice [Beyer 1998; Cockton *et al.* 1995; Dey *et al.* 2001; Johnson *et al.* 1993; Lim and Long 1994] as well as consideration of technical constraints. For example, environmental noise should be considered in relation to audio feedback. Lighting condition is an issue when it can influence the reliability of a computer vision-based tracking system [Crowley *et al.* 2000].

3.2.2. MULTI-TARGET USER INTERFACES AND PLASTIC USER INTERFACES

A multi-target user interface is capable of supporting multiple targets. A plastic user interface is a multi-target user interface that preserves usability across the targets. Usability is not intrinsic to a system. Usability must be validated against a set of properties elicited in the early phases of the development process. A multi-target user interface is plastic if these usability-related properties are kept within the predefined range of values as adaptation occurs to different targets. Although the properties developed so far in HCI [Gram and Cockton 1996] provide a sound basis for characterizing usability, they do not cover all aspects of plasticity. In [Calvary *et al.* 2001a] we propose additional metrics for evaluating the plasticity of user interfaces.

Whereas multi-target user interfaces ensure technical adaptation to different contexts of use, plastic user interfaces ensure both technical adaptation and usability. Typically, portability of Java user interfaces supports technical adaptation to different platforms but may not guarantee consistent behaviour across these platforms.

3.2.3. TERMINOLOGY: SUMMARY

In summary, for the purpose of our analysis:

- A target is defined as a triple 'user, platform, environment' envisioned by the designers of the system.
- A context of use is a triple 'user, platform, environment' that is effective at run-time.
- A multi-target user interface supports multiple targets, i.e., multiple types of users, platforms and environments. Multi-platform and multi-environment user interfaces are sub-classes of multi-target user interfaces:
- A multi-platform user interface is sensitive to multiple classes of platforms but supports a single class of users and environments.
- Similarly, a multi-environment user interface is sensitive to multiple classes of environments, but supports a single class of platforms and users. Multi-environment user interfaces are often likened to context-aware user interfaces [Moran and Dourish 2001].
- A plastic user interface is a multi-target user interface that preserves usability as adaptation occurs.

Having defined the notions of context of use, multi-target and plastic user interfaces, we are now able to present a taxonomic space that covers both multi-targeting and plasticity. The goal of this taxonomy is to identify the core issues that software tools aimed at multi-targeting and plasticity should address.

3.3. THE "PLASTIC UI SNOWFLAKE"

Figure 3.1 is a graphical representation of the problem space for reasoning about user interface plasticity. The plastic UI snowflake can be used to characterize existing tools or to express requirements for future tools. Each branch of the snowflake presents a number of issues relevant to UI plasticity. These include: the classes of targets that the tool supports (adaptation to platforms, environments and users), the stages of the development process that the tool covers (design, implementation or run-time), the actors that perform the adaptation of the user interface to the target (human or system intervention) and the dynamism of user interfaces that the tools are able to produce (static pre-computed or dynamic on-fly computed user interfaces). When considering adaptation to multiple platforms, we also need to discuss the way the user interface is migrated across platforms.

In the following sub-sections, we present each dimension of the snowflake in detail, illustrated with state-of-the-art examples. In particular, we develop multi-platform targeting. Although multi-user targeting is just as important, we are not yet in a position to provide a sound analysis for it. For adaptation to multi-environment targeting, please refer to [Moran and Dourish 2001] and [Coutaz and Rey 2002].

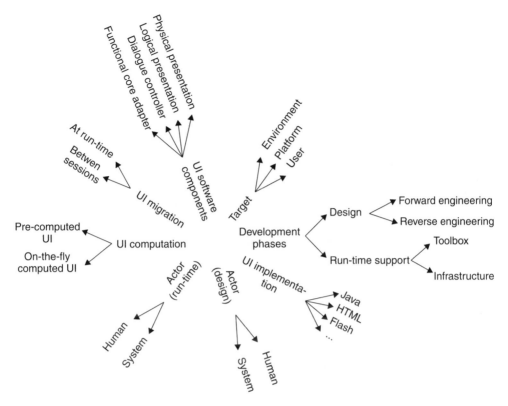

Figure 3.1. The *Plastic UI Snowflake*: a problem space for characterizing software tools, and for expressing requirements for software tools aimed at plastic user interfaces.

3.3.1. TARGET SENSITIVITY

In software tools for plasticity, the first issue to consider is the kind of targets a particular tool addresses or is supposed to address. Are we concerned with multi-platform or multi-environment only? Do we need adaptation to multiple classes of users? Or is it a combination of platforms, environments and users?

For example, ARTStudio [Thevenin 2001] addresses the problem of multi-platform targeting whereas the Context Toolkit [Dey *et al.* 2001] is concerned with environment sensitivity only. AVANTI, which can support visually impaired users, addresses adaptation to end-users [Stephanidis *et al.* 2001]. There is currently no tool (or combination of tools) that supports all three dimensions of plasticity, i.e. users, platforms and environments.

3.3.2. CLASSES OF SOFTWARE TOOLS

As with any software tool, we must distinguish between tools that support the design phases of a system versus implementation tools and mechanisms used at run-time.

Design phases are primarily concerned with forward engineering and reverse engineering of legacy systems. Forward engineering is supported by specification tools for modelling, for configuration management and versioning, as well as for code generation:

- Modelling is a fundamental activity in system design. In HCI, model-based tools such as Humanoid [Szekely 1996], ADEPT [Johnson *et al.* 1993] and TRIDENT [Vanderdonckt 1995] have shown significant promise, not only as conceptualization tools, but also as generators. If these approaches have failed in the past because of their high learning curve [Myers *et al.* 2000], they are being reconsidered for multi-target generation as in MOBI-D [Eisenstein *et al.* 2001] and USE-IT [Akoumianakis and Stephanidis 1997].
- Configuration management and versioning have been initiated with the emergence of large-scale software. They apply equally to multi-targeting and plasticity for two reasons. First, the code that supports a particular target can be derived from the high-level specification of a configuration. Secondly, the iterative nature of user interface development calls for versioning support. In particular, consistency must be maintained between the configurations that support a particular target.
- Generation has long been viewed as a reification process from high-level abstract description to executable code. For the purpose of multi-targeting and plasticity, we suggest generation by reification, as well as by translation where transformations are applied to descriptions while preserving their level of abstraction. The Process Reference framework described in Section 3.4 shows how to combine reification and translation.
- Tools for reverse engineering, that is eliciting software architecture from source code, are recent. In Section 3.4, we will see how tools such as Vaquita [Bouillon *et al.* 2002] can support the process of abstracting in order to plastify existing user interfaces.

Implementation phases are concerned with coding. Implementation may rely on infrastructure frameworks and toolkits. Infrastructure frameworks, such as the Internet or the X window protocol, provide implementers with a basic reusable structure that acts as a foundation for other system components such as toolkits. BEACH is an infrastructure that supports any number of display screens each connected to a PC [Tandler 2001]. MID is an infrastructure that extends Windows to support any number of mice to control a single display [Hourcade and Bederson 1999]. We are currently developing I-AM (Interaction Abstract Machine), an infrastructure aimed at supporting any number of displays and input devices, which from the programmer's perspective will offer a uniform and dynamic interaction space [Coutaz *et al.* 2002]. Similar requirements motivate the blackboard-based architecture developed for iRoom [Winograd 2001]. The Context Toolkit is a toolkit for developing user interfaces that are sensitive to the environment [Dey *et al.* 2001].

3.3.3. ACTORS IN CHARGE OF ADAPTATION

The actors in charge of adaptation depend on the phase of the development process:

- At the design stage, multi-targeting and plasticising can be performed explicitly by humans such as system designers and implementers, or it can rely on dedicated tools.

- At run-time, the user interface is adaptable or adaptive. It is adaptable when it adapts at the user's request, typically by providing preference menus. It is adaptive when the user interface adapts on its own initiative. The right balance between adaptability and adaptivity is a tricky problem. For example, in context-aware computing, Cheverst *et al.* [2001] report that using location and time to simplify users' tasks sometimes makes users feel that they are being pre-empted by the system. Similarly, adaptivity to users has been widely attempted with limited success [Browne *et al.* 1990].

3.3.4. COMPUTATION OF MULTI-TARGET AND PLASTIC USER INTERFACES

The phases that designers and developers elicit for multi-targeting and plasticity have a direct impact on the types of user interfaces produced for the run-time phase. Multi-target and plastic user interfaces may be pre-computed, or they may be computed on the fly:

- Pre-computed user interfaces result from adaptation performed during the design or implementation phases of the development process: given a functional core (i.e., an application), a specific user interface is generated for every envisioned target.
- Dynamic multi-target and plastic user interfaces are computed on the fly based on run-time mechanisms. Examples of run-time mechanisms include the Multimodal Toolkit [Crease *et al.* 2000], which supports dynamic adaptation to interactive devices. Flex-Clock [Grolaux 2000], which dynamically adapts to window sizes, is another example.
- The generated user interface can be a combination of static pre-computed components with on-the-fly adaptation. In this case, we have a hybrid multi-target plastic user interface. As a general rule of thumb, pre-computation is used for the overall structure of the user interface to ensure that the system runs quickly. However since this approach does not always provide an ideal adaptation to the situation, dynamic computation is added for fine-grain adjustments.

3.3.5. USER INTERFACE SOFTWARE COMPONENTS

A number of software components are affected when adapting an interface for multi-targeting and plasticity. There is a large body of literature on this issue. However, because the software perspective is often mixed with the user's perception of adaptation, the state of the art does not provide a clear, unambiguous picture. For example, Dieterich *et al.* introduce five levels of adaptation: the lexical, syntactic, semantic, task and goal levels [Dieterich *et al.* 1993]. More recently, Stephanidis *et al.* define the lexical, syntactic and semantic levels of adaptation using examples [Stephanidis and Savidis 2001]. We propose to use Arch [Bass *et al.* 1992], a reference software architecture model, as a sound basis for characterizing software adaptation to target changes.

As shown in Figure 3.2, the Functional Core (FC) covers the domain-dependent concepts and functions. At the other extreme is the Physical Presentation Component (PPC), which is dependent on the toolkit used for implementing the look and feel of the interactive system. The PPC is in charge of presenting the domain concepts and functions in terms of physical interactive objects (also known as widgets or interactors). The keystone of the arch structure is the Dialog Control (DC) whose role consists of regulating task sequencing. For example, the Dialog Control ensures that the user executes the task open

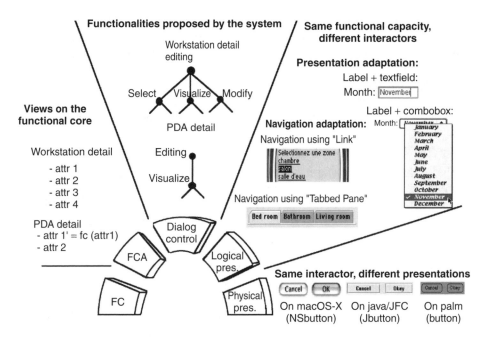

Figure 3.2. Arch architecture model.

document before performing any editing task. The FC, DC and PPC do not exchange data directly. Instead, they mediate through adaptors: the Functional Core Adaptor (FCA) and the Logical Presentation Component (LPC). The FCA is intended to accommodate various forms of mismatch between the Functional Core and the user interface. The Logical Presentation Component insulates the rendering of domain objects from the interaction toolkit of the target platform.

Using Arch as a structuring framework, the software components affected by multi-targeting and plasticity are the FCA, the DC, the LPC, the PPC, or a combination of them. In particular:

- At the Physical Presentation Component level, physical interactor classes used for implementing the user interface are kept unchanged but their rendering and behaviour may change across platforms. For example, if a concept is rendered as a button class, this concept will be represented as a button regardless of the target platform. However, the look and feel of the button may vary. This type of adaptation is used in the Tk graphical user interface toolkit as well as in Java/AWT with the notion of peers.
- At the Logical Presentation Component level, adaptation consists of changing the representation of the domain concepts. For example, the concept of month can be rendered as a Label + TextField, or as a Label + ComboBox, or as a dedicated physical interactor. In an LPC adaptation, physical interactors may change across platforms provided that their representational and interactional capabilities are equivalent. The implementation of an LPC level adaptation can usefully rely on the distinction between abstract

interactive objects and concrete interactive objects as presented in [Vanderdonckt and Bodard 1993].

- At the Dialogue Control level, the tasks that can be executed with the system are kept unchanged but their organization is modified. As a result, the structure of the dialogue is changed. AVANTI's polymorphic tasks [Stephanidis *et al.* 2001] are an example of a DC level adaptation.
- At the Functional Core Adaptor level, the nature of the entities as well as the functions exported by the functional core are changed. Zizi's semantic zoom is an example of an FCA level adaptation [Zizi and Beaudouin-Lafon 1994].

As illustrated by the above examples, Arch offers a clear analysis of the impact of a particular adaptation on the software components of a user interface.

3.3.6. USER INTERFACE MIGRATION

User interface migration corresponds to the transfer of the user interface between different platforms. It may be possible either at run-time or only between sessions:

- On-fly migration requires that the state of the functional core be saved as well as that of the user interface. The state of the user interface can be saved at multiple levels of granularity: when saved at the Dialogue Control level, the user can pursue the task from the beginning of the current task; when saved at the Logical Presentation or at the Physical Presentation levels, the user is able to carry on the current task at the exact point where they left off, and there is no discontinuity.
- When migration is possible only between sessions, the user has to quit the application, and then restart the application from the saved state of the functional core. In this case, the interaction process is interrupted. More research is required to determine how to minimize this disruption.

User interface migration between platforms places a high demand on the underlying infrastructure and toolkits. It also raises interesting user-centred design issues that should be addressed within the design process. Design phases are addressed next with the presentation of the Process Reference Framework.

3.4. THE PROCESS REFERENCE FRAMEWORK FOR MULTI-TARGET AND PLASTIC UIs

The Process Reference Framework provides designers and developers with generic principles for structuring and understanding the development process of multi-target and plastic user interfaces. We present an overall description of the framework in Section 3.4.1 followed by a more detailed expression of the framework applied to the design stage in Section 3.4.2. Different instantiations of the framework are presented in Section 3.4.3. Run-time architecture, which can be found in [Crease *et al.* 2000] and [Calvary *et al.* 2001b], is not discussed in this chapter.

3.4.1. GENERAL DESCRIPTION

As shown in Figure 3.3, the framework stresses a model-based approach coupled with a software development lifecycle.

3.4.1.1. Models and Lifecycle

Model-based approaches, which rely on high-level specifications, provide the foundations for code generation and code abstraction. This process of code generation and code abstraction reduces the cost of code production and code reusability while improving code quality.

The Process Reference Framework uses three types of models, where each type corresponds to a step of the lifecycle:

- *Ontological models* are meta-models that define the key dimensions of plasticity. They are independent from any domain and interactive system but are conveyed in the tools used for developing multi-target and plastic user interfaces. They are useful for the tool developer. When instantiated with tool support, ontological models give rise to archetypal models.
- *Archetypal models* depend on the domain and the interactive system being developed. They serve as input specifications for the design phase of an interactive system.
- *Observed models* are executable models that support the adaptation process at run-time.

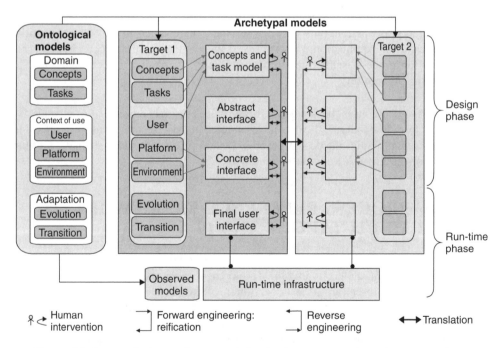

Figure 3.3. Process Reference Framework for the development of plastic user interfaces.

As shown in Figure 3.3, the *design phase* complies with a structured development process whose end result is a set of executable user interfaces (Final User Interfaces) each aimed at a particular archetypal target.

3.4.1.2. Coverage of the Models

As shown in Figure 3.3, the Process Reference Framework uses the following classes:

- *Domain models* cover the domain concepts and user tasks. Domain concepts denote the entities that users manipulate in their tasks. Tasks refer to the activities users undertake in order to attain their goals with the system.
- *Context of use models* describe a target in terms of user, platform and environment.
- *Adaptation models* specify how to adapt the system when the context of use and/or the target change. They include rules for selecting interactors, building user interface dialogues, etc.

These three classes of models (i.e., domain, context of use and adaptation models) may be ontological, archetypal or observed. As an illustration, in ARTStudio, the ontological task model is similar to the ConcurTaskTree concept [Breedvelt-Schouten *et al.* 1997], but is enhanced with decorations that specify the target audience. When instantiated as an archetypal task model, the ontological model can indicate that a given task does not make sense with a specific device and context, for example on a PDA in a train.

Having introduced the principles of the Process Reference Framework, we now present the framework as it is used in the design phase of multi-target and plastic user interfaces.

3.4.2. THE PROCESS REFERENCE FRAMEWORK IN THE DESIGN PHASE

In the design phase, the Process Reference Framework provides designers and developers with generic principles for structuring and understanding the development process of multi-target and plastic user interfaces. The design phase employs domain, context of use and adaptation models that are instantiations of the same models in the ontological domain. Archetypal models are referenced as well in the development process. As shown in Figure 3.3, the process is a combination of vertical reification and horizontal translation. Vertical reification is applied for a particular target while translation is used to create bridges between the descriptions for different targets. Reification and translation are discussed next.

3.4.2.1. Reification and Translation

Reification covers the inference process from high-level abstract descriptions to runtime code. As shown in Figure 3.3, the framework uses a four-step reification process: a Concept and Task Model is reified into an Abstract User Interface which, in turn, leads to a Concrete User Interface. The Concrete User Interface is then turned into a Final User Interface.

At the highest level, the *Concept and Task Model* brings together the concepts and task descriptions produced by the designers for that particular interactive system and that particular target.

An *Abstract User Interface* (Abstract UI) is a canonical expression of the rendering of the domain concepts and functions in a way that is independent of the interactors available for the target. For example, in ARTStudio, an Abstract UI is a collection of related workspaces. The relations between the workspaces are inferred from (i) the task relationships expressed in the Concept and Task model and (ii) the structure of the concepts described in the Concept model. Similarly, connectedness between concepts and tasks is inferred from the Concept and Task model. The canonical structure of navigation within the user interface is defined in this model as access links between workspaces.

A *Concrete User Interface* (Concrete UI) turns an Abstract UI into an interactor-dependent expression. Although a Concrete UI makes explicit the final look and feel of the Final User Interface, it is still a mock-up that runs only within the development environment.

A *Final User Interface* (Final UI), generated from a Concrete UI, is expressed in source code, such as Java and HTML. It can then be interpreted or compiled as a pre-computed user interface and plugged into a run-time infrastructure that supports dynamic adaptation to multiple targets.

A *translation* is an operation that transforms a description intended for a particular target into a description of the same class but aimed at a different target. As shown in Figure 3.3, translation can be applied between tasks and concepts for different targets, and/or between Abstract UIs, and/or Concrete UIs, and/or Final UIs.

Although high-level specifications are powerful tools, they have a cost. As observed by Myers *et al.* concerning the problem of 'threshold and ceiling effects' [Myers *et al.* 2000], powerful tools require steep learning curves. Conversely, tools that are easy to master do not necessarily provide the required support. Human intervention, decoration and factorisation, discussed next, can solve this dual problem.

3.4.2.2. Human Intervention

In the absence of tool support, reification and translation are performed manually by human experts. At the other extreme, tools can perform them automatically. However, full automation has a price: either the tool produces common-denominator solutions (e.g., standard WIMP UIs produced by model-based UI generators) or the designer has to specify an overwhelming number of details to get the desired results.

As shown in Figure 3.3, the Process Reference Framework addresses cooperation between human and tool as follows: the development environment infers descriptions that the designer can then adjust to specific requirements. For example, in ARTStudio, the designer can modify the relationships between workspaces, can change the layouts of the interactors, or even replace interactors. Decorations, presented next, provide another way to perform adjustments.

3.4.2.3. Decorations

A decoration is a type of information attached to description elements. Although a decoration does not modify the description per se, it provides information that modifies the interpretation of the description.

Applied to the design of multi-target and plastic UIs, decorations are used to express exceptions to standard or default conditions. Designers focus on the representative target, and then express deviations from the reference case study. We propose three types of decorations:

- *Directive decorations* correspond to rules that cannot be easily expressed in terms of general-purpose inference rules. For example, suppose the development environment includes the following generation rule: 'Any domain concept of type Integer must be represented as a Label in the Concrete UI'. If the designer wishes to represent the temperature domain concept with a gauge, then a directive decoration can be attached to that concept. Directive decorations are proactive. Corrective decorations are reactive.
- *Corrective decorations* can be used by designers to override standard options of the development environment. For example, suppose that, for workstations, the development environment generates the Final UI shown in Figure 3.5b. The designer can override the use of the thumbnail interactor, which gives access to one room at a time, by decorating the Concrete UI so as to obtain the presentation shown in Figure 3.5a, where all of the rooms are shown at the same time.
- *Factorisation decorations* express exceptions to the standard case. They are useful for expressing local differences without modifying the model. They are complementary to the factorisation function presented next, which has a global result on the model.

3.4.2.4. Factorisation

Figure 3.4 illustrates factorisation applied to task models. At the top are three task models for Java/AWT, HTML and WML-enabled target platforms. At the bottom left is the task model obtained from factorisation. At the bottom right is the task model obtained with factorisation decorations. Whereas factorisation produces three specific parts linked to a shared part, the use of factorisation description leads to a single shared description where exceptions are expressed as decorations.

Given a set of descriptions of the same class (e.g. a set of task models, of platform models, etc.), each aimed at a particular target, *factorisation* produces a new description composed of a description shared by all the targets and descriptions specific to each target. When applied to task models, factorisation corresponds to AVANTI's polymorphic tasks. Figure 3.4 illustrates the principles of factorisation.

Factorisation supports multi-targeting and plasticity in a 'one-target-at-a-time' approach. It allows designers to focus on one target at a time, and then to combine the descriptions produced independently for each target into a single description where common parts are factored out. Decoration supports a different approach where designers focus on a reference target and then define exceptions.

Having presented the general principles of the Process Reference Framework, we need to analyse how it can be instantiated.

3.4.3. INSTANTIATIONS OF THE PROCESS REFERENCE FRAMEWORK

The generic representation of the Process Reference Framework shown in Figure 3.3 can be instantiated in ways that reflect distinct practices and design cultures:

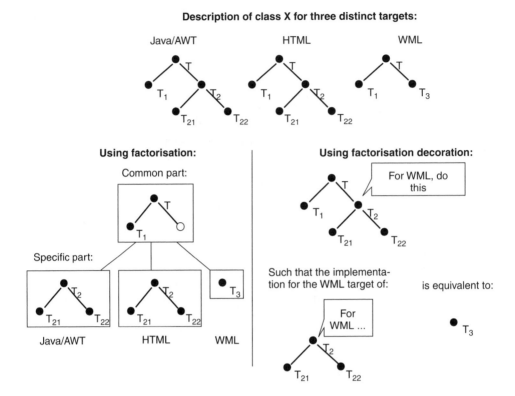

Figure 3.4. Factorisation function and factorisation decorations applied to task models.

- Reification and translation can be combined in different patterns. For example ART-Studio uses reification all the way through the final user interface (Figure 3.7).
- The reification process can start from any level of abstraction. Designers choose the entry point that best fits their practice. Typically, designers initiate the design by prototyping, i.e., by producing a concrete UI. If necessary, the missing abstractions higher in the reification process can be retrieved through reverse engineering as in Vaquita [Bouillon *et al.* 2002].
- The Process Reference Framework can be used to perform multi-targeting and plasticising through reverse engineering. Legacy systems, which have been designed for a particular target, must be redeveloped from scratch to support different targets. Alternatively, legacy systems can be reverse-engineered, then forward engineered. For example, an abstract UI can be inferred from a concrete UI. In turn, a concept and task model can be retrieved from an abstract UI. This is the approach adopted in WebRevEnge in combination with Teresa [Mori *et al.* 2002].

The following section describes how our software tool ARTStudio implements a subset of the framework.

3.5. ARTSTUDIO: AN APPLICATION OF THE PROCESS REFERENCE FRAMEWORK

ARTStudio (Adaptation through Reification and Translation Studio) is a software environment for the development of pre-computed multi-platform UIs. It supports elementary single screen platforms whose resources are known at the design stage. Therefore, distributed user interfaces are not addressed. It complies with the precepts of the Process Reference Framework, but implements a subset of its principles. Before discussing ARTStudio in more detail, we first present a case study: the EDF home heating control system.

3.5.1. THE EDF HOME HEATING CONTROL SYSTEM

The heating control system planned by EDF (the French Electricity Company) will be controlled by users in diverse contexts of use. These include:

- at home: through a dedicated wall-mounted device or through a PDA connected to a wireless home network;
- in the office: through the Web, using a standard workstation;
- anywhere: using a WAP-enabled mobile phone.

Target users are archetypal adult family members, and the target environments are implicit.

A typical user's task consists of consulting and modifying the temperature of a room. Figures 3.5 and 3.6 show the final UIs of the system for the target platforms. In Figure 3.5a, the system displays the temperature settings for each of the rooms of the house (the bedroom, bathroom and living room). Here, 24-hour temperature settings are available at a glance for every room of the house. The screen size is comfortable enough to make the entire system state observable. In 5b, the system shows the temperature settings of one room at a time. A thumbnail allows users to switch between rooms, and 24-hour temperature settings are displayed for a single room at a time. In contrast with 5a, the system state is not observable, but is browsable [Gram and Cockton 1996]. Additional navigational tasks, such as selecting the appropriate room, must be performed to access the desired information.

Figure 3.6 shows the interaction sequence for setting the temperature of a room with a WAP-enabled mobile phone. In 6a, the user selects the room (e.g., le salon – the living room). In 6b, the system shows the current temperature of the living room. By selecting the editing function ('Donner ordre'), users can modify the temperature settings of the selected room (6c).

In comparison to the scenario depicted in Figure 3.5a, two navigation tasks (i.e., selecting the room then selecting the edit function) must be performed in order to reach the desired state. In addition, a title has been added to every deck (i.e. WML page) to remind the user of their current location in the interaction space.

3.5.2. ARTSTUDIO

In its current implementation, ARTStudio supports the four-step reification process of the Process Reference Framework, as well as human intervention and factorisation

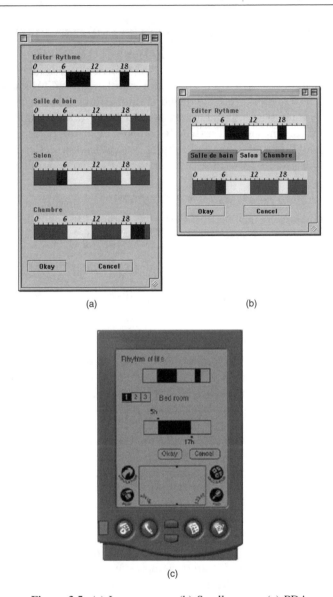

(a)   (b)

(c)

Figure 3.5. (a) Large screen; (b) Small screen; (c) PDA.

Figure 3.6. Modifying temperature settings using a WAP-enabled mobile phone.

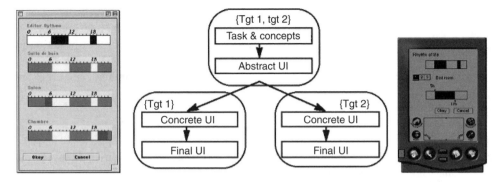

Figure 3.7. Instantiation of the Process Reference framework in ARTStudio.

(Figure 3.7). It does not support translation or decoration. Multi-targeting is limited to Java-enabled single-screen platforms and Web pages using a Java HTTP server. ARTStudio is implemented in Java and uses JESS, a rule-based language, for generating abstract and concrete UIs. All of the descriptions used and produced by ARTStudio are saved as XML files.

In Figure 3.7, the first and second design steps are common to all targets. The difference appears when generating the concrete user interface. ARTStudio does not use translation.

As illustrated in Figure 3.8, the development of a multi-platform plastic UI is called a project. A project includes the concept and task model, the context of use model, the concrete UI and the final UI.

3.5.2.1. Domain Concept Model

Domain concepts are modelled as UML objects using form-filling or XML schema. In addition to the standard UML specification, a concept description includes the specification

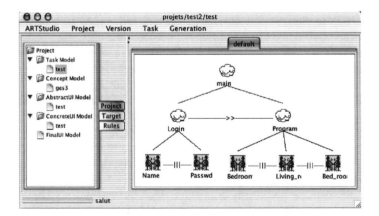

Figure 3.8. The task editor in ARTStudio. The picture shows the task model for a simplified version of the EDF Heating Control System.

by extension of its domain of values. For example, the value of the attribute *name* of type *string* of the *room* concept may be one value among others including *Living room, Bed room*, etc. The type and the domain of values of a concept are useful information for identifying the candidate interactors in the Concrete UI. In our case of interest, the *room* concept may be represented as a set of strings (Figure 3.5a), as a thumbnail (Figure 3.5b), or as dedicated icons.

3.5.2.2. Task Model

Task modelling is based on ConcurTaskTree [Breedvelt-Schouten *et al.* 1997]. The ART-Studio task editor allows the designer to add, cut and paste tasks by direct manipulation. Additional parameters are specified through form-filling. These include:

- Specifying the name and the type of a task. For interaction tasks, the designer chooses from a predefined set of universal interaction tasks such as 'selection', 'specification' and 'activation'. This set may be extended to fit the work domain (e.g., 'Set to anti-freeze mode').
- Specifying a prologue and an epilogue. This involves providing function names whose execution will be launched before and after the execution of the task. These functions are used when generating the final UI. They serve as gateways between the dialogue controller and the functional core adaptor of the interactive system. For example, for the task 'setting the temperature of the bedroom', a prologue function is used to get the current value of the room temperature from the functional core. An epilogue function is specified to notify the functional core of the temperature change.
- Referencing the concepts involved in the task and ranking them according to their level of importance in the task. This ordering is useful for generating the abstract and concrete user interfaces: typically, first class objects should be visible whereas second class objects may be browsable if visibility cannot be guaranteed due to the lack of physical resources.

3.5.2.3. Abstract UI Model

An abstract UI is modelled as a structured set of workspaces isomorphic to the task model: there is a one-to-one correspondence between a workspace and a task. In addition, a workspace that corresponds to an abstract task includes the workspaces that correspond to the subtasks of the abstract task: it is a compound workspace. Conversely, a leaf workspace is elementary. For example, Figure 3.9 shows three elementary workspaces (the bedroom, bathroom and living room) encapsulated in a common compound workspace. This parent workspace results from the Prog task of the task model. In turn, this workspace and the Rhythm elementary workspace are parts of the top-level workspace.

By direct manipulation, the designer can reconfigure the default arrangements. For example, given the task model shown in Figure 3.10a, the designer may decide to group the Rhythm workspace with the *Room* workspaces (Figure 3.10b) or, at the other extreme, suppress the intermediate structuring compound workspace (Figure 3.10c).

ARTStudio completes the workspace structure with a *navigation scheme* based on the logical and temporal relationships between the tasks. The navigation scheme expresses the user's ability to migrate between workspaces at run-time.

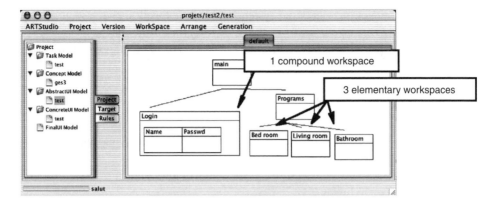

Figure 3.9. The Abstract UI generated by ARTStudio from the Task Model of Figure 3.8. Thick line rectangles represent compound workspaces whereas thin line rectangles correspond to elementary workspaces.

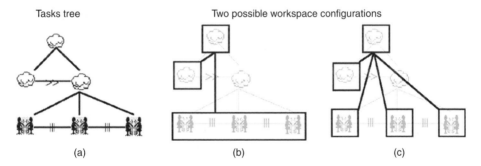

Figure 3.10. ARTStudio supports human intervention. (a) Task tree; (b), (c) Two possible workspace configurations. Workspaces can be reconfigured at will.

3.5.2.4. The Platform and Interactor Model

The platform model is a UML description that captures the size and depth of the screen, and the programming language supported by the platform (e.g., Java).

For each interactor available on the target platform, the interactor model specifies the representational capacity, interactional capacity, and the usage cost:

- Representational capacity: an interactor can either serve as a mechanism for switching between workspaces (e.g., a button, a thumbnail), or can be used to represent domain concepts (e.g., a multi-valued scale as in Figure 3.5). In the latter case, the interactor model includes the specification of the data type it is able to render.
- The interactional capacity denotes the tasks that the interactor is able to support (e.g., specification, selection, navigation).

• The usage cost, which measures the system resources as well as the human resources the interactor requires, is expressed as the 'x,y' footprint of the interactor on a display screen and its proactivity or reactivity (i.e. whether it avoids users making mistakes a priori or a posteriori) [Calvary 1998].

3.5.2.5. The Concrete UI Model

The generation of the concrete UI uses the abstract user interface, the platform and the interactor models, as well as heuristics. It consists of a set of mapping functions:

• between workspaces and display surfaces such as windows and canvases;
• between concepts and interactors;
• between the navigation scheme and navigation interactors.

The root workspace of the Abstract User Interface is mapped into a window. Any other workspace is mapped either as a window or as a canvas depending on the navigation interactor used for entering the workspace. Typically, a button navigation interactor opens a new window whereas a thumbnail leads to a new canvas. Mapping concepts to interactors is based on a constraint resolution system. For each of the concepts, ARTStudio matches the type and the domain of values of the concepts with the interactors' representational capacity, their interactional capacity and their usage cost.

Figure 3.11 illustrates the concrete UIs that correspond to the final UIs shown in Figure 3.5a for large screen targets (e.g., Mac and PC workstations) and in Figure 3.5b for PDA targets.

Concrete UIs, like abstract UIs, are editable by the designer. The layout of the interactors can be modified by direct manipulation. In addition, the designer can override the default navigation scheme. Pre-computed UIs can now be linked to a run-time environment that supports dynamic adaptation.

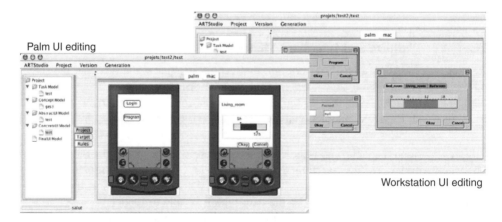

Figure 3.11. The concrete UI editor in ARTStudio.

3.6. CONCLUSION

This chapter covers the problem of multi-target and plastic user interfaces from the software development perspective. A taxonomic space makes explicit the core issues that need to be addressed in the development and run-time stages of multi-target and plastic user interfaces. This taxonomy is complemented with a conceptual framework, the Process Reference Framework, which helps in structuring the development process of multi-target and plastic UIs as well as supporting run-time execution. The Process Reference Framework helps to clearly characterize the functional coverage of current tools and identify requirements for future tools. Considerable research and development are needed to address the run-time infrastructure.

ACKNOWLEDGEMENT

This work has been partly supported by the EC project IST CAMELEON (IST-2000-30104). The notions of entry point and abstraction have been suggested by Jean Vanderdonckt and Fabio Paternò, members of the CAMELEON project.

REFERENCES

Akoumianakis, D. and Stephanidis, C. (1997) Supporting user-adapted interface design: The USE-IT system. *Interacting with Computers*, 9(1), 73–104, Elsevier.

Bass, L., Little, R., Pellegrino, R., *et al.* (1992) The Arch model: Seeheim revisited (version 1.0). The UIMS Developers Workshop (April 1991), in *SIGCHI Bulletin*, 24(1).

Beyer, H. and Holtzblatt K. (eds) (1998) *Contextual Design*. Morgan Kaufmann.

Bouillon, L., Vanderdonckt, J. and Souchon, N. (2002) Recovering Alternative Presentation Models of a Web Page with VAQUITA. Chapter 27 *Proceedings of 4th International Conference on Computer-Aided Design of User Interfaces CADUI, May 15–17, 2002, Valenciennes, France*. Dordrecht, Kluwer Academics.

Breedvelt-Schouten, I.M., Paternò, F.D. and Severijns, C.A. (1997) Reusable Structure in Task Models. *Proceedings of Eurographics Workshop on Design, Specification and Verification of Interactive System DSVIS, June 4–6, 1997, Granada, Spain*. Springer Verlag.

Browne, D., Totterdell, P. and Norman, M. (eds) (1990) *Adaptive User Interface*. Academic Press, Computer and People Series.

Calvary, G. (1998) Proactivité et Réactivité: de l'Assignation à la complémentarité en conception et evaluation d'interfaces homme-machine. Phd of the University Joseph-Fourier-Grenoble I, Speciality Computer Science.

Calvary, G., Coutaz, J. and Thevenin, D. (2001a) A Unifying Reference Framework for the Development of Plastic User Interfaces. *Proceedings of 8th IFIP International Conference on Engineering for Human-Computer Interaction EHCI, May 11–13, 2001, Toronto, Canada*. Lecture Notes in Computer Science, 2254. Springer-Verlag.

Calvary, G., Coutaz, J. and Thevenin, D. (2001b) Supporting Context Changes for Plastic User Interfaces: a process and a mechanism. *Proceedings of the joint AFHIM conference on Interaction Homme Machine and British HCI Group conference on Human Computer Interaction HCI* (eds A. Blandford, J. Vanderdonckt, P. Gray). BCS conference series. Springer.

Cheverst, K., Davies, N., Mitchell, K. and Efstratiou, C. (2001) Using Context as a Crystal Ball: Rewards and pitfalls. *Personal and Ubiquitous Computing*, 5(1), 8–11.

Cockton, G., Clarke S., Gray, P. and Johnson, C. (1995) Literate Development: Weaving human context into design specifications in *Critical Issues in User Interface Engineering* (eds P. Palanque and D. Benyon). London, Springer-Verlag.

Coutaz, J., Lachenal, C. and Rey, G. (2002) Initial Reference Framework for Interaction Surfaces, Version V1.0, GLOSS Deliverable D17.

Coutaz, J. and Rey, G. (2002) Foundations for a Theory of Contextors. *Proceedings of 4th International Conference on Computer-Aided Design of User Interfaces CADUI, May 15–17, 2002, Valenciennes, France*, Dordrecht, Kluwer Academics.

Crease, M., Gray, P. and Brewster, S. (2000) A Toolkit Mechanism and Context Independent Widgets. *Proceedings of ISCE Workshop on Design, Specification and Verification of Interactive System DSVIS, June 5–6, 2000, Limerick, Ireland*. Springer Verlag.

Crowley, J., Coutaz, J. and Bérard, F. (2000) Things that See. *Communication of the ACM*, 43(3), 54–64.

Dey, A., Abowd, G. and Salber, D. (2001) A Conceptual Framework and a Toolkit for Supporting the Rapid Prototyping of Context-Aware Applications. *Human-Computer Interaction*, 16(2–4), 97–166, Lawrence Erlbaum.

Dieterich, H., Malinowski, U., Kühme, T. and Schneider-Hufschmidt, M. (1993) State of the Art in Adaptive User Interfaces in *Adaptive User Interfaces, Principles and Practice* (eds H.J. Bullinger, P.G. Polson) Human Factors in Information Technology series, Elsevier.

Eisenstein J., Vanderdonckt J. and Puerta A. (2001) Applying Model-Based Techniques to the Development of UIs for Mobile Computers. *Proceedings of ACM Conference on Intelligent User Interfaces IUI, 14–17, 2001, Santa Fe, New Mexico, USA*. ACM Press.

Gram, C. and Cockton, G. (eds) (1996) *Design Principles for Interactive Software*. Chapman & Hall.

Grolaux, D. (2000) FlexClock. Available at: http://www.info.ucl.ac.be/people/ned/flexclock/

Hourcade, J. and Bederson, B. (1999) *Architecture and Implementation of a Java Package for Multiple Input Devices (MID)*. Available at: http://www.cs.umd.edu/hcil/mid/

Johnson, P., Wilson, S., Markopoulos, P. and Pycock, Y. (1993) ADEPT–Advanced Design Environment for Prototyping with Task Models. *Proceedings of the joint ACM Conference on Human Factors in Computing Systems CHI and IFIP Conference on Human Computer Interaction INTERACT, April 24–29, 1993, Amsterdam, The Netherlands*. ACM Press.

Lim, K.Y. and Long, J. (eds) (1994) *The MUSE Method for Usability Engineering*. Cambridge University Press.

Moran, T. and Dourish, P. (eds) (2001) Context-Aware Computing. Special Issue of *Human Computer Interaction* 16(2–4), Lawrence Erlbaum.

Mori, G., Paganelli, L., Paternò, F., *et al.* (2002) *Tools for Model-Based Design of Multi-Context Applications, September 5, 2002*. CAMELEON Document, Deliverable 2.1.

Myers, B., Hudson, S. and Pausch, R. (2000) Past, Present, Future of User Interface Tools. *Transactions on Computer-Human Interaction*, 7(1), 3–28.

Rekimoto, J. and Saitoh, M. (1999) Augmented Surfaces: A spatially continuous workspace for hybrid computing environments. *Proceedings of the ACM conference on Human Factors in Computing Systems CHI, May 15–20, 1999, Pittsburgh, PA, USA*. ACM Press.

Stephanidis, C., Paramythis, A., Sfyrakis, M. and Savidis, A. (2001) A Case Study in Unified User Interface Development: *The AVANTI web browser in User Interface for All: Concepts, methods and tools* (ed. C. Stephanidis). Lawrence Erlbaum.

Stephanidis, C. and Savidis, A. (2001) Universal Access in the Information Society: Methods, tools, and interaction technologies. *Universal Access in the Information Society UAIS*, 1(1), 40–55.

Streitz, N., Geißler, J., Holmer, T. *et al.* (1999) I-LAND: An interactive landscape for creativity and innovation. *Proceedings of the ACM conference on Human Factors in Computing Systems CHI, May 15–20, 1999, Pittsburgh, PA, USA*. ACM Press.

Szekely, P. (1996) Retrospective and Challenges for Model-Based Interface Development. *Proceedings of the 2nd International Workshop on Computer-Aided Design of User Interfaces CADUI, June 5–7, 1996, Namur, Belgium* (ed. J. Vanderdonckt). Presses Universitaires de Namur.

Tandler, P. (2001) Software Infrastructure for Ubiquitous Computing Environments: Supporting synchronous collaboration with heterogeneous devices. *Proceedings of 3rd ACM conference on Ubiquitous Computing Ubicomp, September 30 – October 2, 2001, Atlanta, Georgia, USA.* Springer Verlag.

Thevenin, D. (2001) *Adaptation en Interaction Homme-Machine: le cas de la Plasticité.* Phd Thesis of the University Joseph-Fourier Grenoble I, Speciality Computer Science.

Thevenin, D. and Coutaz, J. (1999) Plasticity of User Interfaces: Framework and Research Agenda. *Proceedings of IFIP Conference on Human Computer Interaction INTERACT.* August 30 – September 3, 1999, Edinburgh, Scotland (eds Sasse, A. and Johnson, C.). IFIP IOS Press.

Vanderdonckt, J. (1995) *Knowledge-Based Systems for Automated User Interface Generation: The TRIDENT Experience.* RP-95-010, University Notre-Dame de la Paix, Computer Science Institute, Namur, Belgium.

Vanderdonckt, J. and Bodard, F. (1993) Encapsulating Knowledge for Intelligent Automatic Interaction Objects Selection. *Proceedings of the joint ACM Conference on Human Factors in Computing Systems CHI and IFIP Conference on Human Computer Interaction INTERACT, April 24–29, 1993, Amsterdam, The Netherlands.* ACM Press.

Winograd, T. (2001) Architecture for Context. *Human Computer Interaction,* 16(2–4), 401–419. Lawrence Erlbaum.

Zizi, M. and Beaudouin-Lafon, M. (1994) Accessing Hyperdocuments through Interactive Dynamic Maps. *Proceedings of the European Conference on Hypertext Technology ECHT, September 19–23, 1994, Edinburgh, Scotland.* ACM Press.

Temporal Aspects of Multi-Platform Interaction

David England and Min Du

School of Computing and Mathematical Sciences, Liverpool John Moores University, UK

4.1. INTRODUCTION

Throughout the history of Human-Computer Interaction research there have been many attempts to develop common design methods and deployment technologies to cope with the range of possible interaction platforms. The earliest of these attempts was [Hayes *et al.* 1985], which provided the first executable user interface description language. Their tool, COUSIN, provided rules for controlling the automatic layout of interfaces from application descriptions. COUSIN finds modern echoes in recent research based on XML and XSL technology. In this research, interfaces are described in XML. Style-sheets, written in XSL, are used to transform the generic XML description into a device-specific layout. For example, the work of [Luyten and Coninx 2001] uses XML to create interfaces for mobile platforms. The chief limitation of this approach is that it focuses on layout issues, or more generally, visual presentation issues. That is, it uses rules, usually expressed in XSL, to trade-off presentation issues between the possible display areas. The main technique of presentation adaptation has been to degrade the presentation from designs for the highest-level device to the lower-level platforms. For example, the lower-level

Multiple User Interfaces. Edited by A. Seffah and H. Javahery
© 2004 John Wiley & Sons, Ltd ISBN: 0-470-85444-8

platform may use reduced image sizes, or the breadth and depth of navigation structures would be altered. In the latter case we would have traded-off display space for the time it takes to access menu items lower down the menu structure. This kind of presentation adaptation work has until recently ignored the issues of user task support. More recent work by [Pribeanu *et al.* 2001] has attempted to model the task structures that users employ and match these to appropriate user interface components. Again the main focus has been on the presentation aspects of adapting the design to the display limitations of the target platform. In theory, such work should allow the user's full task structure to be properly adapted to a range of platforms without too many compromises in end-user performance. However, such work is in its early stages and currently the majority of cross-platform adaptation has to be done manually, in the design and implementation stages of interaction development.

In our own work we have focused on task modelling using the eXecutable User Action Notation or XUAN [Gray *et al.* 1994]. XUAN is based on the User Action Notation [Hartson and Gray 1992]. XUAN permits the modelling of temporal aspects of interaction such as the durations of user tasks and the relations between user tasks. Other researchers have used Petri nets [Turnell *et al.* 2001], UML [Navarre *et al.* 2001] and other systems-based, concurrency notations, such as CSP [Hoare 1984], to address temporal issues of interaction. However, such approaches move the focus of interaction design too far towards the systems level, and away from pure user interaction concerns. The advantage of XUAN, for user interface designers, is that it is based on Allen's temporal logic [Allen 1984]. Allen's model is an artificial intelligence model of the human planning of temporal intervals. As such it is a better level of abstraction for reasoning about user tasks, than systems level models that deal with system-level or computational artefacts. We have found XUAN to be a useful tool for reasoning about such issues as shared interaction in virtual environments [England and Gray 1998] and multi-user interaction in medical settings [England and Du 2002]. As an executable notation, XUAN also supports the specification–prototype–evaluation loop which is an important part of the interactive systems development cycle. More recently we have realised some of the limitations of XUAN in terms of allowing the repeated description of common interaction scenarios. We have adopted the concept of patterns [Gamma *et al.* 1995] to develop a Pattern-User Action Notation [Du and England 2001], which allows us to specify re-usable, temporal specifications of interaction. Common interaction scenarios are specified in PUAN and then parameterised to adopt the scenario to a specific temporal context. The term 'pattern' has been adopted in different ways by researchers in Human-Computer Interaction. In our case, we follow the software engineering tradition of Gamma and our patterns are formal, hierarchical descriptions of interaction scenarios which concentrate on the temporal aspects of interaction. To be clear, our concept of 'pattern' is not that derived from [Alexander *et al.* 1977] which are informal templates for designers and end-users to use together.

It is this aspect of PUAN that we will concentrate on in the rest of this paper, namely, the ability to parameterise scenarios across different platforms by making use of the knowledge of the different temporal contexts of multiple platforms. In the next section we shall informally introduce some of the temporal issues of cross-platform interaction. We shall then go on to their more formal descriptions in PUAN. Finally, we will discuss the implications for our work for user interface designers, the designers of the platforms and future HCI research.

4.2. TEMPORAL CONTEXTS OF MULTIPLE PLATFORMS

Let us now, informally, look at the temporal characteristics of interactive platforms that can affect their users' performance. We have divided these into temporal characteristics of:

- Input/output devices;
- Computational power;
- Operating system capabilities and the associated user interface layer;
- Communications capabilities.

Initially we shall consider each of these factors separately but as the discussion progresses it will become obvious that there are overlap and interactions between these factors. In addition, some of these factors are static to the platform while others can change with platform usage.

4.2.1. FITTS' LAW AND THE CONTROL:DISPLAY RATIO

If we firstly consider the characteristics of input and output devices, the most important relationship is that between the chief visual output mode and the main input mode. In all cases this relationship is governed by two main factors, Fitts' Law [Fitts 1954] and the control:display ratio of the input device. Fitts' Law is a fundamental tenet of graphical human-computer interaction and states that, the time to acquire a target is a function of the distance to and size of the target. What this means for screen design, is that the less movement a user has to make to acquire a user interface component and the larger the component is, the more likely it is that the user will hit the component without error. This also has implications for the way that menus and sub-menus are designed. Walker and Smelcer [1990] show how attention to the design of menus and sub-menus can ease menu item selection. However, for our current discussion the main point of interest is the comparison of different sizes of components on different display devices.

A simplistic interpretation of Fitts' Law would be to re-scale interface components when we move applications between interaction platforms. For a smaller display we would need less distance to move and could have smaller targets (down to some limiting readable size) and still retain the same performance as the larger platform. However, this would ignore the second important factor, the control:display ratio. A typical, desktop graphical interface is driven by a mouse, which is an indirect, relative positioning device. A typical control ratio between mouse movement and on-screen cursor movement is 1:3. A typical PDA is driven by a touch pen, which is a direct, absolute positioning device. Thus a user could move 10 cm with either input device, and move across the complete screen. The PDA user would be faced with smaller user interface components and thus, Fitts' Law would predict that they would commit more interaction errors and take longer to carry out operations, than the desktop user. An additional factor that comes into play is which muscle group the user is employing to control the device. Different muscle groups have different levels of accuracy. For example, fingers, wrists and arm joints are each increasingly less accurate and are used to control pens, mice and whiteboard pointers respectively. So we would expect more errors and a slower task time depending on which muscle group was used to control the input device. Thus we can see that when we are designing an application for a number of platforms, there are a number of factors that have to be borne in mind. If we were designing a development tool to take account of the effects of these factors on the temporal aspects of interaction, it would need to be able to express and control these factors during design and prototyping.

4.2.2. COMPUTATION SPEED OF THE PLATFORM

The computational power of platforms has an obvious effect on the functionality that can be supported and the response time that users can expect from a platform. This is true even when we are moving from platforms of the same kind (e.g. desktop PCs) of differing CPU and/or GPU (Graphical Processor Unit) speeds. This fact is often overlooked by designers who specify progress indicators and busy signals to run at a speed relative to the machine's speed rather than in the user's real-time [Dix 1987]. Feedback on the successful or unsuccessful completion of a task has also been recognised since [Norman 1988] as an important aspect of interaction. One example is the email delivery indicator on Microsoft Outlook 98. The pop-up signal for successful connection to a mail server, which was visible to the user on a 200 MHz machine, flashed by unseen on an 800 Mhz machine. Thus the user would be unsure whether the task had completed successfully unless there were some other indicators. When we are designing any form of user interface feedback we have to take into account the speed at which the underlying platform can deliver the feedback and that information about task progress and completion is delivered in the user's time frame. We can model the interaction effects of Fitts' Law and computation power in our notation, XUAN, and we will show examples of this below.

4.2.3. SUPPORT FOR TASK SWITCHING ON PLATFORMS

Whereas response time and feedback can be modelled by temporal durations in XUAN, the scheduling capabilities of operating systems and their effects on user tasks can be modelled by temporal relations. At the highest end of capability we have operating systems that support true concurrency and multi-user interaction, e.g. UNIX and its descendants. Next, we have operating systems which support single-user task switching but not true concurrency, e.g. Pocket PC 2002, and lastly, single task operating systems, such as those in the most basic mobile phones.

If we have a complex task model for an application (or set of applications) we can see that users would have to employ different strategies to complete their application depending on the support for task interleaving on their platform's operating system. Imagine a user who is creating a document with downloaded images which is then emailed to a colleague. In the more advanced operating system the user can begin all tasks simultaneously and continue with the document while the images are downloading. Once the images are downloaded and the document complete, they can send their message. Under an operating system such as Pocket PC, however, the user must adopt a different strategy; they can switch between (interleave) the three different tasks but background tasks are suspended while the user interacts with the main task. On the simplest platform the user must adopt the strategy of planning what sequence to attempt and complete the tasks, as only one task can be in operation at any one time. Each task must be completed in its entirety before the next is begun. Thus the scheduling capabilities of the operating system force modifications in the task structure [O'Donnell and Draper 1996], and we need to be able to model these modifications in our designs for multi-platform interaction. Java, with its multiple thread support is becoming increasingly available on a range of lower powered platforms and can offer task switching where it is not supported by the native operating system. In our own work we have built a Java engine to execute PUAN descriptions and we will show later how this can be used to support a greater range of temporal relations between tasks across platforms.

Our fourth and final platform capability is the communications layer of the platform. Again there is a vast range of communication speeds ranging from high speed, gigabyte research and corporate networks to slower, kilobit speed mobile phone connections. We need to take account of the communications rate and its effect on the user's interaction with and through the platform to other users and remote applications. The communications rate affects not only interaction but also the communications protocol. For example, most Internet connections work on a best-effort connection trying to make the best use of the available bandwidth. This may not be suitable for all applications, e.g. video serving where a resource reservation protocol may be more appropriate. In addition, available bandwidth may vary with usage and over time, and again we need to model this possibility in our interaction scenarios.

Of course, each of these factors interacts. The platform response time for client-server applications is a combination of the local platform power and its communications rate. The predictability of response time is a combination of the state of the local platform and the state of its network connection. Whether a user can adapt their strategy, and switch tasks while waiting for a delayed task or communication, depends on the operating system. The time it takes for a user to complete any given task is a combination of the time taken to operate the input device, the time it takes to get a response from the platform and/or network and the task management overhead of the user scheduling and planning their tasks on the platform. In the next section we will show how we can model interaction scenarios taking these factors into account. We will show how we can parameterise platforms so that we can adopt user task models to different platforms.

4.3. MODELLING TEMPORAL CONTEXTS

In the previous section we gave a partial, informal picture of the temporal aspects of multiple platform interaction. This is similar to the majority of the guideline-driven advice that many HCI practitioners seek out and which some HCI researchers provide. However, if we wish to move HCI forward as an engineering discipline we need to continue to codify and validate good designs. Adopting PUAN, and similar notations, is one way of codifying designs. In this section we show how we can represent some of the previous informal descriptions in PUAN. PUAN borrows the table format of UAN for part of its representation. The tables have three columns for user actions, system feedback and system (or agent) state. PUAN, like UAN, tries to be neutral about the cognitive state of the user(s) and the application logic of the system. These separate concerns are dealt with in the task (or user object) modelling phase and the systems development phases, respectively. This separation of concerns allows the designer to concentrate on human-computer interaction issues. In the tables there is a partial ordering of events from left-to-right, and top-to-bottom. Additionally, with PUAN, we can add temporal constraints to further specify the temporal behaviour of the interface.

In the following descriptions we can parameterise the device-independent patterns in order to adapt them to the temporal context of use of a specific platform. Thus, we can re-use platform-independent interaction patterns across a range of platforms. However, we still need to assess the validity of the derived pattern. We can do this by inspecting the parameterised pattern and by executing the pattern and evaluating the resulting user interface.

4.3.1. ACTION SELECTION PATTERN

We first describe our platform-independent patterns. These are rather simple but will prove useful when we come to our high-level patterns. Firstly, we have our display driver pattern that models the interaction between the display device and the input device. This is the base pattern that supports all user input face component selection operations (e.g. buttons, menu items, or text insertion points). Here we are modelling the low level Fitts' Law and the control:display ratio for inclusion in higher-level patterns. Our Action Selection Pattern (Table 4.1) describes the simple steps of moving an input device, getting feedback from the display and acquiring a displayed object:

Here x and y are input control coordinates and sx and sy are the initial screen coordinates, and x', y', sx' and sy' the final coordinates on hitting the object. Adding additional temporal constraints we can say, firstly that we have a sequence of steps (separated by',') where the `move_input_1` and `move_cursor` steps are repeated until an activation event (e.g. pressing a mouse button, tapping a pen) is received:

```
(Move input_1(x,y), Move_Cursor(sx,sy))*, Activate Input_1(x',y'),
    System_action(sx',sy'), Feedback_action(sx',sy')
```

Table 4.1. Action selection pattern.

User	System Feedback	System State
Move input_1(x,y)		
	Move_Cursor(sx,sy)	
Activate input_1(x',y')		If (sx,sy) intersects UI_object { System_action(sx',sy')
	Feedback_action(sx',sy') }	

Secondly we can have an indication of the control:display ratio from

```
Control movement = sqrt((x'-x)^2 + (y'-y)^2)
Screen movement = sqrt((sx'-sx)^2 + (sy'-sy)^2)
C:D = Control movement/Screen movement
```

We can now compare device interactions according to Fitts' Law by stating that the time to select an object is proportional to the control distance moved and the size of the `UI_Object`, i.e.

```
End(Activate Input_1) - Start(move_input_1) ∝ Control Movement &
    UI_Object.size()
```

So when we are designing displays for different platforms we can look at the Control movement distances and the size of the target objects, and by making estimates for the end and start times, we can derive estimates for the possible degrees of error between usages of different platforms. Or more properly, we can make hypotheses about possible differing error rates between platforms and test these empirically during the evaluation phase. For a given platform, inspection of the above formula and empirical testing will give us a set of values for Control Movement and `UI Object.size()` which will minimise object selection times.

Now this pattern is not entirely device-independent as it expects a pointing device (mouse, pen, foot pedal) and corresponding display. We could quite easily replace these steps with, say, voice-activated control of the platform, as in the 'Speech Action Selection' (Table 4.2) pattern. The temporal issues then would be the time to speak the command, the time for the platform to recognise the command and the time taken to execute the recognised command. We can go further and specify concurrent control of operations by pointing device and voice-activation.

4.3.2. PROGRESS MONITORING PATTERN

As Dix [1987] pointed out, there are no infinitely fast machines, even though some user interface designers have built interfaces on that assumption. Instead we need to assume the worst-case scenario that any interaction step can be subjected to delay. We need to inform the user about the delay status of the system. For designs across interaction platforms, we also need a base description of delayed response that can be adapted to

Table 4.2. Speech action selection pattern.

User	System Feedback	System State
(Issue_Command(text)		If Recognise (text)
	Then {Affirm_recognition	System_action (text)}
	Else Issue(repeat prompt)	
	Feedback_action(text))*	

differing delay contexts. The Progress Monitoring Pattern (Table 4.3) deals with three states of possible delay;

1. The task succeeds almost immediately;
2. The task is delayed and then succeeds;
3. The task is subjected to an abnormal delay.

We can represent this in PUAN as follows:

Table 4.3. Progress monitoring pattern.

User	System Feedback	System State
`Action Selection`		`Begin = end(Action Selection.Activate input_1)` `End = start(Action Selection.Feedback_Action)`
	`If (End-Begin) < S` `Show Success` `If (End-Begin) > S &&` `(End-Begin) < M` `Show Progress Indicator` `Else Show Progress Unknown`	
`Cancel Action`	`Show Cancellation`	`System Action.End()`

Here we begin with the complete pattern for Action Selection. Part of the system state is to record the end of the user's activation of the task and the start of the display of feedback. We react to our three possible states as follows:

1. The task succeeds almost immediately:
   If the beginning of feedback occurs within some time, S, from the end of task activation, show success.
2. The task is delayed and then succeeds:
   If the beginning of feedback does not occur within the time interval, S to M, from the end of task activation, show a progress indicator.
3. The task is subjected to an abnormal delay:
   If the beginning of feedback does not occur within the time, M, from the end of task activation, show a busy indicator.

There are some further temporal constraints we wish to specify. Firstly, that the calculation of *End* is concurrent with *Show Progress Indicator* and *Progress Unknown*, i.e.

```
(Show Progress Indicator, Progress Unknown) || End = start(Action
    Selection.Feedback_Action)
```

Secondly, that *Show Progress Indicator* and *Progress Unknown* can be interrupted by Cancel Action:

```
(Show Progress Indicator, Progress Unknown) <= Cancel Action.
```

The latter step is often missed out in interaction design and users lose control of the interaction due to blocking delays, even when there are no issues for data integrity by cancelling an action.

The choice of values for S and M is where we parameterise the pattern for different platforms and contexts. However, S and M are not simple, static values. They are a combination of sub-values which may themselves vary over time. Each of S and M is composed of the sub-values:

$S_{display}$, $M_{display}$ the time to display the resulting information

$S_{compute}$, $M_{compute}$ the time to perform the computation for the selected action

$S_{transfer}$, $M_{transfer}$ the time to get a result from a remote host

$S = S_{display} + S_{compute} + S_{transfer}$

$M = M_{display} + M_{compute} + M_{transfer}$

Just as there are no infinitely fast machines, so there are no networks of infinite speed or infinite bandwidth so $S_{transfer}$ and $M_{transfer}$ become important for networked applications. And, the choice of values for $S_{transfer}$ and $M_{transfer}$ also depend on the size and nature of the information (static images, streamed media) that is being transferred.

4.3.3. TASK MANAGEMENT PATTERN

Our final pattern example considers the different strategies that need to be supported for the management of task ordering and switching on different platforms. In our foregoing discussion we presented three different platforms offering full- to no task switching. In the general case a user may be performing, or attempting to complete, N tasks at any one time. The most fully supported case, ignoring data dependencies between tasks, is that all tasks can be performed in parallel.

$$A_1 \| A_2 \dots \| A_N$$

The next level of support is for task interleaving without true concurrency, i.e.

$$A_1 \Leftrightarrow A_2 \dots \Leftrightarrow A_N$$

Finally the lowest level of support is for strict sequence-only

$$A_1, A_2 \dots, A_N$$

These temporal relational constraints represent the base level of task management on particular platforms. If we begin to look at particular sets of tasks, we can see how the temporal model of the tasks can be mapped onto the temporal constraints of the platform. For example, consider our earlier discussion of writing a document that includes downloaded images which we are then going to send by email. The task sequence for the fully supported case is

```
(write document || download images), send email
```

With the next level of support the user loses the ability to operate concurrently on downloading and document writing, i.e.

```
(write document ⇔ download images), send email
```

In the least supported case, it is up to the user to identify the task data dependencies (e.g. the images required for the document) in order to complete the sequential tasks in the correct order of

```
download images, write document, send email
```

For the designer we have a parallel here to the problems of degrading image size and quality when moving to platforms of lower display presentation capabilities. In the case of task management we have a situation of degrading flexibility, as task-switching support is reduced. How can we help users in the more degraded context? One solution would be for the applications on the platform and the platform operating system to support common data dependencies, i.e. as we move from one application to another, the common data from one application is carried over to the next.

We represent the degradation of temporal relations in the three different contexts in Table 4.4. For some relations the degradation to a relation of flexibility is straightforward. For others it involves knowledge of the data dependencies between the tasks.

Thus, when we are designing applications we can use the above table to transform task management into the appropriate context.

4.3.4. PLATFORM INTERACTION PATTERN

Finally we can represent the overlapping of the issues of the different factors affecting temporal interaction with an overall pattern, Platform Interaction.

```
Task Management ⇔mapping (Action Selection ||mapping Progress Monitoring)*
```

Table 4.4. Mapping temporal relations to different task switching contexts.

Temporal relation	Full concurrency	Task switching	Sequential only
Concurrent ||	||	⇔	, data dependent
Interleavable ⇔	⇔	⇔	, data dependent
Order independent &	&	&	,
Interruptible ->	->	->	,
Strict Sequence ,	,	,	,

Where $\Leftrightarrow_{mapping}$ and $\|_{mapping}$ are the mappings of the temporal relations into the platform task switching context. That is, we have a repeated loop of action selection and monitoring of the progress of the actions under the control of the task management context of the platform.

4.4. THE TEMPORAL CONSTRAINT ENGINE

Notations for user interface design, as with PUAN presented above, are useful in themselves as tools for thinking about interaction design issues. However, in our current state of knowledge about user interface design, much of what is known and represented is still either informal or too generic in nature to give formal guidance to the designer. Thus, we still need to perform empirical evaluations of interfaces with users, and this process is shown in Figure 4.1. Using notational representations with evaluation help us to focus the issues for the evaluation phase. In addition, we can use notations as a means of capturing knowledge from evaluation to reuse designs and avoid mistakes in future projects. In our work we use a prototype, Java-based temporal constraint engine to validate our PUAN descriptions and to support the evaluation of interface prototypes with users. The Java. PUAN engine is compiled with the candidate application and parses and exercises the PUAN temporal constraint descriptions at run-time.

So for our examples above, e.g. Progress Monitoring, the values of S and M would be set in the PUAN text which is interpreted by the Java.PUAN engine. The values of S and M are then evaluated at run-time and are used to control the threads which support the tasks of the Java application. The constraint engine checks the start and end times of the relevant tasks to see if they are within the intervals specified by S and M, and executes the corresponding conditional arm of the constraint accordingly. We can evaluate an application with different temporal conditions by changing the PUAN text and re-interpreting it. The Java application itself does not need to be changed. In addition to changing simple values we can also change temporal constraints and relations. We could in fact simulate the temporal characteristics of multiple platforms simply by changing the PUAN description of an application, i.e. by supplying the appropriate temporal relation mapping to the overall platform interaction pattern.

In our work so far [Du and England 2001] we have used the Java.PUAN engine to instantiate an example of a user employing a word processor, file transfer program and email program with different cases of temporal constraints on task switching between the

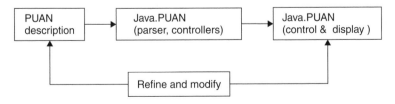

Figure 4.1. Process of PUAN evaluation and refinement.

tasks. We are currently working on a more substantial example which models multiple tasks amongst multiple users in an A&E (Accident and Emergency or Emergency Room) setting. Here we have modelled the different tasks and their interactions between different users. The next stage is to support this model on multiple platforms, namely, desktops for reception staff, PDAs for doctors and a whiteboard for patient information [England and Du 2002].

We make use of lightweight, Java threads to express concurrency in the candidate application. The use of the Java virtual machine, JVM, means we avoid some of the problems of task switching context between platforms, as most JVMs fully support Java threads. Even the CLDC (connection limited device configuration) [Sun 2002] supports threads, all be it with some restrictions on their use. However, the majority of today's applications are not built using Java so we still need the ability to map task-switching contexts when we are modelling the majority of applications. Our research looks forward to the time when most operating systems support lightweight threads or processes. Our constraint engine has some limitations on what it can capture. It cannot capture events that are not reported by the underlying system. For example, Java threads may have different priorities on different virtual machines and some events may go unreported if they occur between machine cycles. Our constraint engine works in 'soft' real-time, that is, it can only report the expiration of time intervals expressed in PUAN; it cannot enforce them. It is left to the designer to have a method for coping with broken constraints. Finally, our constraint engine does not offer any magical solution to the standard problems of concurrent processes, such as deadlock and mutual exclusion. Again, it is currently left to the designer to analyse such situations and avoid them.

In our future work with the Java.PUAN engine we are looking at supporting interaction with networked appliances, which means focusing more on CLDC-style platforms. More particularly we are looking at mobile control platforms for networked appliances. As part of this we are considering the user interface 'handshaking' between the control platforms and the appliance so that the appliance can parameterise the temporal parameters of the control platform, as they meet on an ad hoc basis.

4.5. DISCUSSION

We have presented some representational solutions to dealing with the temporal issues of interaction across different interaction platforms. We have presented some basic interaction patterns written in PUAN, which can be parameterised to set the platform interaction context for an application which migrates across different platforms. These parameters can be static for a platform for all applications or they can be dynamic and adjusted according to the platform and application contexts. We have used the word 'task' throughout our discussion without much definition. This has been deliberate as, from our engine's point of view, a user task can be mapped on to any computational task that can be represented in a Java thread. Thus, the actions set off by a button, menu item or other user interface component could run concurrently in the appropriate task-switching context. However, for most designers, their control over user tasks is limited to the process level of the particular operating system, and by the permitted level of application interrupts. For most designers, this means they cannot fully exploit the potential of concurrency in allowing users to choose their own strategies in task switching. However, as machines become increasingly concurrent and multi-modal at the user interface level, user interface designers will face greater challenges to approaching and dealing with concurrency-enabled interfaces in a disciplined way. We believe that PUAN, and similar future notations, offer designers a disciplined framework with which to approach user interface concurrency and temporal interface design.

A common question about our work is why we did not start with notation 'X' instead of XUAN? We hope we have justified our use of XUAN and its foundation in temporal logic, as presenting the most appropriate level of abstraction for dealing with the issues discussed here. We would like to dismiss discussions of using XML and XSL combinations, as, from a research point of view, these languages are just a manifestation of decades-old parsing and compilation techniques that go back to Lex and YACC. In other words, they may make our work slightly more accessible but they do not address any of the conceptual issues we are trying to investigate.

4.6. CONCLUSIONS

Temporal issues of interaction are an important, but sadly neglected, aspect of user interface design. Presentation and input/output issues have dominated user interface research and practice for many years. However, with the growth of concurrent user interfaces, multi-user interaction and multi-model I/O, designers will be faced with many challenges in the coming decade. We believe it is necessary to develop executable notations and associated tools, like PUAN and the Java.PUAN engine, both to help current designers of complex, multiple platform interfaces and to set the research agenda for the future exploitation of multiple platform interaction.

REFERENCES

Alexander, C., Ishikawa. S. and Silverstein, M. (eds) (1977) *A Pattern Language: Towns, Buildings, Construction*. Oxford University Press.

Allen, J.F. (1984) Towards a General Theory of Action and Time. *Artificial Intelligence*, 23, 123–54.

Dix, A.J. (1987) The Myth of the Infinitely Fast Machine. *People and Computers III: Proceedings of HCI'87*, 215–28 D. Diaper and R. Winder (eds.). Cambridge University Press.

Du, M. and England, D. (2001) Temporal Patterns for Complex Interaction Design. *Proceedings of Design, Specification and Verification of Interactive Systems DSVIS 2001*, C Johnson (ed.). Lecture Notes in Computer Science 2220, Springer-Verlag.

England, D. and Gray, P.D. (1998) Temporal aspects of interaction in shared virtual worlds. *Interacting with Computers*, 11 87–105.

England, D. and Du, M. (2002) *Modelling Multiple and Collaborative Tasks in XUAN: A&E Scenarios* (under review ACM ToCHI 2002).

Fitts, P.M. (1954) The Information Capacity of the Human Motor System in Controlling the Amplitude of Movement. *Experimental Psychology*, 47, 381–91.

Gamma, E., Helm,R., Johnson, R. and Vlissides, J. (1995) *Design Patterns: Elements of Reusable Object- Oriented Software*. Addison-Wesley.

Gray, P.D., England, D. and McGowan, S. (1994) XUAN: Enhancing the UAN to capture temporal relationships among actions. *Proceedings of BCS HCI '94*, 1(3), 26–49. Cambridge University Press.

Hartson, H.R. and Gray, P.D. (1992) Temporal Aspects of Tasks in the User Action Notation. *Human Computer Interaction*, 7(92), 1–45.

Hartson, H.R., Siochi, A.C., and Hix, D. (1990) The UAN: A user oriented representation for direct manipulation interface designs. *ACM Transactions on Information Systems*, 8(3): 181–203.

Hayes, P.J., Szekely, P.A. and Lerner, R.A. (1985) Design Alternatives for User Interface Management Systems Based on Experience with COUSIN. *Proceedings of the CHI '85 conference on Human factors in computing systems*, 169–175.

Hoare, C.A.R. (1984) *Communicating Sequential Processes*. Prentice Hall.

Luyten, K. and Coninx, K. (2001) An XML-Based Runtime User Interface Description Language for Mobile Computing Devices, in *Proceedings of Design, Specification and Verification of Interactive Systems DSVIS 2001*, C Johnson (ed.). Lecture Notes in Computer Science 2220, Springer-Verlag.

Navarre, D., Palanque, P., Paternò, F., Santoro, C. and Bastide, R. (2001) A Tool Suite for Integrating Task and System Models through Scenarios. *Proceedings of Design, Specification and Verification of Interactive Systems DSVIS 2001*, C Johnson (ed.). Lecture Notes in Computer Science 2220, Springer-Verlag

Norman, D.A. (1988) *The Psychology of Everyday Things*. Basic Books.

O'Donnell, P. and Draper, S.W. (1996) Temporal Aspect of Usability, How Machine Delays Change User Strategies. *SIGCHI*, 28(2), 39–46.

Pribeanu, C., Limbourg, Q. and Vanderdonckt, J. (2001) Task Modelling for Context-Sensitive User Interfaces, in *Proceedings of Design, Specification and Verification of Interactive Systems DSVIS 2001*, C Johnson (ed.). Lecture Notes in Computer Science 2220, Springer-Verlag.

Sun Microsystems (2002), *CLDC Specification*, available at http://jcp.org/aboutJava/community-process/final/jsr030/index.html, last accessed August 2002.

Turnell, M., Scaico, A., de Sousa, M.R.F. and Perkusich, A. (2001) Industrial User Interface Evaluation Based on Coloured Petri Nets Modelling and Analysis, in *Proceedings of Design, Specification and Verification of Interactive Systems DSVIS 2001*, C Johnson (ed.). Lecture Notes in Computer Science 2220, Springer-Verlag.

Walker, N. and Smelcer, J. (1990) A Comparison of Selection Time from Walking and Bar Menus. *Proceedings of CHI'90*, 221–5. Addison-Wesley, Reading, Mass.

A. THE PUAN NOTATION

PUAN (Pattern User Action Notation) is a variant of the User Action Notation (UAN) [Hartson *et al.* 1990] developed as part of the Temporal Aspects of Usability (TAU) project [Gray *et al.* 1994] to support investigation of temporal issues in interaction.

In the Notation, tasks consist of a set of temporally-related user actions. The temporal ordering among elements in the action set is specified in the PUAN *action language* (Table A1). For example, if task T contains the action set {$A1$, $A2$}, the relationship of strict sequence would be expressed by:

A1 , A2(usually shown on separate lines)

Order independent execution of the set (i.e., all must be executed, but in any order) is shown with the operator '&':

A1 & A2

The full set of relations is shown below:

Table A1. PUAN action language.

	XUAN	note
Sequence	A1 , A2	
Order independence	A1 & A2	
Optionality	A1 | A2	
Interruptibility	A1 -> A2	also: A2 <- A1
Concurrent	A1 || A2	
Interleavability	A1 <|> A2	
Iteration	A* or A+	also: while (condition) A
Conditionality	if condition then A	
Waiting	various alternatives	

User actions are either primitive actions, typically manipulations of physical input devices (pressing a key, moving a mouse) or tasks:

```
<user action> ::= <primitive user action> | <task>
```

Additionally, an action specification may be annotated with information about system feedback (perceivable changes in system state), non-perceivable changes to user interface state and application-significant operations. Syntactically, a UAN specification places its user actions in a vertically organised list, with annotations in columns to the right. Thus, consider a specification of clicking a typical screen button widget (Table A2).

PUAN is primarily concerned with expressing temporal relationships of sequence among the actions forming a task. The tabular display is a syntactic device for showing strict sequence simply and effectively. Actions and their annotations are read from left to right and from top to bottom. However, certain interactive sequences demand that the ordering imposed by the tabular format be relaxed.

In dealing with time-critical tasks, it is often necessary to express temporal constraints based on the actual duration of actions. PUAN includes several functions for this purpose, including the following time functions:

```
start(a:ACTION), stop(a:ACTION)
```

Table A2. UAN task description for click button.

User Actions	Feedback	User Interface State/ Application Operations
move to screen button	cursor tracks	
mouse button down	screen button highlighted	
mouse button up	button unhighlighted	execute button action

These primitive time functions return a value indicating the start and stop times of a particular action. These two primitives can be built upon to derive more specific temporal relationships (see below).

```
duration(a:ACTION)
```

This function returns a value which is the length of a in seconds. `duration()` is defined in terms of `start()` and `stop()` as:

```
duration(a) = stop(a) - start(a)
<, < =, =, > =, >
```

In time-comparison relations, comparison operators evaluate temporal relationships. For example, `start(a1) < start(a2)` assesses whether the absolute start time of action a1 was less than (earlier than) the absolute start time of action a2.

PUAN has a special operator for conditionality. In order to improve the readability of temporal constraints, we have found it helpful to introduce a more conventional conditional structure.

```
if (condition) then a : ACTION
```

The PALIO Framework for Adaptive Information Services

Constantine Stephanidis,[1,2] Alexandros Paramythis,[1] Vasilios Zarikas,[1] and Anthony Savidis[1]

[1] *Institute of Computer Science Foundation for Research and Technology-Hellas, Greece*
[2] *Department of Computer Science University of Crete, Greece*

5.1. INTRODUCTION

In recent years, the concept of adaptation has been investigated with the perspective of providing built-in accessibility and high interaction quality in applications and services in the emerging information society [Stephanidis 2001a; Stephanidis 2001b]. Adaptation characterizes software products that automatically configure their parameters according to the given attributes of individual users (e.g., mental, motor and sensory characteristics, requirements and preferences) and to the particular context of use (e.g., hardware and software platform, environment of use).

In the context of this chapter, adaptation concerns the interactive behaviour of the User Interface (UI), as well as the content of applications and services. Adaptation implies the capability, on the part of the system, of capturing and representing knowledge concerning

Multiple User Interfaces. Edited by A. Seffah and H. Javahery
© 2004 John Wiley & Sons, Ltd ISBN: 0-470-85444-8

alternative instantiations suitable for different users, contexts, purposes, etc., as well as for reasoning about those alternatives to arrive at adaptation decisions. Furthermore, adaptation implies the capability of assembling, coherently presenting, and managing at run-time, the appropriate alternatives for the current user, purpose and context of use [Savidis and Stephanidis 2001].

The PALIO project[1] addresses the issue of universal access to community-wide services, based on content and UI adaptation beyond desktop access. The main challenge of the PALIO project is the creation of an open system for the unconstrained access and retrieval of information (i.e., not limited by space, time, access technology, etc.). Under this scenario, mobile communication systems play an essential role, because they enable access to services from anywhere and at anytime. One important aspect of the PALIO system is the support for a wide range of communication technologies (mobile or wired) to facilitate access to services.

The PALIO project is mainly based on the results of three research projects that have preceded it: TIDE-ACCESS, ACTS-AVANTI, and ESPRIT-HIPS. In all of these projects, a primary research target has been the support for alternative incarnations of the interactive part of applications and services, according to user and usage context characteristics. As such, these projects are directly related to the concept of Multiple User Interfaces (MUIs), and have addressed several related aspects from both a methodological and an implementation point of view.

The ACCESS project[2] developed new technological solutions for supporting the concept of *User Interfaces for all* (i.e., universal accessibility of computer-based applications), by facilitating the development of UIs adaptable to individual user abilities, skills, requirements, and preferences. The project addressed the need for innovation in this field and proposed a new development methodology called Unified User Interface development. The project also developed a set of tools enabling designers to deal with problems encountered in the provision of access to technology in a consistent, systematic and unified manner [Stephanidis 2001c; Savidis and Stephanidis 2001].

The AVANTI project[3] addressed the interaction requirements of disabled individuals who were using Web-based multimedia applications and services. One of the main objectives of the work undertaken was the design and development of a UI that would provide equitable access and quality in use to all potential end users, including disabled and elderly people. This was achieved by employing adaptability and adaptivity techniques at both the content and the UI levels. A unique characteristic of the AVANTI project was that it addressed adaptation both at the client-side UI, through a dedicated, adaptive Web browser, and at the server-side, through presentation and content adaptation [Stephanidis *et al.* 2001; Fink *et al.* 1998].

The HIPS project[4] was aimed at developing new interaction paradigms for navigating physical spaces. The objective of the project was to enrich a user's experience of a city by overlapping the physical space with contextual and personalized information on the human environment. HIPS developed ALIAS (Adaptive Location aware Information Assistance for nomadic activitieS), a new interaction paradigm for navigation. ALIAS allowed people to simultaneously navigate both the physical space and the related information space. The gap between the two was minimized by delivering contextualized and personalized information on the human environment through a multimodal presentation, according to the user's movements. ALIAS was implemented as a telecommunication system taking advantage of an extensive electronic database containing information about the particular place. Users interacted with the system using mobile palmtops with wireless connections or computers with wired connections [Oppermann and Specht 1998].

PALIO, building on the results of these projects, proposes a new software framework that supports the provision of tourist services in an integrated, open structure. It is capable of providing information from local databases in a personalized way. This framework is based on the concurrent adoption of the following concepts: (a) integration of different wireless and wired telecommunication technologies to offer services through fixed terminals in public places and mobile personal terminals (e.g., mobile phones, PDAs, laptops); (b) location awareness to allow the dynamic modification of information presented (according to user position); (c) adaptation of the contents to automatically provide different presentations depending on user requirements, needs and preferences; (d) scalability of the information to different communication technologies and terminals; (e) interoperability between different service providers in both the envisaged wireless network and the World Wide Web.

The framework presented above exhibits several features that bear direct relevance to the concept of MUIs. Specifically, the framework supports adaptation not only on the basis of user characteristics and interests, but also on the basis of the interaction context. The latter includes (amongst other things) the capabilities and features of the access terminals, and the user's current location. On this basis, the PALIO framework is capable of adapting the content and presentation of services for use on a wide range of devices, with particular emphasis on nomadic interaction from wireless network devices.

The rest of this chapter is structured as follows: Section 5.2 provides an overview of the PALIO system architecture and its adaptation infrastructure. Section 5.3 discusses the PALIO system under the Adaptive Hypermedia Systems perspective. Section 5.4 goes into more depth on those characteristics of the framework that are of particular interest regarding MUIs and presents a brief example of the framework in operation. The chapter concludes with a summary and a brief overview of ongoing work.

5.2. THE PALIO SYSTEM ARCHITECTURE

5.2.1. OVERVIEW

The *Augmented Virtual City Centre* (AVC) constitutes the core of the PALIO system. Users perceive the AVC as a system that groups together all information and services available in the city. It serves as an augmented, virtual facilitation point from which different types of information and services can be accessed. Context- and location- awareness, as well as the adaptation capabilities of the AVC, enable users to experience their interaction with services as *contextually grounded* dialogue. For example, the system always knows the user's location and can correctly infer what is near the user, without the user having to explicitly provide related information.

The main building blocks of the AVC are depicted in Figure 5.1, and can be broadly categorized as follows:

The **Service Control Centre** (SCC) is the central component of the PALIO system. It serves as the access point and the run-time platform for the system's information services.

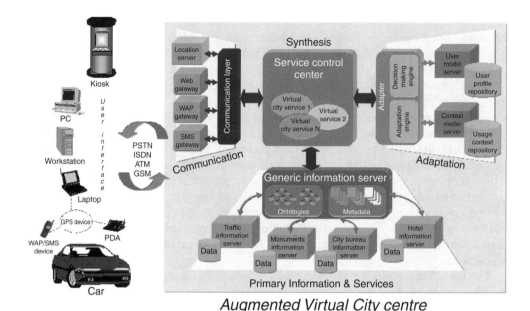

Figure 5.1. Overall architecture of the PALIO system.

The SCC is the framework upon which other services are built. It provides the generic building blocks required to compose services. Examples include the maintenance of the service state control, the creation of basic information retrieval mechanisms (through which service-specific modules can communicate with, and retrieve information from, various distributed information sources/servers in PALIO), etc. Seen from a different perspective, the SCC acts as a central server that supports multi-user access to integrated, primary information and services, appropriately adapted to the user, the context of use, the access terminal and the telecommunications infrastructure.

The **Communication Layer** (CL)[5] encapsulates the individual communication servers (Web gateway, WAP gateway, SMS gateway, etc.) and provides transparent communication independent of the server characteristics. This component unifies and abstracts the different communication protocols (e.g., WAP, http) and terminal platforms (e.g., mobile phone, PC, Internet kiosk). Specifically, the CL transforms incoming communication from the user into a common format, so that the rest of the system does not need to handle the peculiarities of the underlying communication networks and protocols. Symmetrically, the CL transforms information expressed in the aforementioned common format into a format appropriate for transmission and presentation on the user's terminal. In addition to the above, information regarding the capabilities and characteristics of the access terminal propagates across the PALIO system. This information is used to adapt the content and presentation of data transmitted to the user, so that it is appropriate for the user's terminal (e.g., in terms of media, modalities and bandwidth consumption).

The **Generic Information Server** (IS) integrates and manages existing information and services (which are distributed over the network). In this respect, the IS acts as a two-way facilitator. Firstly, it combines appropriate content and data models (in the form of an information ontology and its associated metadata), upon which it acts as a mediator for the retrieval of information and the utilization of existing services by the Service Control Centre. Secondly, it communicates directly with the distributed servers that contain the respective data or realize the services. The existing information and services that are being used in PALIO are termed *primary*, in the sense that they already exist and constitute the building blocks for the PALIO services. The PALIO (virtual city) services, on the other hand, are synthesized on top of the primary ones and reside within the SCC.

The **Adaptation Infrastructure** is responsible for content- and interface- adaptation in the PALIO System. Its major building blocks are the Adapter, the User Model Server and the Context Model Server. These are described in more detail in the next section.

From the user's point of view, a PALIO service is an application that can get information on an area of interest. From the developer's point of view, a PALIO service is a collection of dynamically generated and static template files, expressed in an XML-based, device-independent language, which are used to generate information pages.

A PALIO service is implemented using eXtensible Server Pages (XSPs). XSPs are template files written using an XML-based language and processed by the Cocoon[6] publishing framework (used as the ground platform in the implementation of the SCC) to generate information pages that are delivered to the users in a format supported by their terminal devices. If, for example, a user is using an HTML browser, then the information pages are delivered to that user as HTML pages, while if the user is using WAP, then the same information is delivered as WML stacks.

A PALIO XSP page may consist of (a) static content expressed in XHTML, an XML-compatible version of HTML 4.01; (b) custom XML used to generate dynamic content, including data retrieval queries needed to generate dynamic IS content; (c) custom XML tags used to specify which parts of the generated information page should be adapted for a particular user.

In brief, services in PALIO are collections of:

- Pages containing static content expressed in XHTML, dynamic content expressed in the PALIO content language, information retrieval queries expressed in the PALIO query and ontology languages, and embedded adaptation rules.
- External files containing adaptation logic and actions (including files that express arbitrary document transformations in XSLT format).
- Configuration files specifying the mappings between adaptation logic and service pages.
- Other service configuration files, including the site map (a term used by Cocoon to refer to mappings between request patterns and actual service pages).

An alternative view of the PALIO architecture is presented in Figure 5.2. Therein, one can better observe the interconnections of the various components of the framework, as well as their communication protocols. Furthermore, the figure shows the relation between the Cocoon and PALIO frameworks.

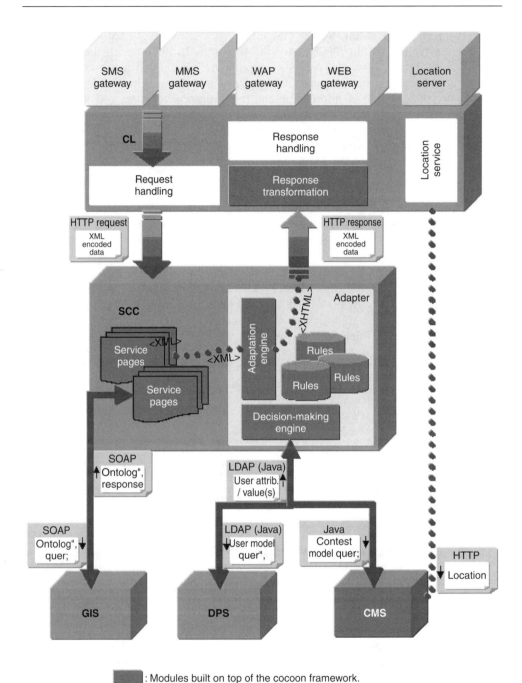

Figure 5.2. Components and communication protocols in the PALIO framework.

5.2.2. THE PALIO ADAPTATION INFRASTRUCTURE

In the PALIO system, the Adaptation Infrastructure is responsible for content and presentation adaptation. As already mentioned, its major components include the User Model Server, the Context Model Server, and the Adapter.

In PALIO, user modelling is carried out by *humanIt's*[7] **Dynamic Personalization Server** (DPS). The DPS maintains four models: a *user* model, a *usage* model, a *system* model, and a *service* model. In general, user models consist of a part dedicated to users' interests and preferences, as well as a demographic part. In PALIO's current version of the DPS, the principal part of a user model is devoted to representing users' interests and preferences. This part's structure is compliant with the information ontology, providing PALIO with a domain taxonomy. This domain taxonomy is mirrored in the DPS-hosted system model (see below).

User models also incorporate information derived from group modelling, by providing the following distinct probability estimates: *individual probability*, an assumption about a user's interests, derived solely from the user's interaction history (including information explicitly provided by the user); *predicted probability*, a prediction about a user's interests based on a set of similar users, which is dynamically computed according to known and inferred user characteristics, preferences, etc.; and *normalized probability*, which compares an individual's interests with those of the whole user population.

The DPS *usage* model is a persistent storage space for all of the DPS's usage-related data. It is comprised of interaction data communicated by the PALIO SCC (which monitors user activity within individual services) and information related to processing these data in user modelling components. These data are subsequently used to infer users' interests in specific items in PALIO's information ontology and/or domain taxonomy.

The system model encompasses information about the domain that is relevant for all user modelling components of the DPS. The most important example of system model contents is the domain taxonomy.

In contrast, the *service* model contains information that is relevant for single components only. Specifically, the service model contains information that is required for establishing communication between the DPS core and its user modelling components.

The **Context Model Server** (CMS), as its name suggests, maintains information about the usage context. A usage context is defined, in PALIO, to include all information relating to an interactive episode that is not directly related to an individual user. PALIO follows the definition of context given in [Dey and Abowd 2000], but diverges in that users engaged in direct interaction with a system are considered (and modelled) separately from other dimensions of context. Along these lines, a context model may contain information such as: characteristics of the access terminal (including capabilities, supported mark-up language, etc.), characteristics of the network connection, current date and time, etc.

In addition, the CMS also maintains information about (a) the user's current location (which is communicated to the CMS by the Location Server, in the case of GSM-based localization, or the access device itself, in the case of GPS) and (b) information related to *push* services to which users are subscribed. It should be noted that in order to collect information about the current context of use, the CMS communicates, directly or indirectly, with several other components of the PALIO system. These other components are the *primary carriers* of the information. These first-level data collected by the CMS then undergo further analysis, with the intention of identifying and characterizing the current context of use. Like the DPS, the CMS responds to queries made by the Adapter regarding the context and relays notifications to the Adapter about important modifications (which may trigger specific adaptations) to the current context of use.

One of the innovative characteristics of the PALIO CMS is its ability to make context information available at different levels of abstraction. For example, the current time is available in a fully qualified form, but also as a day period constant (e.g., morning); the target device can be described in general terms (e.g., for a simple WAP terminal: tiny screen device, graphics not supported, links supported), etc. These abstraction capabilities also characterize aspects of the usage context that relate to the user's location. For instance, it is possible to identify the user's current location by geographical longitude and latitude, but also by the type of building the user may be in (e.g., a museum), the characteristics of the environment (e.g., noisy), and so on. The adaptation logic (that is based on these usage context abstractions) has the advantage that it is general enough to be applied in several related contexts. This makes it possible to define adaptation logic that addresses specific, semantically unambiguous characteristics of the usage context, in addition to addressing the context as a whole. Section 5.4.1 below discusses this issue in more detail.

The third major component of the adaptation infrastructure is the **Adapter**, which is the basic adaptation component of the system. It integrates information concerning the user, the context of use, the access environment and the interaction history, and adapts the information content and presentation accordingly. Adaptations are performed on the basis of the following parameters: user interests (when available in the DPS or established from the ongoing interaction), user characteristics (when available in the DPS), user behavior during interaction (provided by the DPS, or derived from ongoing interaction), type of telecommunication technology and terminal (provided by the CMS), location of the user in the city (provided by the CMS), etc.

The Adapter is comprised of two main modules, the **Decision Making Engine** (DME) and the **Adaptation Engine** (AE). The DME is responsible for deciding upon the need for adaptations, based on (a) the information available about the user, the context of use, the access terminal, etc. and (b) a knowledge base that interrelates this information with adaptations (i.e., the adaptation logic). Combining the two, the DME makes decisions about the most appropriate adaptation for any particular setting and user/technology combination addressed by the project.

The AE instantiates the decisions communicated to it by the DME. The DME and AE are kept as two distinct functional entities in order to decouple the adaptation decision logic from the adaptation implementation. In our view, this approach allows for a high level of flexibility. New types of adaptations can be introduced into the system very easily. At the same time, the rationale for arriving at an adaptation decision and the functional steps required to carry it out can be expressed and modified separately.

5.3. PALIO AS AN ADAPTIVE HYPERMEDIA SYSTEM

Adaptive Hypermedia Systems (AH systems) are a relatively new area of research in adaptive systems. They have drawn considerable attention since the advent of the World

Wide Web. Today, there exist numerous AH systems in a variety of application domains, with a great variety of capabilities [Ardissono and Goy 1999; Balabanovic and Shoham 1997; Brusilovsky *et al.* 1998; Henze 2001; Schwab *et al.* 2000; Kobsa 2001]. While a full review of all AH systems related to PALIO is beyond the scope of this chapter, this section will discuss the most important characteristics of PALIO and attempt to position it within the general space of AH systems.

Major categories of AH systems include educational hypermedia, on-line information systems, on-line help systems, information retrieval systems, and institutional hypermedia. A closer inspection of the main categories[8] gives rise to further classification, based on the goal or scope of the system's adaptation, the methods used to achieve that goal, etc. These are some of the sub-categories:

- On-line information systems can be further categorized into classic on-line information systems, electronic encyclopedias, information kiosks, virtual museums, handheld guides, e-commerce systems, and performance support systems.
- Information retrieval (IR) systems can be categorized into search-oriented adaptive IR hypermedia systems (classic IR in Web context, with search filters), browsing-oriented adaptive IR hypermedia systems (adaptive guidance, adaptive annotation, adaptive recommendation), adaptive bookmark systems (systems for managing personalized views), and information services (search services, filtering services).

The PALIO framework is not constrained to any one of the above categories of AH systems. Being a comprehensive adaptation framework, PALIO can be used to develop AH services and systems in several of these categories. For instance, the PALIO framework could be employed in the development of an adaptive educational system equally as well as in the development of an on-line help system. However, PALIO is more targeted towards facilitating the development of on-line information systems (and sub-categories therein) and would require significant extensions to be rendered suitable for implementing all aspects of adaptive IR systems.

The rest of this section provides a brief overview of the adaptation tools available in PALIO. The discussion will be split into three parts. The first part will discuss the type of information that is taken into account when adaptations are decided upon. The second part will discuss the facilities available for expressing decisions in a computable form. Finally, the third part will address the types of adaptation actions that can be affected by the system.

5.3.1. ADAPTATION DETERMINANTS

The term *adaptation determinant* refers to any piece of information that is explicitly represented in the system and which can serve as input for adaptation logic. Seen from a different perspective, adaptation determinants are the facts known by the system, which can be used to decide upon the need for, and the appropriate type of, adaptation at any point in time. A partial taxonomy of typical adaptation determinants is depicted in Figure 5.3 [adapted from Brusilovsky 1996]. Determinants appearing in the figure in normal text are modelled in PALIO, while those that appear in *italics* are not currently supported. In

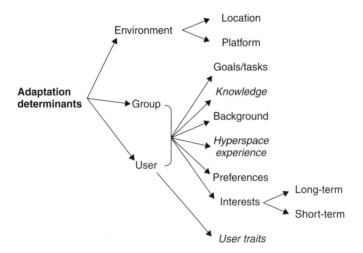

Figure 5.3. Adaptation determinants.

general, three main categories of determinants can be identified: information about users themselves; information about groups to which users may belong; and, information about the environment of execution.[9]

An important fact that is already evident from the taxonomy is that information about the group to which a user belongs can be used alongside (or, even better, in combination with) information about the individual user. This approach is actively pursued in PALIO, where, for instance, the interest of a particular person in a specific type of information can be inferred (with some degree of probability), from the system's knowledge about the related interests of other members in the user's group. Naturally, this approach cannot be extended to all user-related information: user traits (affecting a person's personality, cognitive aptitudes, learning style, etc.), for instance, cannot be inferred from group modelling.

A second important point concerns the fact that knowledge about the user is not always sufficient to identify user needs. One should also consider the more general context in which the user interacts (including the user's location, the characteristics of the access terminal/device, etc.). In fact, with respect to context-related factors, users with varying characteristics can be expected to have similar or identical needs. As already discussed, the PALIO framework takes full account of this fact and incorporates a Context Modelling Component, which undertakes the monitoring of observable context-related parameters. These are available to the adaptation designer to be used as explicit adaptation determinants.

5.3.2. DECISIONS ON THE BASIS OF ADAPTATION DETERMINANTS

The primary decision engine implemented for PALIO is rule-based. Although first order logic rule engines can easily be plugged into the framework (the latter having been specifically designed to allow for that), it was decided that, to facilitate the wide adoption of the framework, a simpler and more accessible approach was in order. Along these lines, a new rule language was created, borrowing from control structures that were commonly supported in functional programming languages. The premise was that such structures were much more familiar to designers of adaptive systems, while at the same time they afforded lower degrees of complexity when it came to understanding the interrelations and dependencies between distinct pieces of adaptation logic. This approach was first applied in the AVANTI project with very good results [Stephanidis *et al.* 2001].

An XML binding was developed for the aforementioned rule language, while a rule interpreter and a corresponding rule engine supported the run-time operation of the system. Adaptation rules expressed in such a rule language may be either defined in external files or embedded in the document to be adapted. Rules embedded in pages are constrained in that they can only be applied within the specific document in which they reside and therefore are not reusable.

Rules external to documents are organized into rule-sets; typically, each rule-set resides in a different file, but this is a matter of organization, rather than a constraint imposed by the framework. Rule-sets have a specific *name* and *scope*. The name may be used to refer to the rule-set within configuration files and can be used, for example, to enable/disable a whole set of rules, by referring to it via the name of the enclosing rule-set. The possible values of the scope attribute are *global, service* and *local*. These denote that the rules defined in a rule-set may apply to all of the pages served by PALIO, to pages belonging to a specific service, or to specific pages indicated through configuration files, respectively. The use of names and scopes is complementary to the use of the service configuration files and is intended to add more flexibility to the definition of relationships between rule-sets and service documents.

Every rule, whether internal or external to the document, has the following attributes (a) the rule *name*, which is an identifier for the rule; (b) the rule *class*, which is optional and may be used as an alternative way to group rules; (c) the rule *stage*, which defines whether the rule should be applied during the first or the second adaptation phase; and (d) the rule *priority*, which provides a partial ordering scheme for rule evaluation and application and may take the values of *high, medium* or *low*. The stage property defines whether the adaptation rule should be applied before or after querying the IS (if this is required to process the document). Adaptations performed during the first phase (before querying the IS) are either unconcerned with IS-ontology data or adapt (i.e. apply modifications to) IS queries. Rules applied during the second phase (after querying the IS) are concerned with IS-ontology data.

The framework currently supports three types of rules: *if-then−else* rules, *switch-case−default* rules, and *prologue-actions−epilogue* rules. If-then−else rules are the simplest type of conditional rules; they bind sets of adaptation actions with the truth-value of a conditional expression. Following the example made by functional languages, an if-then−else rule in PALIO is composed of a *condition* (containing an expression based on adaptation determinants), a *then* part (containing the adaptation actions to be taken whenever the condition is satisfied), and an optional *else* part (containing the actions to be taken when the condition fails).

Switch-case−default rules can be used to relate multiple values (outcomes of run-time expression evaluation) to sets of adaptation actions. In this case, adaptation actions are executed if the value of a *variant* (expression) is equal to a value, or within the range of values, specified as a selection *case*. The switch-case–default rule construct supports the definition of multiple cases and requires the additional definition of an (optionally empty) set of *default* adaptation actions to be performed if none of the defined cases apply.

Prologue-actions–epilogue rules are intended mainly for the definition of unconditional rules. In other words, the specific rule construct is provided to support the definition of (sets of) actions to be performed at a particular stage in the sequence of adaptations. Although the construct can be simulated with an if-then–else construct (where the condition is always true), the prologue-actions–epilogue structure (through *action-sets*) provides an explicit separation between preparatory actions, the adaptations themselves, and clean-up actions. This separation allows for better rule structuring and improves the maintainability of the rule definition. A very common use of the *prologue* and *epilogue* parts is the creation/assignment and retraction of variables that are used by the adaptation actions (e.g., in order to determine the point of application of an action).

The conditional parts of the first two rule constructs presented above, as well as the definition of variants and variant ranges in the case of the switch-case–default construct, are composed of expressions. These consist, as one would expect, of operands and operators applied on the operands. The main supported operand types in PALIO are: *string, number, boolean, date, location* and *null*. The *string* type is used for character sequences. The *number* type is used for all kinds of numerals including integers, floating point numbers, etc. The *boolean* type is used to express truth values and can only assume the values *true* or *false*. The *date* type is used for absolute, relative and recurring dates and temporal ranges. The *location* type is used to express absolute and relative geographical locations. Finally, the *null* type is a utility type with restricted use; authors may use it to determine whether a variable exists in the scope of a particular data source, and whether its type has been set. The operators supported by the PALIO framework can be classified into the following main categories: comparison operators ($>$, $>=$, $<$, $<=$, $<>$), mathematical operators ($+$, $-$, $*$, $/$, $\%$), logical operators (*and, or, not*), string operators (*concatenate, contains*, and *substring*), date-specific operators (*get-year, get-month*, etc.), and location-specific operators (*near* and *distance*).

5.3.3. ADAPTATION ACTIONS

The PALIO framework supports the following adaptation action categories: inserting document fragments, removing document fragments, replacing document elements/fragments, sorting document fragments, setting and removing element attributes, selecting among alternative document fragments, and applying arbitrary document transformations expressed in XSLT. In addition to the above basic adaptation actions, the PALIO framework also supports the manipulation of variables (i.e. storing and retrieving values from different run-time scopes).

Adaptation actions can be (and usually are) defined in external files, but can also be embedded within documents. The *target* or *reference* element of an adaptation action, which can also be a dummy element acting as a placeholder for the action's target, needs to be identified. When the adaptation action is specified externally, such an identification can be done using (a) the element's *class* attribute, which is better suited for the identification of multiple insertion points in a document, as several elements can share the same *class* attribute; (b) the element's *id* attribute, which is better suited for the identification of single reference points in the document – the *id* attribute of any element should be unique within a document; (c) an XPath expression, to be used whenever more complex constraints need to be expressed with respect to the insertion point. When the adaptation action is embedded in a document, the default reference element is the rule (context) node itself.

In the following descriptions, constraints related to maintaining XML document validity are omitted for brevity, but should be intuitively understood in most cases (e.g., when adding an element as a parent to another element, or changing the value of an attribute).

Inserting Document Elements/Fragments

This adaptation action enables authors to insert new elements or document fragments into the document. For the insertion to be performed, the author needs to specify the reference element (or elements) at which the insertion will take place, as well as the relative position of the new element/fragment with respect to the reference element. The relative position of the inserted element/fragment with respect to the reference element can be (a) immediately *before* the reference element; (b) immediately *after* the reference element; (c) *inside* the reference element, as its *first child*; (d) *inside* the reference element, as its *last child*; (e) *outside* the reference element, as its *parent*.

Removing Document Elements/Fragments

This adaptation action enables authors to remove individual or multiple elements, or document fragments contained within elements, from the document. For the removal to be performed, the author needs to specify the element (or elements) to be removed, as well as whether any child elements should also be removed. The identification of the element to be removed can be done using the same approach as in the case of insertion, with the following important exception: when the action is part of a rule embedded in a document, authors cannot make use of the 'current context node' idiom, because then the removal would affect the rule node itself, which is impossible. When removing an element, authors have the option to *remove* or *retain* the element's children.

Replacing Document Elements/Fragments

This adaptation action enables authors to replace single or multiple elements in the document, or document fragments contained within elements. The elements to be removed, along with any child elements that should also be removed, and the elements that are to be inserted instead, need to be specified by the author. This action can be treated, from a semantic point of view, as a removal action combined with a subsequent insertion action. This action improves on the aforementioned combination by making it possible to easily transfer the child elements of the elements being removed to the elements being inserted.

Manipulating Element Attributes

This adaptation action enables authors to set the value of specific element attributes or remove element attributes entirely. To set an attribute value, the author needs to specify the elements affected, as well as the name of the attribute to be set and the value that it will assume. To remove an element attribute entirely, the author only needs to specify the elements affected and the name of the attribute to be removed. When setting an attribute value, authors can use either static values, or (more typically) dynamically calculated/retrieved values (relating to the current user/session, context of use, etc.).

Selecting Amongst Alternative Document Elements/Fragments

This adaptation action facilitates the selection of fragments from alternative fragments declared within the document. The use and selection of alternative content fragments requires authors to first introduce, and identify, the alternative (but not necessarily mutually exclusive) fragments into the document. Secondly the author must select one or more of them through the respective adaptation actions. All alternatives that are not explicitly selected are automatically removed from the document. The selection of alternatives can take place (a) by selecting exactly one of the alternatives and discarding the rest; or (b) by selecting zero or more of the alternatives and discarding the rest. The first case is a special case of the second. However, it is provided as a separate adaptation action in PALIO for reasons of authoring convenience.

Sorting Document Elements/Fragments

This adaptation action facilitates a common technique (that is difficult to define and implement with basic building blocks) in Adaptive Hypermedia Systems: the sorting of document fragments, links, etc. As in the case of the definition and selection of alternatives, sorting adaptations requires two steps of specification on the part of the author. First, the author provides a list of document elements/fragments that are to be sorted within the document. Secondly, a respective sorting adaptation action needs to be authored and introduced into a rule (typically external to the document).

The specification of elements/fragments adheres to the following principles: there exists a generic container termed *list*, which holds the items (elements/fragments) to be sorted (note that the container has been called a *list* to benefit from author familiarity with the container concept and not because the container provides any facilities for traversing its contents); the container holds *items*, each of which must provide a *name* attribute (used to relate the item, directly or indirectly, to a value that will be used subsequently for sorting); the container may also hold additional supporting content that accompanies the items to be sorted.

The specification of each of the items to be sorted is accomplished through the item element and requires only the identification of a name for the item, which will be used during sorting to associate the item with a concrete value. The specification of sorting behavior includes two additional optional components: the identification of the maximum number of items to be retained, and the identification of a threshold value that acts as a cut-off point for items. The sorting order to be applied to the items is defined as an attribute of the sort adaptation action.

Applying Arbitrary XSL Transformations

This type of adaptation action is intended to cover any complex cases that cannot be addressed through the use of the basic adaptation actions listed above. Specifically, this adaptation only defines that an external XSLT file is to be applied to the document, providing the URL of that file as an attribute. The PALIO adaptation framework then loads the file and its application to the (in-memory) representation of the document. The current implementation of the framework contains a number of ready-made XSLT files, which address common non-trivial adaptations, such as the transformation of nested lists into tables and vice versa.

5.4. PALIO IN THE CONTEXT OF MUI

5.4.1. PALIO AS A WEB UI

Following the definitions found in Chapter 1 of this book, PALIO would be classified as a framework for building Web UIs. This means that the framework is specifically intended to support the development of on-line, server-side information services characterized by their capability to dynamically adapt on the basis of user and context characteristics. PALIO goes far beyond *transcoding* approaches [Maglio and Barrett 2000] to transform content and interface for presentation on different devices. In fact, this type of transformation is the last (and, in some ways, the least significant) adaptation step that takes place before documents are served to a user. The rest of this section is devoted to pointing out and discussing the capabilities of the PALIO framework in the context of MUIs and their associated principles.

5.4.1.1. Device- and Platform-Independence

One of the highly innovative characteristics of PALIO is the extensive support it offers for creating applications and services that are truly independent from the target device and computing platform. As the benefits of such independence are well explained in other chapters of this book, we will focus on the ways in which this independence is achieved within the PALIO framework.

The first level of support can be clearly traced to the use of a single format for describing services in a document-oriented fashion – with each document encapsulating portions of the system's interactive functionality. Service developers do not need to concern themselves with the exact characteristics of the access terminals or network connections. Instead, they are free to focus on providing rich service functionality and conveying that in a flexible, XML-based form. This serves as an *abstraction* that transcends the device- and platform- levels, although it may, when necessary, make provisions for them. As an example of the latter, authors can indicate which portions of a document go together naturally. These sections can then be used to sub-divide the document, for example to split it into cards for presentation on a WML-enabled device.

The second level of support concerns the encoding and representation of user actions within the framework. As already discussed, the CL component of the architecture is responsible for converting all incoming user requests (originally expressed in various application-level communication protocols) into a unified representation. This common representation is then propagated into the rest of the framework, thus shielding run-time components and (more importantly) service developers from the specifics of the platforms and communication protocols. Although this approach is clearly oriented toward a request–response system, it has been successfully employed to support settings as diverse as traditional Web-based interaction and mobile-phone SMS-based interaction. Furthermore, the particular model has been employed in the realization of *push* services, where there is no immediate action on the part of the user, but rather a pre-existing request to receive particular types of information as these become available.

5.4.1.2. Device- and Platform-Awareness

The fact that service authors/developers can construct services in a device- and platform-independent manner does not imply that they cannot take specific devices and platforms into account at different stages of the authoring process. As already discussed, the framework offers support for dynamically adapting the system's output to fit the characteristics of the access terminal. More interestingly, this support goes far beyond the mapping or transformation of documents to the desired output format.

When discussing the CMS component of the architecture (see Section 5.2.2), we mentioned that the context characteristics are available at various levels of abstraction. This implies that the content and presentation of documents can be adapted to abstract characteristics of the device-platform combination, without binding adaptations to any specific device or platform. Achieving this involves the employment of generic adaptation rules, which are aware of a subset of the characteristics of devices or supported markup language and are applicable over a multitude of device-platform combinations.

The power of this approach may be better exemplified in the following scenario. A service author wishes to apply different navigation and presentation strategies/schemes, based on the size of the access terminal's screen and its capability to support links. A decision is made to (a) differentiate between three screen sizes: normal (e.g., PC monitor), small (e.g., PDA screen) and tiny (e.g., mobile phone screen); and (b) distinguish between support for direct hyperlinks (e.g., normal hyperlinks in an HTML document), support for indirect links (e.g., for SMS-based interaction, which asks the user to send a particular message to a specified number), and no support for links altogether. The PALIO framework inherently supports these distinctions and enables the author to employ adaptation rules that explicitly address them individually or together (e.g., creating a navigation scheme for small-screen devices with support for direct links). Effectively, this enables authors to automatically reuse these rules in settings or combinations that were not known or anticipated at the time of authoring. Without such support, one would need to explicitly address every single output platform in isolation. Additionally, new or unknown platforms would require the author to repeat the process all over again.

As an example, consider the case of the various versions of the WML markup language and their forthcoming complement, XHTML Mobile Profile 1.1.[10] All these languages and versions differ from each other – differences that are profound in some cases. However, they are all intended for deployment on similar, resource-constrained devices. If one were to address each existing combination of markup language/version and screen size, one would be faced with a daunting task. Following the PALIO approach, supporting a new or updated markup language would only require the appropriate revision of the lowest-level transformations undertaken by the CL component (mapping from the uniform internal document representation to the final output format). In other words, one can define a specific set of transformations for WML 1.3 compliant browsers and a different set for XHTML Mobile Profile browsers. However, *both* sets of transformations could be applied to the same document, which would have already been adapted for presentation on small-screen devices with support for direct links.

5.4.1.3. Uniformity and Cross-Platform Consistency

Beyond enabling the creation of services that are aware of the characteristics and capabilities of different devices and platforms, the principles discussed in the previous section also have a major impact on attaining the goals of interface uniformity and consistency. Specifically, using the described mechanisms and approach, the navigation schemes and presentation templates employed in PALIO are bound to outgoing documents by means of adaptation. As a result, ensuring consistency across platforms can be addressed as a service-level issue, or even seen from the much broader perspective of the service's container, or service center. Furthermore, since presentation and content can be adapted independently in PALIO, achieving consistency in the interface does not in any way affect the generation of the service's information content.

In more practical terms, the two service centers that were developed to demonstrate the PALIO project followed the concept of gradually downsizing a common presentation and navigation theme. This involves (a) the creation of multiple *key templates* intended for major categories of output device configurations (these templates are comprised of both presentation and navigation elements and were derived from a common visual and interactive design); (b) the authoring of adaptation rules which further modify the templates on the basis of device characteristics within each of the categories. Each of the key templates, as well as the adaptations applied to them, created a step-wise simplification of the system's navigation and presentation scheme, by observing a set of principles (i.e., what functionality should be available through the interface) and their coupling to the characteristics of the device at hand (i.e., how users perceive and interact with that functionality on a device or platform with given characteristics).

Using the case of inter-service navigation as an example, the 'gradual downsizing' approach is demonstrated by the following adaptation rules: when the access terminal has a normal-size screen and direct-link support, access to services other than the 'current' one is affected through a 'menu' on the left side of the document (containing hyperlinks to each of the services); for terminals with small-size screens, direct-link support and high-bandwidth network connections, the service menu is moved to the end of the page; for set-ups similar to the latter, but with low-bandwidth connections, the service menu resides in its own, separate 'page' and a link pointing to that page replaces the menu. The screenshots in Figure 5.4 illustrate two examples of the approach described here.

5.4.1.4. General Context-Awareness

The characteristics of devices and network connections are not the only dimensions of context that are taken into account in PALIO. Of at least equal importance are a user's location and the time of use. We have already discussed PALIO's support for context-based adaptation, as well as how certain characteristics of context can be abstracted over, or

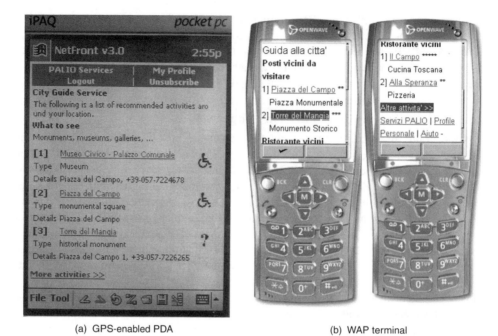

(a) GPS-enabled PDA   (b) WAP terminal

Figure 5.4. Output from the PALIO Information System in Siena, Italy to various devices.

inferred from raw sensor data. Here we will focus on how this enables the delivery of services that can support a user in different contexts of use, or conversely, different users in the same or similar contexts of use.

The premise of context-based (or context-aware) adaptation is that by identifying distinctive characteristics of the context within which a user is interacting, it becomes possible to better understand and address the needs of the user. A basic assumption behind that premise is that context can define, or at least influence, a user's needs in terms of interaction and information-seeking behavior, in a way that is independent of the particular characteristics of any given user.

Residing on the server-side, the PALIO framework has limited access to context-related information. Specifically, the primary sources of raw context information are (a) the access terminal's browsing application[11] (from which one can typically identify the type of device, markup language supported, and network connection at hand); (b) the user's current location with a granularity dependent upon the technology used (i.e., GPS- or GSM-based positioning); and (c) the current date and time (in principle it is possible to identify the user's date and time, if this is different from the system's, through the browsing application).

A first level of abstraction over context data is derived, as discussed in the preceding sections, by generically describing characteristics of the context, instead of specific instance values. For example: specific times are categorized in *zones* (e.g., morning), specific dates are associated with a season (e.g., summer), specific screen sizes are classified as tiny, small, normal and large, etc. Where possible, the same is done for the user's location. For example, specific geographical locations (expressed in terms of latitude and longitude) are associated with landmarks, buildings, etc.

This augmented representation of the user's interaction context is subsequently used to infer further characteristics of the context that are not available through actual sensors. For example, the noise level within a building is a potentially important context attribute, which is not available as a primary source of context information in PALIO. In some cases it is possible to infer the amount of noise (with varying levels of certainty) from semantic information about the user's location, possibly coupled with information about the time of day and season of year. Consider, for instance, the case of a library building: it is a rather safe assumption that the level of ambient noise in this environment is quite low throughout the day and year. Conversely, tourists frequent the archaeological museum in the city of Heraklion, Crete, primarily during summer months and in the mornings and afternoons. It would be a logical assumption (although not as certain an assumption as in the previous example) that during those times the level of noise in the museum is higher than during the rest of the year.

All explicitly derived and inferred context characteristics are made available through the CMS component of the framework as adaptation determinants. This means that service authors can create adaptation logic that is exclusively context-oriented or opt to include context parameters in adaptation rules that relate the characteristics of users with those of the context in which they are immersed. The examples of adaptation on the basis of device or platform characteristics in the previous section belong to the first category. An example belonging in the second category would be the recommendation of restaurants to a user on the basis of the user's culinary preferences, coupled with the user's current location (i.e. which restaurants close to the user, might be to his/her liking). A further example can be found in Section 5.4.2 below.

5.4.1.5. User Awareness

Although not at the center of MUI research, user-oriented adaptation plays a very significant role in the PALIO framework. The intent of adaptation in this respect is to tailor the system to the particular abilities, skills, requirements and preferences of the individual user. A large part of adaptations in this category are directly intended to tailor the service (i.e., the information content) to the needs of the individual user. However, another dimension (arguably more relevant to MUIs) is the adaptation of presentation with the objective of tailoring the interface to accommodate user requirements and preferences.

The number of adaptations possible in this direction is quite large. One could even envisage using rudimentary demographic data (e.g., a user's age group) to perform substantial adaptations in the UI (e.g., by using different color or graphic themes for different age groups). Out of this large number of possibilities, PALIO is focusing on interface adaptations that are intended to facilitate *universal access* [Stephanidis 2001b] of the services. Specifically, users that register with a PALIO service center can identify that they require a specialized interface due to a disability. Disabilities explicitly addressed by PALIO at this point in time include blindness, color blindness, low vision, and motor impairments.

Elements of the interface that are adapted to ensure high-quality interaction for these categories of users include the overall organization of the interface, the manipulation of color schemes and graphics (including replacing graphics with equivalent content), the introduction of specialized navigation facilities, etc.

Furthermore, since user profiles are stored on the server, users can enjoy the accessibility afforded by such specialized adaptations over the full range of devices supported by the system. Additionally, the PALIO framework is capable of selecting among alternative presentation modalities (when these are available), adopting the one that is most suitable for the user and device combination. For example, if the description of a venue is available in both text and audio formats and the user is blind, and the user's device is capable of rendering audio files, PALIO can compose a response that incorporates the audio description, rather than the textual one.

5.4.2. A BRIEF EXAMPLE

To better illustrate the capabilities of the PALIO framework and its relation to MUIs, this section presents an example from the PALIO information system installed for demonstration and testing purposes in the city of Siena, Italy. For brevity, the example will not delve into technical details; instead, it will focus on the results of applying the PALIO framework to the following two fictional interaction scenarios.

1st scenario: An English-speaking wheelchair user is in Piazza del Campo (Siena, Italy) in the morning. She is interested in sightseeing and prefers visiting monuments and museums. She is accessing the system from her palmtop, which she rented from the Siena Tourist Bureau and which comes with a GPS unit. She accesses the **City Guide** service and asks for recommendations about what she can see or do next.

2nd scenario: An Italian-speaking able-bodied user is also in Piazza del Campo (Siena, Italy) around noon. He hasn't used the system before and, therefore, there is no information about what he might prefer. He is accessing the system from his mobile phone through WAP. He also accesses the **City Guide** service and asks for recommendations of what he can do next.

The result of the first user's interaction with the PALIO system is shown in Figure 5.4a. Relevant characteristics include (a) the presentation language is English; (b) the front-end is tailored for a small-screen terminal capable of color and graphics; (c) the system's recommendations are in accordance with the user's preferences; (d) recommended sites are in the immediate vicinity of the user; and (e) accessibility information is provided immediately (at the first level), since this information will impact on whether the user will decide to visit the place or not.

The result of the second user's interaction with the PALIO system is shown in Figure 5.4b. Relevant characteristics include (a) the presentation language is Italian; (b) the front-end is tailored for tiny-screen terminals without assumptions made about color and graphics; (c) the system's recommendations are derived from preferences for the user's group – asterisks next to each recommendation indicate other users' collective assessments of the venue recommended; (d) recommendations include a wider range of activities (e.g., sightseeing and eating); (e) one type of activity (eating) is relevant to the temporal context (it's noon); (f) recommended sites are in the general vicinity of the user; and (g) accessibility information is not provided immediately (at the first level).

5.5. SUMMARY AND ON-GOING WORK

This chapter has presented the PALIO service framework, with a focus on its adaptation infrastructure. PALIO constitutes a substantial extension over previous efforts toward universal access, since it introduces and explicitly accounts for novel types of adaptation and new interactive platforms beyond the desktop. Accordingly, it pursues an architectural model of interaction that is expected to be widely applicable in service sectors other than tourism.

From the perspective of MUIs, PALIO offers an innovative approach to the development of user- and context-aware Web UIs. The framework supports virtually all types of hypermedia adaptation techniques, thus enabling the easy creation of on-line systems. These systems maintain high levels of consistency over different platforms, while at the same time allowing developers to make the best use of available resources. PALIO provides support for the strict separation of content and presentation, allowing each to be manipulated and adapted independently, thus making it possible to cater for the needs of non-traditional user groups (such as disabled users).

For service creators, PALIO is a powerful and easy-to-use development system for many reasons. First, the notions of *user model* and *context model* (including user location) are seamlessly and transparently supported through their incorporation in the PALIO adaptation rules language. Second, adherence to web standards, as well as the use of XML, XSLT and other technologies developed by the W3C, allows Web developers to easily move from other Web development systems to PALIO. Finally, PALIO supports automatic content creation for different user terminal device types, by automatically transforming content format.

The PALIO system can be easily expanded to include new sources of information and it can be ported to different application domains. This is possible because PALIO abstracts the information sources it uses through the use of a domain-specific ontology. Adding new information sources will require the incorporation of the new information databases to the GIS and the modification of the PALIO information ontology. Furthermore, using the Cocoon publishing framework as the base of the PALIO system and Java as the development language means that PALIO can run on any server platform and that it can support a wide range of adaptation using standard Web publishing technology.

The PALIO framework, as well as the developed information systems and demonstration services, were under evaluation during the time of this writing. Preliminary evaluation results were positive, both from the perspective of end-user experiences and from the perspective of service development.

Ongoing work is targeting the provision of better support for *push* services and the incorporation of support for standardized device profiles. In particular, we are currently working on more intelligent push services that can take the full interaction context and previous interaction history into account when deciding if and what type of information should be sent to a user (based on possibly uncertain knowledge). In parallel, we are working to incorporate support for Composite Capability/Preference Profiles (CC/PP),[12] and for UAProf[13] in order to attain a more standardized way to describe, communicate and retrieve device capabilities.

ACKNOWLEDGEMENTS

The PALIO project (IST-1999-20656) is partly funded by the Information Society Technologies Programme of the European Commission – DG Information Society. The partners in the PALIO consortium are: ASSIOMA S.p.A. (Italy) – Prime Contractor; CNR-IROE (Italy); Comune di Firenze (Italy); FORTH-ICS (Greece); GMD (Germany); Telecom Italia Mobile S.p.A. (Italy); University of Sienna (Italy); Comune di Siena (Italy); MA Systems and Control Ltd (UK); FORTHnet (Greece).

The authors would also like to explicitly acknowledge the contributions from the following members of the Human-Computer Interaction Laboratory of FORTH-ICS: Chrisoula Alexandraki, Ioannis Segkos, Napoleon Maou, and Margherita Antona.

REFERENCES

Ardissono, L. and Goy, A. (1999) Tailoring the interaction with users of electronic shops. *Proceedings of 7th International Conference on User Modeling, UM99* (ed. J. Kay), Wien, 35–44. Springer.

Balabanovic, M. and Shoham, Y. (1997) Fab: content-based collaborative recommendation. *Communications of the ACM* 40(3), 66–72.

Brusilovsky, P. (1996) Methods and techniques of adaptive hypermedia. *User Modeling and User-Adapted Interaction*, 6(2–3), 87–129.

Brusilovsky, P. (2001) Adaptive hypermedia. *User Modeling and User Adapted Interaction, Ten Year Anniversary Issue*, 11(1/2), 87–110.

Brusilovsky, P., Kobsa, A. and Vassileva, J. (eds) (1998) *Adaptive Hypertext and Hypermedia*. Dordrecht: Kluwer Academic Publishers.

Dey A.K., and Abowd, G.D. (2000) Towards a Better Understanding of Context and Context Awareness, in *Proceedings of the Workshop on the What, Who, Where, When and How of Context-Awareness, affiliated with the CHI 2000 Conference on Human Factors in Computer Systems*. New York: ACM Press.

Fink, J., Kobsa, A. and Nill, A. (1998) Adaptable and Adaptive Information Provision for All Users, Including Disabled and Elderly People. *The New Review of Hypermedia and Multimedia*, 4, 163–188.

Henze, N. (2001) Open Adaptive Hypermedia: An approach to adaptive information presentation on the Web, in *Proceedings of the 1st International Conference on Universal Access in Human-Computer Interaction (UAHCI 2001), held jointly with HCI International 2001* (ed. C. Stephanidis), 818–21. Mahwah, NJ: Lawrence Erlbaum.

Kobsa, A. (2001) Generic User Modeling Systems. *User Modeling and User Adapted Interaction* 11(1–2), 49–63.

Maglio, P.P. and Barrett, R. (2000) Intermediaries Personalize Information Streams. *Communications of the ACM*, 43(8), 96–101.

Oppermann, R. and Specht, M. (1998) Adaptive Support for a Mobile Museum Guide, in *IMC '98 Workshop 'Interactive Application of Mobile Computing'*, Rostock, November 1998. Available at: http://www.egd.igd.fhg.de/~imc98/Proceedings/imc98-SessionMA3-2.pdf.

Savidis, A., and Stephanidis, C. (2001) The Unified User Interface Software Architecture, in *User Interfaces for All: Concepts, Methods and Tools* (ed. C. Stephanidis), 389–415. Mahwah, NJ: Lawrence Erlbaum.

Schwab, I., Pohl, W., and Koychev, I. (2000) Learning to Recommend from Positive Evidence, in *Proceedings of 2000 Int. Conf. on Intelligent User Interfaces*, 241–7. New York: ACM Press.

Stephanidis, C. (2001a) Adaptive Techniques for Universal Access. *User Modelling and User Adapted Interaction International Journal*, 11(1–2), 159–79. Kluwer Academic.

Stephanidis, C. (2001b) User Interfaces for All: New perspectives into Human-Computer Interaction, in *User Interfaces for All: Concepts, Methods and Tools* (ed. C. Stephanidis), 3–7. Mahwah, NJ: Lawrence Erlbaum.

Stephanidis, C. (2001c) The concept of Unified User Interfaces, in *User Interfaces for All: Concepts, Methods and Tools* (ed. C. Stephanidis), 371–88. Mahwah, NJ: Lawrence Erlbaum.

Stephanidis, C., Paramythis, A., Sfyrakis, M., and Savidis, A. (2001) A Case Study in Unified User Interface Development: The AVANTI Web Browser, in *User Interfaces for All: Concepts, Methods and Tools* (ed. C. Stephanidis), 525–68. Mahwah, NJ: Lawrence Erlbaum.

FOOTNOTES

1. Personalized Access to Local Information and services for tourists (see Acknowledgments section).
2. TIDE TP1001 – ACCESS 'Development platform for unified ACCESS to enabling environments', funded by the European Commission (1994–96). The partners of the ACCESS consortium are: CNR-IROE (Italy), Prime Contractor; FORTH-ICS (Greece); University of Hertforshire (UK); University of Athens (Greece); NAWH (Finland); VTT (Finland); Hereward College (UK); RNIB (UK); Seleco (Italy); MA Systems & Control (UK); PIKOMED (Finland).
3. ACTS – AC042 AVANTI 'Adaptable and Adaptive Interaction in Multimedia Telecommunications Applications', funded by the European Commission (1995–98). The partners of the ACTS-AVANTI consortium are: ALCATEL Italia, Siette division (Italy), Prime Contractor; CNR-IROE (Italy); FORTH-ICS (Greece); GMD (Germany), VTT (Finland); University of Sienna (Italy); TECO Systems (Italy); STUDIO ADR (Italy); MA Systems and Control (UK).
4. ESPRIT-25574 HIPS 'Hyper-Interaction within Physical Spaces' funded by the European Commission (1997–2000). The partners of the HIPS consortium are: Universita degli Studi di Siena (Italy), Prime Contractor; University of Edinburgh (UK); Alcatel Italia SpA (Italy); Istituto Trentino di Cultura (Italy); Cara Broadbent & Jegher Associes (France); GMD (Germany); SINTEF (Norway). The project's website is http://www.media.unisi.it/hips.
5. The term 'layer' is used in the PALIO project for historical reasons; the CL is not a layer in the sense of layered software architecture, but rather a component.
6. Apache Cocoon is an XML publishing framework that facilitates the usage of XML and XSLT technologies for server applications. Designed around pipelined SAX processing, to benefit performance and scalability, Cocoon offers a flexible environment based on the separation of concerns between content, logic and style – the so-called

pyramid model of web contracts. More information can be obtained at the project's homepage, at http://xml.apache.org/cocoon/index.html.

7. http://www.humanit.de/

8. The detailed description of the different types of AHS is beyond the scope of this chapter. Interested readers are referred to the excellent reviews, [Brusilovsky 1996] and [Brusilovsky 2001], which have long served as reference points in the area.

9. The term *usage context* (or *context of use*, often abbreviated to *context*) is used throughout this chapter to denote all factors characterizing an interactive situation that are external to the user (including the user's geographical location). This encompasses the space of factors denoted by *environment* in the taxonomy.

10. For more information on WML and XHTML Mobile Profile, refer to the Web site of the Open Mobile Alliance: http://www.openmobilealliance.org/.

11. Here, the term 'browsing' does not necessarily refer to hypermedia browsing. Instead, a user's 'browsing application' is understood to be that part of the device's software environment which enables a user to access the PALIO information systems.

12. For more information on CC/PP refer to http://www.w3.org/Mobile/CCPP/.

13. For more information on UAProf refer to http://www.openmobilealliance.org/documents.asp.

Part III

Development Technology
and Languages

6

Building Multi-Platform User Interfaces with UIML

Mir Farooq Ali,[1] Manuel A. Pérez-Quiñones,[1] and Marc Abrams[2]

[1] *Department of Computer Science, Virginia Tech, USA*
[2] *Harmonia, Inc., USA*

6.1. INTRODUCTION

Developers now face the daunting task of building UIs that must work on multiple devices. Two of the problems we encountered in our own work [Ali and Abrams 2001; Ali *et al.* 2002] in building UIs for different platforms were the different layout features and screen sizes associated with each platform and device. Dan Olsen [1999], Peter Johnson [1998], and Stephen Brewster *et al.* [1998] all talk about problems in interaction due to the diversity of interactive platforms, devices, network services and applications. They also talk about the problems associated with the small screen size of hand-held devices. In general, it is difficult to develop a multi-platform UI without duplicating development effort.

There have been some approaches (including XWeb [Olsen *et al.* 2000]) proposed for solving this problem of multi-platform UI development. Building 'plastic interfaces' [Calvary *et al.* 2000; Thevenin *et al.* 2001; Thevenin and Coutaz 1999] is one such method in which the UIs are designed to 'withstand variations of context of use while preserving usability' (see Chapter 3). Their approach follows a traditional lifecycle

Multiple User Interfaces. Edited by A. Seffah and H. Javahery
© 2004 John Wiley & Sons, Ltd ISBN: 0-470-85444-8

development approach in which a UI is transformed into a more suitable solution for a particular platform or user. This transformation is performed in a step-wise manner by human intervention, automatic adaptation or a mixture of both. Eisenstein *et al.* [2000; 2001] discuss the use of a UI model to create UIs for mobile platforms.

The User Interface Markup Language (UIML) was developed to address the need for a uniform language for building multi-platform applications. UIML is a user interface implementation language for multiple devices. That is, it includes features that are particularly suited to describing interactive software while at the same time remaining platform independent. It emphasizes the separation of concerns of an interactive application in such a way that either: moving a program from one platform to another (see our definition of platform, later) is relatively easy, or the changes are somewhat localized. Also, the language was designed to provide access to all the functionality of the run-time platform.

UIML supports more than just a common denominator of all platforms; it provides access to the full power of each platform. Furthermore, because it is based on XML, it is easy to write transformations using a tool like XSLT [Clark 1999] that converts the language from an abstract representation into a more concrete version. The tools built around UIML extend the language through the use of transformations. These transformations allow the developer to create UIs with a single language that executes on multiple platforms. In this chapter, we discuss some of the design decisions of the language, describe our approach toward creating a new UI design methodology to build multi-platform UIs, and reveal a development environment we created to support this methodology.

As can be seen in the literature review that follows and in the other chapters in this book, there are many different approaches to solving the problem of building UIs for multiple platforms. We view the process of developing a multi-platform UI as a series of steps where the developer begins by working at a very abstract level that is common to all platforms and then moves to more concrete designs that are more specific to each platform. We have made a concerted effort to match these development stages to traditional usability engineering stages, so that developers can easily accept the methodology.

We are using a task model representation [Paternò 1999; Paternò *et al.* 2001] as the starting point for the development process. A typical task model is a hierarchical tree with different sub-trees indicating the different tasks that the user can perform in the application domain. This representation is independent of the platform. A transformation algorithm then converts the task model into a UIML program described using a generic vocabulary.

A UIML program, with its generic vocabulary, is specific to a *family of devices* (see the definition of family of devices in the next section). There is a transformation algorithm for each family of devices. For example, using a generic vocabulary for desktop applications, the developer can write a program in UIML only once and have it rendered for Java or HTML. Moreover, it can be rendered into different HTML versions and/or browser versions. Later, we present some details of the generic vocabulary and discuss the advantages of this approach.

The rest of the chapter is organized as follows: we offer some terminology explanation in Section 6.2; related work is presented in Section 6.3; Section 6.4 introduces the language features of UIML that are relevant to the discussion; the general framework that is used for multi-platform UI development is presented in Section 6.5 along with a few examples; Section 6.6 presents our development environment, TIDE, which provides developer support for the framework we propose; and Section 6.7 concludes the paper with a brief description of ongoing and future work.

6.2. TERMINOLOGY

An *application* is defined as the back-end logic behind a UI that implements the interaction supported by the user interface.

A *device* is a physical object with which an end-user interacts using a UI, such as a Personal Computer (PC), a hand-held computer (e.g. a Palm), a cell-phone, an ordinary desktop telephone or a pager.

A *toolkit* is the library or markup language used by the application program to build its UI. A toolkit typically describes the behaviour of widgets like menus, buttons and scrolling bars. In the context of this chapter, the term toolkit includes both markup languages like WML, XHTML, and VoiceXML (with their sets of tags) and more traditional APIs for imperative languages like Java Swing, Java AWT and Microsoft Foundation Classes for C++.

A *platform* is the combination of a device, an operating system (OS) and a toolkit. An example of a platform is a PC running Windows 2000 on which applications use the Java Swing toolkit. In the case of HTML, this definition has to be expanded to include the version of HTML and the particular web browser being used. For example, Internet Explorer 5.5 running on a PC with Windows NT would be a different platform than Netscape Communicator 6.0 running on the same PC with the same OS since they implement HTML differently. A family of platforms is a group of platforms that have similar layout features.

Rendering is the process of converting a UIML document into a form that can be presented in a meaningful way (e.g. through sight or sound) to an end-user, and with which the user can interact. Rendering can be accomplished in two forms: by compiling UIML into another language (e.g. WML or VoiceXML) or by interpreting UIML. An interpreter is a program that reads UIML code and makes calls to an API that displays the UI and allows interaction.

A UI *element* or *widget* is a primitive building block provided by the toolkit for creating UIs. A *vocabulary* is the set of names, properties and associated behaviours for UI elements. A *generic vocabulary* is the vocabulary shared by all platforms of a family.

Figure 6.1 clarifies some of the terminology presented. Based on our definitions, using HTML with Internet Explorer 5.5 on a Windows 2000 PC is one platform within the desktop family. This family also includes the following platforms: the Java Swing 1.3 toolkit on a Windows 2000 PC, and HTML with Netscape 6.1 on a Windows 2000 PC. Each version of Internet Explorer or Netscape (on the same device and OS) is a separate platform within the same family. Our definition of the desktop family is based on the similar layouts and screen sizes provided by the platforms that make it up. Netscape, Internet Explorer and Java Swing are all designed for devices that have ample screen real estate and similar layout facilities. On the other hand, the WAP and PDA families have significantly different layout capabilities and their devices have smaller screen sizes.

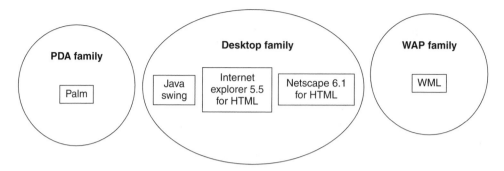

Figure 6.1. Sample families and platforms.

6.3. RELATED WORK

Multi-platform UI development falls under the umbrella of 'variety challenges' [Wiecha and Szekely 2001]. Today, there are new challenges for application developers due to the emergence of a variety of new users, devices, channels, roles and functions. In order to help surmount these challenges, we will categorize the problem of multi-platform UI development.

Transcoding [Asakawa and Takagi 2000; Hori *et al.* 2000; Huang and Sundaresan 2000] is a technique used in the World Wide Web to adaptively convert web-content for the increasingly diverse devices that are being used to access web pages. This technique converts an HTML web page into the desired format. Transcoding assumes multiple forms: in its simplest form, semantic meaning is inferred from the structure of the web page and the page is transformed using this semantic information. A more sophisticated version of transcoding associates annotations with the structural elements of the web page and the transformation occurs based on these annotations. Another version infers semantics based on a group of web pages. Although these approaches work within a limited context of use, they are not very extensible since it is not always possible to infer semantic information from the structural elements (i.e. syntactic structure) of a web page.

It is useful to revisit some of the concepts behind model-based UI development tools since some of these concepts have to be used to generate multi-platform UIs [Myers *et al.* 2000]. Eisenstein *et al.* [2000; 2001] state that for mobile computing a UI model is an essential step in the design and development of the UI. This makes the process less time-consuming and error-prone. Model-based UI development tools use high-level specifications of different aspects of the UI and automatically generate some or all of the user interface [Myers 1995]. One of the central ideas behind model-based tools is to achieve balance between detailed control over the design of the UI and automation. Some of the model-based tools built in the late 80s and early 90s include UIDE [Sukaviriya and Foley 1993; Sukaviriya *et al.* 1993], Interactive UIDE [Frank and Foley 1993], HUMANOID [Luo *et al.* 1993], MASTERMIND [Szekely *et al.* 1995], and ITS [Wiecha *et al.* 1990].

The primary component in a model-based tool is the *model* that is used to represent the UI in an abstract fashion. Different types of models have been used in different systems, including task models, dialogue models, user models, domain models, and application models. All of these models represent the UI at a higher level of abstraction than what is possible with a more concrete representation. The UI developer builds these models and they are automatically or semi-automatically transformed into the final UI.

Some recent approaches to using models for building web sites include WebML [Bonifati *et al.* 2000; Ceri *et al.* 2000] and the AutoWeb system [Fraternali 1999; Fraternali and Paolini 2000]. The basic idea behind WebML is to develop a conceptual specification for a web site using four different models: structural, hypertext, presentation and personalization. These models are adapted from the E/R model and UML class diagrams. The AutoWeb system uses a similar notation called Hypermedia Design Model-lite (HDM-lite) that utilizes concepts from database modelling to develop conceptual models for web site development. Once these models are built, an automatic transformation is done to yield the final web pages. One other recent approach that uses a model-based system for multi-device UIs is presented in [Vanderdonckt *et al.* 2001]. A couple of other approaches to developing multi-device collaborative applications are Manifold [Marsic 2001] and WebSplitter [Han *et al.* 2000].

Since the advent of the World Wide Web in the mid-90s and the emergence of the eXtensible Markup Language (XML) as a standard meta-language, a number of different markup languages have surfaced for creating UIs for different devices. The foremost among these are HTML for desktop machines, Wireless Markup Language (WML) [WAPForum] for small hand-held devices, and VoiceXML [McGlashan *et al.* 2001] for voice-enabled devices. HTML can be considered a language for multi-platform development, but it only supports platforms that are in the same family (primarily desktop computers). Some people have tried to make HTML files available on other devices (e.g. Transcoding, above) but in general HTML has remained tied to desktop computers. XForms [Dubinko *et al.* 2002] is the next generation of HTML; it greatly enhances the capabilities of the current forms available in HTML. One of XForms' goals is to provide support for hand-held, television and desktop browsers.

While all of these markup languages have almost entirely removed the UI developer's need to know the specifics about the toolkit, hardware, and operating system, they have not made a significant contribution toward multi-platform development. Model-based tools were primarily designed for a single platform and do not consider multi-platform UIs. The other shortcoming of these tools is that some of them incorporate layout information within their transformation algorithms, thus making them inextensible. Most of the languages that we have mentioned, such as WML and VoiceXML, are geared toward a single platform.

UIML shares some aspects of its development with earlier attempts at researching model-based tools. However, one must not discount UIML's contribution to UI research and development on the basis of its similarity to these earlier approaches. A brief look at the history of UI research explains the difference.

According to Myers, Hudson and Pausch [Myers *et al.* 2000], UI toolkits that have made an impact share the following characteristics (among others): a low threshold, a high ceiling, and predictability. According to the authors, other model-based tools fail because they have a high threshold, a low ceiling, and unpredictability. A *high threshold* means that the toolkit often requires the developer to learn specialized languages in order to use it. A *low ceiling* indicates that the toolkit only works for a small class of UI applications (e.g. a Web-based UI tool that will not work with other interface styles). Developers quickly run into the toolkit's limitations. Finally, a toolkit's *unpredictability* is due in large part to its approach. Most unpredictable tools apply sophisticated artificial intelligence algorithms to generate their interface. As a result, it is difficult for the developer to know what to modify in the high-level model in order to produce a desired change in the UI. UIML, while similar in nature to some of the other model-based tools, has a few new design twists that make it interesting from a UI research and development point of view.

First, the language is designed for multiple platforms and families of devices. This is done without attempting to define a lowest common denominator of device functionality. Instead, UIML uses a generic vocabulary and other techniques to produce interfaces for the different platforms. The advantage of this approach is that while developers will still need to learn a new language (namely, UIML), this language is all they will need to know to develop UIs for multiple platforms. This helps lower the threshold of using UIML.

Secondly, UIML provides mapping to a platform's toolkit. Thus, UIML in and of itself does not restrict the types of applications that can be developed for different platforms. Therefore, UIML has a high ceiling.

Finally, predictability is not an issue because UIML does not use sophisticated artificial intelligence algorithms to generate UIs. Instead, it relies on a set of simple transformations (taking advantage of XML's capabilities) that produce the resulting interface. From the developer's point of view, it is clear which part of the UIML specification generates a specific part of the UI. Furthermore, the tools we are building attempt to make this relationship between different levels of specification more clear to the developer.

6.4. UIML

UIML [Abrams and Phanouriou 1999; Phanouriou 2000] is a declarative XML-based language that can be used to define user interfaces. One of the original design goals of UIML was to 'reduce the time to develop user interfaces for multiple device families' [Abrams *et al.* 1999]. A related design rationale behind UIML was to 'allow a family of interfaces to be created in which the common features are factored out' [Abrams and Phanouriou 1999]. This indicates that the capability to create multi-platform UIs was inherent in the design of UIML.

Although UIML allows a multi-platform description of UIs, there is limited commonality between the platform-specific descriptions when platform-specific vocabularies are used. This means that the UI designer will have to create separate UIs for each platform using its own vocabulary. Recall that a vocabulary is defined to be a set of UI elements with associated properties and behaviour. Limited commonality is not a shortcoming of UIML itself, but a result of the inherent differences between platforms with varying form factors.

One of the primary design goals of UIML is to provide a single, canonical format for describing UIs that map to multiple devices. Phanouriou [2000] lists some of the criteria used in designing UIML:

1. UIML should map the canonical UI description to a particular device/platform.
2. UIML should separately describe the content, structure, behaviour and style of a UI.
3. UIML should describe the UI's behaviour in a device-independent fashion.
4. UIML should give as much power to a UI implementer as a native toolkit.

6.4.1. LANGUAGE OVERVIEW

Since UIML is XML-based, the different components of a UI are represented through a set of tags. The language itself does not contain any platform-specific or metaphor-dependent tags. For example, there is no tag like `<window>` that is directly linked to the desktop metaphor of interaction. Platform-specific renderers have to be built in order to render the interface defined in UIML for that particular platform. Associated with each platform-specific renderer is a vocabulary of the language widget-set or tags that are used to define the interface on the target platform.

Below, we see a UIML document skeleton:

```
<?xml version="1.0"?>
<!DOCTYPE uiml PUBLIC "-//UIT//DTD
UIML 2.0 Draft//EN" UIML2_Of.dtd">

<uiml>
  <head>...</head>
  <interface>...</interface>
  <peers>...</peers>
  <template>...</template>
</uiml>
```

Figure 6.2. Skeleton of a UIML document.

At its highest level, a UIML document is comprised of four components: `<head>`, `<interface>`, `<peers>` and `<template>`. The `<interface>` and the `<peers>` are the only components that are relevant to this discussion; information on the others can be found elsewhere [Phanouriou 2000].

6.4.2. THE `<INTERFACE>` COMPONENT

This is the heart of the UIML document in that it represents the actual UI. All of the UIML elements that describe the UI are present within this tag. Its four main components are:

`<structure>`: The physical organization of the interface, including the relationships between the various UI elements within the interface, is represented with this tag. Each `<structure>` is comprised of `<part>` tags. Each `<part>` represents an actual platform-specific UI element and is associated with a single class (i.e. category) of UI elements. One may nest `<part>` tags to represent a hierarchical relationship. There might be more than one `<structure>` root in a UIML document, each representing different organizations of the same UI. This allows one to support multiple families or platforms.

`<style>`: The `<style>` tag contains a list of properties and values used to render the UI. The properties are usually associated with individual parts within the UIML document through the part-names. Properties can also be associated with particular classes of parts. Typical properties associated with parts for Graphical User Interfaces (GUIs) could be the background colour, foreground colour, font, etc. It is also possible to have multiple styles within a single UIML document associated with multiple structures or even the same structure. This facilitates the use of different styles for different contexts.

`<content>`: This tag holds the subject matter associated with the various parts of the UI. A clean separation of the content from the structure is useful when different content is needed under different contexts. This feature of UIML is very helpful when creating UIs that might be displayed in multiple languages. An example of this is a UI in French and English, for which different content is needed in each language.

`<behavior>`: Enumerating a set of conditions and associated actions within rules specifies the behaviour of a UI. UIML permits two types of conditions: the first condition is true when an event occurs, while the second condition is true when an event occurs and an associated datum is equal to a particular value. There are four kinds of actions that occur: the first action assigns a value to a property, the second action calls an external function or method, the third action launches an event and the fourth action restructures the UI.

6.4.3. THE `<PEERS>` COMPONENT

UIML provides a `<peers>` element to allow the mapping of class names and events (within a UIML document) to external entities. There are two child elements within a `<peers>` element:

The `<presentation>` element contains mappings of part and event classes, property names, and event names to a UI toolkit. This mapping defines a vocabulary to be used with a UIML document, such as a vocabulary of classes and names for VoiceXML or WML.

The `<logic>` element provides the glue between UIML and other code. It describes the calling conventions for methods that are invoked by the UIML code. An extremely detailed discussion of the language design issues can be found in Phanouriou's dissertation [Phanouriou 2000].

6.4.4. A SAMPLE UI

To better understand the features of the language, consider the sample UI displayed in Figure 6.3. A UIML renderer for Java produced this UI. The UIML code corresponding to this interface is presented in Figure 6.4. The UI itself is pretty simple. As indicated in Figure 6.3, the UI displays the string 'Hello World!' Clicking on the button changes the string's content and colour.

An important point to be observed here is that the UIML code in Figure 6.4 is platform-specific for the Java AWT/Swing platform. Hence, we observe the use of Java Swing-specific UIML part-names like JFrame, JButton and JLabel in the UIML code. The UI

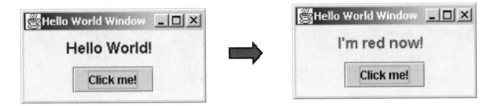

Figure 6.3. Sample interface.

is comprised of the label for the string and the button, both of which are enclosed in a container. This relationship is indicated in the structure part of the UIML code. The other presentation and layout characteristics of the parts are indicated in UIML through various properties. All these properties can be grouped together in the style section. Note that each property for a part is indicated through a name. What actually happens when a user interacts with the UI is indicated in the <behavior> section of the UIML document. In this example, two actions are triggered when the user clicks the button: 'Hello World' changes to 'I'm red now', and the text's colour changes to red. As indicated in Figure 6.4, this is presented in UIML in the form of a rule that in turn is composed of two parts: a condition and an action.

Currently, there are platform-specific renderers available for UIML for a number of different platforms. These include Java, HTML, WML, and VoiceXML. Each of these renderers has a platform-specific vocabulary associated with it to describe its UI elements, behaviour and layout. The UI developer uses the platform-specific vocabulary to create a UIML document that is rendered for the target platform. The example presented in Figure 6.4 is an example of UIML used with a Java Swing vocabulary. The renderers are available from http://www.harmonia.com/.

There is a great deal of difference between the vocabularies associated with each platform. Consequently, a UI developer will have to learn each vocabulary in order to build UIs that will work across multiple platforms. Using UIML as the underlying language for cross-platform UIs reduces the amount of effort required in comparison with the effort that would be required if the UIs had to be built independently using each platform's native language and toolkit.

Unfortunately, UIML alone cannot solve the problem of creating multi-platform UIs. The differences between platforms are too significant to create one UIML file for one particular platform and expect it to be rendered on a different platform with a simple change in the vocabulary. In the past, when building UIs for platforms belonging to different families, we have had to redesign the entire UI due to the differences between the platform vocabularies and layouts. In order to solve this problem, we have found that more abstract representations of the UI are necessary, based on our experience with creating a variety of UIs for different platforms. The abstractions in our approach include using a task model for all families and a generic vocabulary for one particular family. These approaches are discussed in detail in the following sections.

```
<?xml version="1.0" encoding="ISO-8859-1" ?>
<!DOCTYPE uiml PUBLIC "-//Harmonia//DTD UIML 2.0 Draft//EN"
 "UIML2_0g.dtd">
<uiml>
 <head>
   <meta name="Purpose" content="Hello World UIML example"/>
 </head>
 <interface>
  <structure>
    <part id="HWF" class="JFrame">
      <part id="HWL" class="JLabel"/>
      <part id="HWB" class="JButton"/>
    </part>
  </structure>
  <style>
    <property part-name="HWF" name="title">Hello World Window
    </property>
    <property part-name="HWF" name="layout">java.awt.FlowLayout
    </property>
    <property part-name="HWF" name="resizable">true</property>
    <property part-name="HWF" name="background">CCFFFF</property>
    <property part-name="HWF" name="foreground">black</property>
    <property part-name="HWF" name="size">200,100</property>
    <property part-name="HWF" name="location">100,100</property>
    <property part-name="HWL" name="font">ProportionalSpaced-Bold-16
    </property>
    <property part-name="HWL" name="text">Hello World!</property>
    <property part-name="HWB" name="text">Click me!</property>
  </style>
  <behavior>
    <rule>
      <condition>
        <event class="actionPerformed" part-name="HWB"/>
      </condition>
      <action>
       <property part-name="HWL" name="foreground">FF0000</property>
       <property part-name="HWL" name="text">I'm red now!</property>
      </action>
    </rule>
  </behavior>
 </interface>
 <peers>
  <presentation base="Java_1.3_Harmonia_1.0"
   source="Java_1.3_Harmonia_1.0.uiml#vocab"/>
 </peers>
</uiml>
```

Figure 6.4. UIML code for sample UI in Figure 6.3.

6.5. A FRAMEWORK FOR MULTI-PLATFORM UI DEVELOPMENT

The concept of building multi-platform UIs is relatively new. To envision the development process, we consider an existing, traditional approach from the Usability Engineering (UE) literature. One such approach, [Hix and Hartson 1993], identifies three different phases in the UI development process: interaction design, interaction software design and interaction software implementation.

Interaction design is the phase of the usability engineering cycle in which the 'look and feel' and behaviour of a UI is designed in response to what a user hears, sees or does. In current UE practices, this phase is highly platform-specific. Once the interaction design is complete, the interaction software design is created. This involves making decisions about UI toolkit(s), widgets, positioning of widgets, colours, etc. Once the interaction software design is finished, the software is implemented.

The above paragraph describes the traditional view of interaction design. This view is highly platform-specific and works well when designing for a single platform. However, when working with multiple platforms, interaction design has to be split into two distinct phases: platform-independent interaction design and platform-dependent interaction design. These phases lead to different, platform-specific interaction software designs that in turn lead to platform-specific UIs. Figure 6.5 illustrates this process.

We have developed a framework that is very closely related to the traditional UE process (our framework is illustrated in Figure 6.5). The main building blocks of this framework are the *task model*, the *family model* and the *platform-specific UI*. Each building block has a link to the traditional UE process. The three building blocks are interconnected via a process of transformation. More specifically, the task model is transformed into the family model, and the family model is transformed into the platform-specific UI (which is represented by UIML). Next, each of these building blocks will be described, and the transformation process will be explained.

6.5.1. TASK MODEL

Task analysis is an important step in the process of interaction design. It is one of the steps of system analysis, and it is performed to capture the requirements of typical tasks

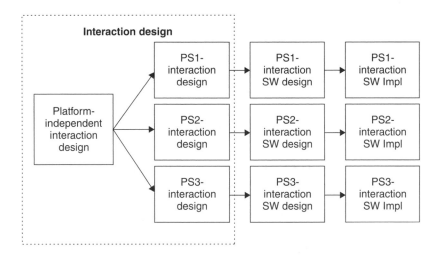

Figure 6.5. Usability Engineering process for multiple platforms.

performed by users. Task analysis is a user-centred process that helps define UI features in terms of the tasks performed by users. It helps to provide a correspondence between user tasks and system features. The task model is an interesting product of task analysis. In its simplest form, the task model is a directed graph that indicates the dependencies between different tasks. It describes the tasks that users perform with the system. Task models have been a central component of many model-based systems including MAS-TERMIND [Szekely *et al.* 1995], ADEPT [Johnson *et al.* 1995], TRIDENT [Bodart *et al.* 1995] and MECANO [Puerta *et al.* 1994].

Recently, Paternò [2001], Eisenstein *et al.* [2000; 2001] and Puerta and Eisenstein [2001] each discussed the use of a task model in conjunction with other UI models in order to create UIs for mobile devices. Depending on the complexity of the application, there are different ways that a task model can be used to generate multi-platform UIs. When an application must be deployed in the same fashion across several platforms, the task model will be the same for all target platforms. This indicates that the user wants to perform the same set of tasks regardless of the platform or device. On the other hand, there might be applications where certain tasks are not suited for certain platforms. Eisenstein *et al.* [2000; 2001] provide a good example of an application where individual tasks are better suited for certain platforms. From the point of view of the task model, this means that some portions of the graph are not applicable for some platforms.

We use a task model in conjunction with UIML to facilitate the development of multi-platform UIs. The task model is developed at a higher level of abstraction than what is currently possible with UIML. The main objective of the task model is to capture enough information about the UI to be able to map it to multiple platforms. An added rationale behind using a task model is that it is already a well-accepted notation in the process of interaction design. Hence, we are not using a notation that is alien to the UI design community.

Our notation is partly based on the Concurrent Task Tree (CTT) notation developed by Fabio Paternò [1999]. The original CTT notation used four types of tasks: *abstraction*, *user*, *application* and *interaction*. We do not use the user task type in our notation.

In our notation, the task model is transformed into a family model, which corresponds to generic UIML guided by the developer. We envision our system providing a set of preferences to facilitate the transformation of each task in the model into one or more elements in the generic UIML. The task model is also used to generate the navigation structure on the target platforms. This is particularly important for platforms like WML and VoiceXML, where information is provided to the user in small blocks. This helps the end-user to navigate easily between blocks of information.

6.5.2. GENERIC DESCRIPTION OF DEVICE FAMILIES

Within our framework, the family model is a generic description of a UI (in UIML) that will function on multiple platforms. As indicated in Figure 6.6, there can be more than one family model. Each family model represents a group of platforms that have similar characteristics.

In distinguishing family models, we use the physical layout of the UI elements as the defining characteristic. For example, different HTML browsers and the Java Swing

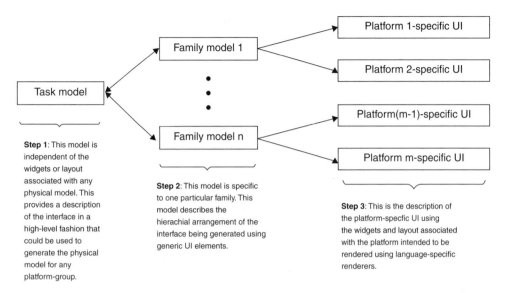

Figure 6.6. The framework for building multi-platform UIs using UIML.

platform can all be considered part of one family model based on their similar layout facilities. Some platforms might require a family model of their own. The VoiceXML platform is one such example, since it is used for voice-based UIs and there is no other analogous platform for either auditory or graphical UIs.

An additional factor that comes up while defining a family is the navigation capabilities provided by the platforms within the family. For example, WML 1.2 [WAPForum] uses the metaphor of a deck of cards. Information is presented on each card and the end-user navigates between the different cards.

Building a family model requires one to build a generic vocabulary of UI elements. These elements are used in conjunction with UIML in order to describe the UI for any platform in the family. The advantage of using UIML is apparent since it allows any vocabulary to be attached to it. In our framework, we use a generic vocabulary that can be used in the family model. Recall that a generic vocabulary is defined to be one vocabulary for all platforms within a family. Creating a generic vocabulary can solve some of the problems outlined above. The family models that can currently be built are for the desktop platform (Java Swing and HTML) and the phone (WML). These family models are based on the available renderers. The specification for the family model is already built.

From Section 6.2, we recall that the definition of family refers to multiple platforms that share common layout capabilities. Different platforms within a family often differ on the toolkit used to build the interface. Consider, for example, a Windows OS machine capable of displaying HTML using some browser and capable of running Java applications. HTML and Java use different toolkits. This makes it impossible to write an application for one and have it execute on the other, even though they both run on the same hardware device using the same operating system. For these particular cases, we have built support for generic vocabularies into UIML.

UIML Vocabularies available August 2001 (from http://www.uiml.org/toolkits):

- W3C's Cascading Style Sheets (CSS)
- W3C's Hypertext Markup Language (HTML) v4.01 with the frameset DTD_ and CSS Level 1
- Java™ 2 SDK (J2SE) v1.3, specifying AWT and Swing toolkits
- A single, generic (or multi-platform) vocabulary for creating Java *and* HTML user interfaces
- VoiceXML Forum's VoiceXML v1.0
- WAP Forum's Wireless Markup Language (WML) v1.3

A generic vocabulary of UI elements, used in conjunction with UIML, can describe any UI for any platform within its family. The vocabulary has two objectives: first, to be powerful enough to accommodate a family of devices, and second, to be generic enough to be used without requiring expertise in all the various platforms and toolkits within the family.

As a first step in creating a generic vocabulary, a set of elements has to be selected from the platform-specific element sets. Secondly, several generic names, representing UI elements on different platforms, must be selected. Thirdly, properties and events have to be assigned to the generic elements. We have identified and defined a set of generic UI elements (including their properties and events). Ali and Abrams [2001] provide a more detailed description of the generic vocabulary.

Table 6.1 shows some of this vocabulary's part classes for the desktop family (which includes HTML 4 and Java Swing).

The mechanism that is currently employed for creating UIs with UIML is one where the UI developer uses the platform-specific vocabulary to create a UIML document that is rendered for the target platform. These renderers can be downloaded from http://www.harmonia.com.

The platform-specific vocabulary for Java uses AWT and Swing class names as UIML part names. The platform-specific vocabularies for HTML, WML, and VoiceXML use

Table 6.1. A generic vocabulary.

Generic Part	UIML Class Name	Generic Part	UIML Class Name
Generic top container	G:TopContainer	Generic Label	G:Label
Generic area	G:Area	Generic Button	G:Button
Generic Internal Frame	G:InternalFrame	Generic Icon	G:Icon
Generic Menu Item	G:Menu	Generic Radio Button	G:RadioButton
Generic Menubar	G:MenuBar	Generic File Chooser	G:FileChooser

HTML, WML, and VoiceXML tags as UIML part names. This enables the UIML author to create a UI that is equivalent to what is possible in Java, HTML, WML, or VoiceXML. However, the platform-specific vocabularies are not suitable for a UI author who wants to create UIML documents that map to *multiple* target platforms. For this, a generic vocabulary is needed. To date, one *generic* vocabulary has been defined, *GenericJH*, which maps to both Java Swing and HTML 4.0. The next section describes how a generic vocabulary is used with UIML.

6.5.3. ABSTRACT TO CONCRETE TRANSFORMATIONS

We can see from Figure 6.6 that there needs to be a transition between the different representations in order to arrive at the final platform-specific UI. There are two different types of transformations needed here. The first type of transformation is the mapping from the task model to the family model. This type of transformation has to be developer-guided and cannot be fully automated. By allowing the UI developer to intervene in the transformation and mapping process, it is possible to ensure usability.

One of the main problems of some of the earlier model-based systems was that a large part of the UI generation process from the abstract models was fully automated, removing user control of the process. This dilemma is also known as the 'mapping problem', as described by Puerta and Eisenstein [1999]. We want to eliminate this problem by having the user guide the mapping process. Once the user has identified the mappings, the system generates the family models based on the target platforms and the user mappings. The task model in the CTT notation is used to generate generic UIML. The task categories and the temporal properties between the tasks are used to generate the `<structure>`, partial `<style>` and the `<behavior>` in the generic UIML for each family.

The second type of transformation occurs between the family model and the platform-specific UI. This is a conversion from generic UIML to platform-specific UIML, both of which can be represented as trees since they are XML-based. This process can be largely automated. However, there are certain aspects of the transformation that need to be guided by the user. For example, there are certain UI elements in our generic vocabulary that could be mapped to more than one element on the target platform. The developer has to select what the mapping will be for the target platform. Currently, the developer's selection of the mapping is a special property of the UI element. The platform-specific UIML is then rendered using an existing UIML renderer. There are several types of transformations that are performed:

- Map a generic class name to one or more parts on the target platform. For example, in HTML a G:TopContainer is mapped to the following sequence of parts:

```
<html>
  <head>...
    <title>...
    <base>...
    <style>...
    <link>...
    <meta>...
  <body>...
```

- Map the properties of the generic part to the correct platform-specific part. In Java a G:TopContainer is mapped to only one part: JFrame.
- Map generic events to the proper platform-specific events.

In order to allow a UI designer to fine-tune the UI to a particular platform, the generic vocabulary contains platform-specific properties. These are used when one platform has a property that has no equivalent on another platform. In the generic vocabulary, these property names are prefixed by J: or H: for mapping to Java or HTML only. The transform engine automatically identifies which target part to associate the property with, in the event that a generic part (e.g. G:TopContainer) maps to several parts (e.g. seven parts for HTML). This is also done for events that are specific to one platform. The resulting interface could be as powerful as the native platform. The multiple style section allows each interface to be as complete as the native platform allows. The generic UIML file will then contain three `<style>` elements. One is for cross-platform style, one for HTML, and one for Java UIs:

```
<uiml>
...
  <style id ="allPlatforms">
     <property id ="g:title">My User Interface</property>
  </style>
  <style id ="onlyHTML" source ="allPlatforms">
     <property id ="h:link-color">red</property>
  </style>
  <style id ="onlyJava" source ="allPlatforms">
     <property id ="j:resizable">red</property>
  </style>
</uiml>
```

In the example above, both a web browser and a Java frame display the title, 'My User Interface'. However, only web browsers can have the colour of their links set, so the property *h:link-color* is used only for HTML UIs. Similarly, only Java UIs can make themselves non-resizable, so the *j:resizable* property applies only to Java UIs. When the UI is rendered, the renderer chooses exactly one `<style>` element. For example, an HTML UI would use *onlyHTML*. This `<style>` element specifies in its *source* attribute the name of the shared *allPlatforms* style, so that the *allPlatforms* style is shared by both the HTML and Java style elements. Figure 6.7 illustrates two interfaces, for Java Swing and HTML, generated from generic UIML thanks to a transformation process.

```
<?xml version ="1.0"?>
<!DOCTYPE uiml PUBLIC "-//Harmonia//DTD UIML 2.0 Draft//EN"
  "UIML2_0g.dtd">
<uiml>
 <head>
  <meta name ="Purpose" content ="Data Collection Form"/>
  <meta name ="Author" content ="Farooq Ali"/>
 </head>
 <interface name ="DataCollectionForm">
```

Figure 6.7. Screenshots of a sample form in Java (left) and HTML (right).

```
<structure>
 <part name ="RequestWindow" class ="G:TopContainer">
   <part name ="EBlock1" class ="G:Area">
     <part name ="TitleLabel" class ="G:Label"/>
     <part name ="FirstName" class ="G:Label"/>
     <part name ="FirstNameField" class ="G:Text"/>
     <part name ="LastName" class ="G:Label"/>
     <part name ="LastNameField" class ="G:Text"/>
     <part name ="StreetAddress" class ="G:Label"/>
     <part name ="StreetAddressField" class ="G:Text"/>
     <part name ="City" class ="G:Label"/>
     <part name ="CityField" class ="G:Text"/>
     <part name ="State" class ="G:Label"/>
     <part name ="StateChoice" class ="G:List"/>
     <part name ="Zip" class ="G:Label"/>
     <part name ="ZipField" class ="G:Text"/>

     <part name ="OKBtn" class ="G:Button"/>
     <part name ="CancelBtn" class ="G:Button"/>
     <part name ="ResetBtn" class ="G:Button"/>
   </part>
 </part>
</structure>
```

6.6. TRANSFORMATION-BASED UI DEVELOPMENT ENVIRONMENT

A transformation-based UI development process places the developer in unfamiliar territory. Developers are accustomed to having total control over the language and the specification of the UI elements. A transformation-based process asks the developer to provide a high-level description of the interface and then to trust the end result. This is one of the cited limitations of code-generators and model-based UI systems [Myers *et al.* 2000].

To address this limitation, we have developed a Transformation-based Integrated Development Environment (TIDE) for UIML. In the first version of TIDE, the developer writes generic UIML code and the interface is rendered using the appropriate UIML renderers. However, the relationship between the UIML code and its resulting interface components are explicitly shown. This section briefly shows how the first version of TIDE operates, TIDE's future design goals, and some screenshots of the redesigned tool (which is currently in the prototype stage).

6.6.1. TIDE VERSION 1

The TIDE application was built on the idea that when developers create an interface in an abstract language (such as UIML) that will be translated into one or more specific languages, they follow a process of trial and error. The developer builds what he thinks will be suitable in UIML, renders his work onto the desired platform, and then makes changes as appropriate. TIDE, an environment designed to help support this process, shows the developer three things: the original UIML source code, the resulting interface after rendering, and the relationship between elements in the two stages. Figure 6.8 shows two screenshots of the TIDE environment.

TIDE uses Harmonia's LiquidUI product suite (version 1.0c) to render from the original generic UIML to Java. The developer may open and close files, view the original UIML source code as plain text or as a tree (using Java's JTree to display it, as shown in Figure 6.8), and make changes from the tree view. The developer may also re-render at any time by pressing the red arrow in the centre of the window.

The relationship between UIML code and the rendered interface is made explicit as shown in Figure 6.8 above. The developer may click on a node in the UIML tree view (the textual view on the left) and the corresponding element on the graphical user interface is highlighted on the right side. The reverse is also true; if the developer clicks on a component of the graphical UI, the corresponding UIML node is highlighted on the left panel. On the right hand side of the bottom frame of Figure 6.8, the developer has clicked on the OK button (the leftmost of the three buttons) and the corresponding code is highlighted on the UIML tree view.

TIDE makes it very easy to explore the different UIML elements and to see the effects they have on the rendering of the UI. For example, a UIML element's property (e.g. the colour of a button) can be directly edited within the tree view. TIDE even supports a history window that keeps track of different changes made to the interface. Each line in the history window (see Figure 6.9) shows a small screen image of the interface at that point in the development cycle. This allows the developer to quickly switch between alternative versions of the interface, thus encouraging more exploration of UIML.

6.6.2. GOALS FOR TIDE 2

The original version of TIDE only had support for UIML with a Java vocabulary. We are currently extending TIDE to provide support for the task model described above and some of the generic vocabularies. The idea is to have four panels that support the

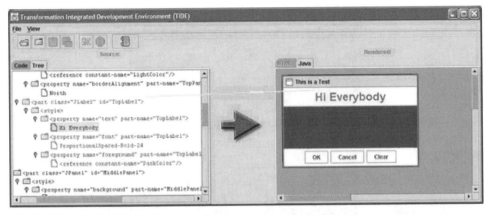

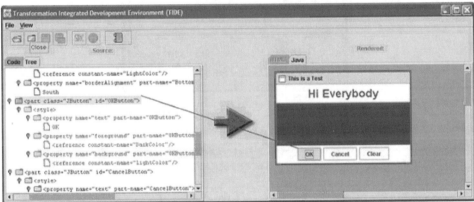

Figure 6.8. UIML code in TIDE.

transformation process, helping the designer understand the nature of each transformation. This way control of the design will not be relinquished to the tool.

We envision that a developer will use TIDE 2 as follows: First, he/she will create a task model. Secondly, this model will be transformed into a series of generic UIML representations (for each of the different families of devices). This generic UIML will require modification, because not all of the UI details are derived from the task model. Thus, at this stage the developer will be able to edit the generic UIML code. We want to support iterative refinement of the UI. To accomplish this we will save the changes the developer makes to the generic UIML code. This will give him/her the ability to edit the UI at any of the different levels of representation without losing the ability to re-generate the UI. The developer's main task is a combination of editing task model details (which apply to all interfaces), editing family-specific UIML, and editing the generated UIML (which is platform-specific).

The initial prototype of TIDE 2 is shown in Figure 6.10. This prototype only supports the desktop family (HTML and Java), but the general idea is clear from the screenshot.

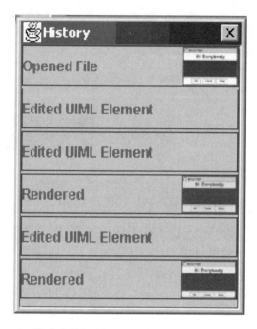

Figure 6.9. History Window in TIDE.

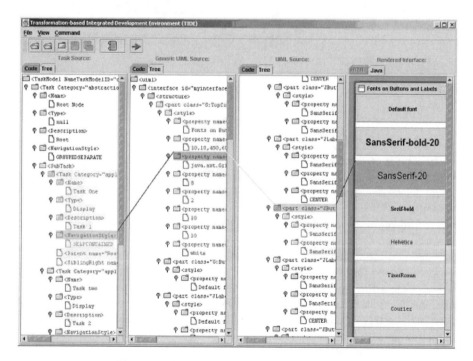

Figure 6.10. TIDE 2, showing different models.

The left-most panel shows the task representation. The second panel from the left shows the result of transforming the representation into a generic UIML for the desktop family. The third panel shows the UIML code for the Java platform. The last panel on the right shows the rendered interface.

One research feature that we are currently implementing is support for the iterative refinement process described earlier. The implementation is straightforward. The transformation algorithm produces a table of mappings between a task representation node and the generated node in the generic UIML. Also, all user actions are already captured in command objects to support Undo/Redo. These command objects are stored in a data structure together with the modified node and the source node where the modified node was generated. When a task model is re-transformed into UIML, the IDE uses this information to do the following:

```
For all command objects representing user actions performed since the last
    transformation
 Find the source node in the mapping table generated by the transformation
    algorithm
 From the mapping table, find the newly generated node and apply the command
    object
```

This simple algorithm supports the maintenance of all changes made to existing UIML parts and properties across multiple transformations. It does not, however, support reinserting new parts into the interface once the transformation algorithm has been executed. We are exploring how to capture that information to better support the iterative development process.

We expect a fully operational version of TIDE to be available upon publication of this book, in 2003. The current version is a high-fidelity prototype that is allowing us to explore how developers accept this highly interactive, exploratory environment.

6.7. CONCLUSIONS

In this paper we have shown some of our research on extending and utilizing UIML to generate multi-platform UIs. We are using a single language, UIML, to provide the multi-platform development support needed. This language is extended via the use of a task model, alternate vocabularies and transformation algorithms.

We have developed a multi-step transformation-based framework, using the UIML language, that can be used to generate multi-platform UIs. The current framework utilizes concepts from the model-based UI development literature and Usability Engineering realm and applies them to this new area of multi-platform UI development. This framework tries to eliminate some of the pitfalls of other model-based approaches by having multiple steps and allowing for developer intervention throughout the UI generation process. Our approach allows the developer to build a single specification for a family of devices. UIML and its associated tools transform this single representation to multiple platform-specific representations that can then be rendered to each device.

We have presented our current research on extending UIML to allow the building of UIs for very different platforms, such as wireless devices and desktop computers. We are currently working on incorporating the task model within TIDE to allow a complete lifecycle-based approach toward developing multi-platform UIs.

ACKNOWLEDGEMENTS

We would like to acknowledge Eric Shell's incredible work in building the TIDE tool. We would like to thank Scott Preddy for his work on the prototype of TIDE 2. This material is based upon work partially supported by the National Science Foundation under Grant No. IIS-0049075.

REFERENCES

Abrams, M. and Phanouriou, C. (1999) *UIML: An XML Language for Building Device-Independent User Interfaces*. Proceedings of the XML'99, Philadelphia.

Abrams, M., Phanouriou, C., Batongbacal, A., and Shuster, J. (1999) *UIML: An Appliance-Independent XML User Interface Language*. Proceedings of the 8th World Wide Web, Toronto.

Ali, M.F. and Abrams, M. (2001) *Simplifying Construction of Multi-Platform User Interfaces Using UIML*. Proceedings of the UIML'2001, Paris, France.

Asakawa, C. and Takagi, H. (2000) *Annotation-Based Transcoding for Nonvisual Web Access*. Proceedings of the Assets'2000, Arlington, Virginia, USA.

Bodart, F., Hennebert, A.-M., Leheureux, J.-M., Provot, I., Sacre, B., and Vanderdonckt, J. (1995) *Towards a Systematic Building of Software Architecture: the TRIDENT Methodological Guide*. Proceedings of the Eurographics Workshop on Design, Specification, Verification of Interactive Systems DSV-IS'95.

Bonifati, A., Ceri, S., Fraternali, P., and Maurino, A. (2000) *Building Multi-device, Content-Centric Applications Using WebML and the W3I3 Tool Suite*. Proceedings of the ER 2000 Workshops on Conceptual Modeling Approaches for E-Business and the World Wide Web and Conceptual Modeling, Salt Lake City, Utah, USA.

Brewster, S., Leplâtre, G., and Crease, M. (1998) *Using Non-Speech Sounds in Mobile Computing Devices*. Proceedings of the First Workshop on Human Computer Interaction of Mobile Devices, Glasgow.

Calvary, G., Coutaz, J., and Thevenin, D. (2000) *Embedding Plasticity in the Development Process of Interactive Systems*. Proceedings of the Sixth ERCIM Workshop 'User Interfaces for All', Florence, Italy.

Ceri, S., Fraternali, P., and Bongio, A. (2000) Web Modeling Language (WebML): A modelling language for designing Web sites. *Computer Networks*, 33.

Clark, J. (1999) XSL Transformations (Version 1.0). http://www.w3.org/TR/xslt.

Dubinko, M., Leigh, L., Klotz, J., Merrick, R., and Raman, T.V. (2002) XForms 1.0: W3C Candidate Recommendation. http://www.w3.org/TR/2002/CR-xforms-20021112/.

Eisenstein, J., Vanderdonckt, J., and Puerta, A. (2000) *Adapting to Mobile Contexts with User-Interface Modeling*. Proceedings of the Third IEEE Workshop on Mobile Computing Systems and Applications.

Eisenstein, J., Vanderdonckt, J., and Puerta, A., (2001) *Applying Model-Based Techniques to the Development of UIs for Mobile Computers*. Proceedings of the Intelligent User Interfaces (IUI'2001), Santa Fe, New Mexico, USA.

Frank, M. and Foley, J. (1993) *Model-Based User Interface Design by Example and by Interview*. Proceedings of the User Interface Software and Tools (UIST).

Fraternali, P. (1999) Tools and Approaches for Developing Data-Intensive Web Applications: A Survey. *ACM Computing Surveys*, vol. 31, pp. 227–263.

Fraternali, P. and Paolini, P. (2000) Model-Driven Development of Web Applications: The Autoweb System. *ACM Transactions on Information Systems*, vol. 28, pp. 323–382.

Han, R., Perret, V., and Nagshineh, M. (2000) *WebSplitter: A Unified XML Framework for Multi-Device Collaborative Web Browsing*. Proceedings of the CSCW 2000, Philadelphia, USA.

Hix, D. and Hartson, R. (1993) *Developing User Interfaces: Ensuring usability through product and process*: John Wiley and Sons.

Hori, M., Kondoh, G., Ono, K., Hirose, S., and Singhal, S. (2000) *Annotation-Based Web Content Transcoding*. Proceedings of the Ninth World Wide Web Conference, Amsterdam, Netherlands.

Huang, A. and Sundaresan, N. (2000) *Aurora: A Conceptual Model for Web-Content Adaptation to Support the Universal Usability of Web-based Services*. Proceedings of the Conference on Universal Usability, CUU 2000, Arlington, VA, USA.

Johnson, P. (1998) *Usability and Mobility: Interactions on the move*. Proceedings of the First Workshop on Human Computer Interaction with Mobile Devices, Glasgow.

Luo, P., Szekely, P., and Neches, R. (1993) *Management of Interface Design in Humanoid*. Proceedings of the Interchi'93.

Marsic, I. (2001) *An Architecture for Heterogenous Groupware Applications*. Proceedings of the 23rd IEEE/ACM International Conference on Software Engineering (ICSE 2001), Toronto, Canada.

McGlashan, S., Burnett, D., Danielsen, P., Ferrans, J., Hunt, A., Karam, G., Ladd, D., Lucas, B., Porter, B., Rehor, K., and Tryphonas, S. (2001) *Voice Extensible Markup Language (VoiceXML) Version 2.0*., http://www.w3.org/TR/2001/WD-voicexml20-20011023/.

Myers, B. (1995) User Interface Software Tools. *ACM Transactions on Computer-Human Interaction*, 2, 64–103.

Myers, B., Hudson, S., and Pausch, R. (2000) Past, Present, and Future of User Interface Software Tools. *ACM Transactions on Computer-Human Interaction*, 7, 3–28.

Olsen, D. (1999) Interacting in Chaos. *Interactions*, 6, 42–54.

Olsen, D., Jefferies, S., Nielsen, T., Moyes, W., and Fredrickson, P. (2000) *Cross-Modal Interaction using XWeb*. Proceedings of the UIST'2000, CA, USA.

Paternò, F. (1999) *Model-Based Design and Evaluation of Interactive Applications*. Springer.

Paternò, F. (2001) *Deriving Multiple Interfaces from Task Models of Nomadic Applications*. Proceedings of the CHI'2001 Workshop: Transforming the UI for Anyone, Anywhere, Seattle, Washington, USA.

Paternò, F., Mori, G., and Galiberti, R. (2001) *CTTE: An Environment for Analysis and Development of Task Models of Cooperative Applications*. Proceedings of the Human Factors in Computing Systems: CHI'2001, Extended Abstracts, Seattle, WA, USA.

Phanouriou, C. (2000) *UIML: An Appliance-Independent XML User Interface Language. Dissertation in Computer Science*, Blacksburg, Virginia Tech.

Puerta, A. and Eisenstein, J. (2001) *A Representational Basis for User Interface Transformations*. Proceedings of the CHI'2001 Workshop: Transforming the UI for Anyone, Anywhere, Seattle, Washington, USA.

Puerta, A., Eriksson, H., Gennari, J.H., and Munsen, M.A. (1994) *Model-Based Automated Generation of User Interfaces*. Proceedings of the National Conference on Artificial Intelligence.

Sukaviriya, P.N. and Foley, J. (1993) Supporting Adaptive Interfaces in a Knowledge-Based User Interface Environment. Proceedings of the Intelligent User Interfaces'93.

Sukaviriya, P.N., Kovacevic, S., Foley, J., Myers, B., Olsen, D., and Schneider-Hufschmidt, M. (1993) *Model-Based User Interfaces: What are they and Why Should We care?* Proceedings of UIST'93.

Szekely, P., Sukaviriya, P.N., Castells, P., Mukthukumarasamy, J., and Salcher, E. (1995) *Declarative Interface Models for User Interface Construction Tools: The MASTERMIND Approach*. Proceedings of the 6th IFIP Working Conference on Engineering for HCI, WY, USA.

Thevenin, D., Calvary, G., and Coutaz, J. (2001) *A Development Process for Plastic User Interfaces*. Proceedings of the CHI'2001 Workshop: Transforming the UI for Anyone, Anywhere, Seattle, Washington, USA.

Thevenin, D. and Coutaz, J. (1999) *Plasticity of User Interfaces: Framework and Research Agenda*. Proceedings of the INTERACT'99.

Vanderdonckt, J., Limbourg, Q., Oger, F., and Macq, B. (2001) *Synchronized Model-Based Design of Multiple User Interfaces*. Proceedings of the Workshop on Multiple User Interfaces over the Internet, Lille, France.

WAPForum, *Wireless Application Protocol: Wireless Markup Language Specification, Version 1.2*, http://www.wapforum.org.

Wiecha, C., Bennett, W., Boies, S., Gould, J., and Greene, S. (1990) ITS: A Tool for Rapidly Developing Interactive Applications. *ACM Transactions on Information Systems*, 8, 204–36.

Wiecha, C. and Szekely, P. (2001) *Transforming the UI for anyone, anywhere*. Proceedings of the CHI'2001, Washington, USA.

XIML: A Multiple User Interface Representation Framework for Industry

Angel Puerta and Jacob Eisenstein

RedWhale Software, USA

7.1. INTRODUCTION

As many chapters of this book testify, developing an efficient and intelligent method for designing and running multiple user interfaces is an important research problem. The challenges are many: automatic adaptation of display to multiple display devices, consistency among interfaces, awareness of context for user tasks, and adaptation to individual users are just some of the research problems to be solved. In the past few years, significant progress has been made in all of these areas and this book reports on many of those achievements.

There is, however, a challenge of a different kind for multiple user interfaces (MUIs). This challenge is that of developing a technology for multiple user interfaces that is acceptable and useful in the software industry. A technology that not only brings efficiency, consistency, and intelligence to the process of building MUIs, but that does so also within an acceptable software engineering framework. This challenge is no doubt

Multiple User Interfaces. Edited by A. Seffah and H. Javahery
© 2004 John Wiley & Sons, Ltd ISBN: 0-470-85444-8

compounded by the fact that throughout the relatively short history of the software industry, the user interface and its engineering have been its poor cousins. Whereas significant engineering advances have been made in databases, applications, algorithms, operating systems, and networking, comparable progress in user interfaces is notable for its absence.

The road to building a solution for MUIs in industry is long. There can be many possible initial paths and in technology development sometimes choosing the wrong one dooms an entire effort. We claim that the essential aspect that such a solution must have is a common representation framework for user interfaces; common from a platform point of view and also from a domain point of view. In this chapter, we report on our process and initial results of our effort to develop an advanced representation framework for MUIs that can be used in the software industry. The eXtensible Interface Markup Language (XIML) is a universal representation for user interfaces that can support multiple user interfaces at design time and at runtime [XIML 2003]. This chapter describes how XIML was conceptualized and developed, and how it was tested for feasibility.

7.1.1. SPECIAL CHALLENGES FOR MUI SOLUTIONS FOR INDUSTRY

Developing a technological framework for MUIs useful to industry imposes a number of special considerations. These requisites, named below, create tradeoffs between purely research goals and practical issues.

- *Common representation.* It is crucial for industry that any key technological solution for MUIs be based on a robust representation mechanism. The representation must be widespread enough to ensure portability. A common representation ensures a computational framework for the technology, which is essential for the development of supporting tools and environments, as well as for interoperability of user interfaces among applications.
- *Requirements engineering.* Definition of the representation must not be attempted without a clear understanding of industry requirements for the technology. In short, the types of applications and features that the representation enables must be in sync with the needs of industry. This may mean that the intended support of the representation may go beyond MUIs if the requirements dictate it.
- *Software engineering support.* Any proposed MUI technological solution for industry must define a methodology that is compatible with acceptable software engineering processes. If that is not the case, even a successful technology will find no acceptance among industry groups.
- *Appropriate foundation technologies.* The software industry is highly reluctant to incorporate any technology that is not based on at least one widely implemented foundation technology. This is the reason why a language like XML is considered an excellent target candidate for MUI representation mechanisms.
- *Feasibility and pilot studies.* MUI technologies for industry must undergo substantial feasibility studies and pilot programs. These naturally go beyond strictly research studies and into realistic application domains.

All of these requirements create a long development cycle. It can be expected that any successful effort towards MUI technology in industry will demand a process stretching over several years.

7.1.2. FOUNDATION TECHNOLOGIES

As we mentioned previously, we state that developing a representation framework for MUIs is the first step in developing a successful MUI technology for industry. To that effect, we have chosen two foundation technologies to build such a framework: model-based interface development and XML. These two technologies combine effectively to allow us to satisfy the industry requirements enumerated in the previous section.

Model-based interface development [Puerta 1997] provides an excellent foundation for the creation of declarative models that capture all relevant elements of a user interface. As such, it provides: (1) organization and structure to the definition of a user interface, (2) an engineering methodology for user interface design, and (3) a software engineering approach to user interface development. These three items take on special importance within our effort since current user interface technologies in industry have considerable shortcomings in all of these areas.

XML has gained wide acceptance within industry in the last few years. It offers a very portable representation mechanism that effectively separates data from content. It is also the preferred technology for implementing interoperability among disparate applications. Many advanced industry efforts, such as Web Services, are using XML as a foundation. In addition, XML representations enjoy the support of various independent organizations that guide the process of their definition and standardization.

7.1.3. SUMMARY OF CHAPTER

The rest of this chapter is divided into two main sections. In Section 7.2, we describe the process that we applied to create the XIML representation framework. We examine the structure of the language and discuss its potential uses for both basic and advanced user interface functionality. We also detail a number of feasibility exercises that we conducted in order to evaluate XIML. In Section 7.3, we present a pilot study in which we build a MUI platform for a realistic domain using XIML. We conclude the chapter with an examination of related work, a proposed plan for future work, and a set of conclusions about the XIML framework.

7.2. THE XIML REPRESENTATION FRAMEWORK

An industry project, especially one dedicated to the development of new infrastructure technologies, must be subdivided into a series of phases. Each phase must have an *exit criterion*, meaning a set of findings and results that justify moving the project into the next phase. Such criteria may include many aspects such as strategic, technological, and financial ones. For the purpose of this chapter, we will focus only on the technological aspects of the project.

With a general goal of creating a representation framework for multiple user interfaces, the logical initial phase of the project is that of a feasibility assessment. In short, we would need to create an initial representation and evaluate whether it can potentially fulfill the requirements that we set at the beginning of the project. This section reports on the feasibility assessment for XIML. The assessment included the following steps:

1. *Industry Computing-Model Evaluation.* This is a study of what computing models are prevalent in industry now and in the near future. Our representation must target one of these models to improve its chances of realizing its potential.
2. *Requirements Elicitation.* An understanding of the functions and features that the framework must enable, and what general objectives it must meet in order to be successful.
3. *Representation Development.* A language development effort based on the target requirements and computing model.
4. *Validation Exercises.* A series of manual and/or automated test exercises that allow us to determine the feasibility of the technology.

7.2.1. TARGET COMPUTING MODEL

The software industry is making a substantial effort to lay the foundation for a new computing model that will enable a standard way for applications to interoperate and interchange data. This is a substantial shift from previous computing models where individual-application capabilities and data manipulation were the main focus of the development process. The model is for now aimed at web-based applications but it is nevertheless extensible to future integration with workstation environments.

Over the past few years, both industry and academia have contributed a number of building blocks to this new computing model. These efforts include, among others, the dissemination and adoption of a common data representation format (XML), the definition of standard protocols for application interoperability (SOAP), and a number of proposed standard definitions for various types of data, such as data for voice-based applications (VoiceXML), and data for directory services (DSML) [OASIS 2003; VoiceXML 2003]. These and many other efforts are being channeled through standards organizations such as the World Wide Web Consortium [W3C 2003] and the Organization for the Advancement of Structured Information Systems [OASIS 2003]. For now, one of the most important examples of this new computing model is the area of *web services*, a platform that enables the building of applications by integrating mostly black-box functional units from multiple providers. All major software companies support the web services platform.

The benefits of the interoperability of software applications and the ease of data interchange among those applications are self-evident. Not only is integration of these applications facilitated in a significant manner, but integrated software support can now be devised for many complex and multi-step workflows and business processes that previously could not be supported.

There is, however, a problem that the user interface software community faces as this new computing model emerges. A standardization effort has not yet emerged for representing and manipulating interaction data – the data that defines and relates all the relevant elements of a user interface. This failure is problematic on at least two fronts. One is that an opportunity is being lost, or delayed, to provide a mechanism to bridge the gaps that exist between the user-interface engineering tasks of design, operation, and evaluation (which are the three critical aspects of the user-interface software cycle). The second is that without a viable solution for interaction-data representation, user-interface engineering will be relegated to the same secondary plane that it has suffered in basically every previous computing model prevalent in industry.

We feel therefore that our effort in building a representation framework is best targeted at this new computing model. By targeting this model, we take advantage of an existing, viable industry model plus we are spared the difficulty of retrofitting a new technology for user interfaces into the limited older computing models.

Admittedly, one key reason why interaction data has not been effectively captured yet is because doing so entails a high level of complexity. Interaction data deals not only with concrete elements, such as the widgets on a screen, but also with abstract elements, such as the context in which the interaction occurs. Therefore, capturing and relating these distinct elements into a cohesive unit presents difficult technical challenges. In turn, solving the abstract-concrete dichotomy becomes one of the key requirements that our representation framework must satisfy.

7.2.2. XIML REQUIREMENTS

In order to effectively define a representation mechanism for interaction data, it is necessary to clearly establish the requirements of such a representation in terms of expressiveness, scope, and underlying support technologies. Figure 7.1 graphically summarizes the major types of requirements that we have found essential for XIML. In this section, we discuss each of those types in detail.

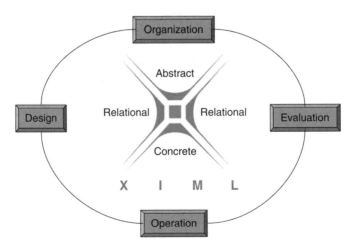

Figure 7.1. XIML represents abstract, concrete and relational data items. It also enables user-interface engineering functions.

- *Central repository of data.* The language must enable a comprehensive, structured storage mechanism for interaction data. These repositories of data may cover in scope one user interface, or a collection of user interfaces. In this manner, purely organizational or knowledge-management functions can be supported by XIML. For example, a cell-phone manufacturer could use XIML to store and manage all the characteristics and design data relevant to the user interfaces for its entire line of products.
- *Comprehensive lifecycle support.* The language must enable support functionality throughout the complete lifecycle of a user interface. This includes design, operation, and evaluation phases. This requirement is critical because it will afford an engineering framework to connect the now disjoint stages in the life of a user interface. For example, an interface-design tool could output an XIML interface specification that can be used at runtime for the management of interaction and that can also be the basis for usability engineering activities.
- *Abstract and concrete elements.* XIML must be able to represent the abstract aspects of a user interface, such as the context in which interaction takes place, and the concrete aspects, such as the specific widgets that are to be displayed on a screen. This requirement is almost a corollary of the previous one as comprehensive lifecycle support would not be possible without it. It is also recognition that interaction decisions – be it in design or operation of a user interface – are dictated in great part by items such as the task flow of a target business process or the characteristics of a specific user type.
- *Relational support.* The language must be able to relate the various elements captured within the scope of its representation. This is particularly important in the case of relating abstract and concrete elements of interaction data. The relational capabilities of the language are what enable the development of knowledge-based support throughout the lifecycle of a user interface [Puerta and Eisenstein 1999; Szekely *et al.* 1995]. For example, model-based interface development tools, interface agents, and intelligent ergonomic critics are some of the technologies that can take advantage of these relational capabilities within their reasoning processes.
- *Underlying technology.* In order to be useful within an industry-based new computing model, XIML must adhere to at least two implementation requirements. First is the use of an underlying technology that is compatible with that computing model. In this case, this points to the use of XML – the representational centerpiece of the new computing model – as the base language for XIML. Second, the language must not impose any particular methodologies or tools on the design, operation, and evaluation of user interfaces. It must be able to co-exist with existing methodologies and tools (limited, of course, by any compatibility issues external to XIML between those tools and methodologies, and the chosen underlying technologies). It should nevertheless be noted that implementation issues are strictly a practical consideration for the language. They impose certain limitations as to what can be achieved in practice, but they do not detract from the theoretical principles of the language and its applicability to different underlying technologies.

7.2.3. STRUCTURE AND ORGANIZATION OF XIML

The XIML language draws mainly from two foundations. One is the study of ontologies and their representations [Neches *et al.* 1991] and the other is the work on interface models [Puerta 1997; Szekely *et al.* 1993; Szekely *et al.* 1995]. From the former, XIML draws the representation principles it uses; from the latter it derives the types and nature of interaction data.

A discussion of the entire XIML schema, or of the specific language constructs would be beyond the scope of this chapter, but is available with the XIML documentation [XIML 2003]. In addition, Section 7.3 includes selected XIML language samples created for our pilot application. For the purpose of this chapter, we focus within this section on describing the organization and structure of the XIML schema. Figure 7.2 shows the basic structure of XIML. In what follows, we examine each of its main representational units.

7.2.3.1. Components

In its most basic sense, XIML is an organized collection of interface elements that are categorized into one or more major interface components. The language does not limit the number and types of components that can be defined; nor is there a theoretical limit on the number and types of elements under each component. In a more practical sense, however, it is to be expected that an XIML specification would support a relatively small number of components with one major type of element defined per component.

In its first version (1.0), XIML predefines five basic interface components, namely *task, domain, user, dialog, and presentation.* The first three of these can be characterized as contextual and abstract while the last two can be described as implementational and concrete. We now examine each of these five components.

- *Task.* The task component captures the business process and/or user tasks that the interface supports. The component defines a hierarchical decomposition of tasks and subtasks that also defines the expected flow among those tasks and the attributes of those tasks. It should be noted that when referring to a business process that is captured by this component, we are referring to that part of the business process that requires interaction with a user. Therefore, this component is not aimed at capturing application

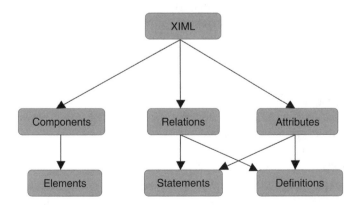

Figure 7.2. The basic representational structure of XIML.

logic. The granularity of tasks is not set by XIML so examples of valid tasks can for example include 'Enter Date', 'View Map', or 'Perform Contract Analysis'.

- *Domain.* The domain component is an organized collection of data objects and classes of objects that is structured into a hierarchy. This hierarchy is similar in nature to that of an ontology [Neches *et al.* 1991] but at a very basic level. Objects are defined via attribute-value pairings. Objects to be included in this component are restricted to those that are viewed or manipulated by a user and can be either simple or complex types. For example, 'Date', 'Map', and 'Contract' can all be domain objects.

- *User.* The user component defines a hierarchy – a tree – of users. A user in the hierarchy can represent a user group or an individual user. Therefore, an element of this component can be a 'Doctor' or can be 'Doctor John Smith'. Attribute-value pairs define the characteristics of these users. As defined today, the user component of XIML does not attempt to capture the mental model (or cognitive states) of users but rather data and features that are relevant in the functions of design, operation and evaluation.

- *Presentation.* The presentation component defines a hierarchy of interaction elements that comprise the concrete objects that communicate with users in an interface. Examples of these are a window, a push button, a slider, or a complex widget such as an ActiveX control to visualize stock data. It is generally intended that the granularity of the elements in the presentation component will be relatively high so that the logic and operation of an interaction element are separated from its definition. In this manner, the rendering of a specific interaction element can be left entirely to the corresponding target display system. We will expand on the practical impact of this separation below when we discuss the issue of cross-platform interface development.

- *Dialog.* The dialog component defines a structured collection of elements that determine the interaction actions that are available to the users of an interface. For example, a 'Click', a 'Voice response', and a 'Gesture' are all types of interaction actions. The dialog component also specifies the flow among the interaction actions that constitute the allowable navigation of the user interface. This component is similar in nature to the Task component but it operates at the concrete levels as opposed to the abstract level of the Task component.

The components predefined in the first version of XIML were selected by studying a large variety of previous efforts in creating interface models [Puerta 1997; Szekely *et al.* 1995]. There are other components that have been identified by researchers in the past as being potentially useful, such as a workstation component (for defining the characteristics of available target displays) or an application component (for defining the links to application logic). We have found that in most practical situations, we have been able to subsume all necessary definitions for a given interface into the existing components. In any event, XIML is extensible so that other components can be added in the future once their presence is justified.

7.2.3.2. Relations

The interaction data elements captured by the various XIML components constitute a body of explicit knowledge about a user interface that can support organization and knowledge-management functions for user interfaces. There is, however, a more extensive body of knowledge that is made up of the relations among the various elements in an XIML specification. A relation in XIML is a definition or a statement that links any two or more XIML elements either within one component or across components. For example, 'Data type A *is displayed with* Presentation Element B or Presentation Element C' is a link between a domain-component element and a presentation-component element.

By capturing relations in an explicit manner, XIML creates a body of knowledge that can support design, operation, and evaluation functions for user interfaces. In particular, the explicit nature of the relations enables knowledge-based support for those interaction functions. In a sense, the set of relations in an XIML specification capture the design knowledge about a user interface. The runtime manipulation of those relations constitutes the operation of the user interface. A more in-depth study of the nature of relations in a declarative interface model can be seen in [Puerta and Eisenstein 1999].

XIML supports relation definitions, which specify the canonical form of a relation, and relation statements, which specify actual instances of relations. It can be expected that there are a number of relations that can be of interest in the design and management of user interfaces. Therefore, XIML includes a number of predefined relations and their semantics. However, it also allows users of the language to define custom relations with semantics residing in the specific applications that utilize XIML.

7.2.3.3. Attributes

In XIML, attributes are features or properties of elements that can be assigned a value. The value of an attribute can be one of a basic set of data types or it can be an instance of another existing element. Multiple values are allowed as well as enumerations and ranges. The basic mechanism in XIML to define the properties of an element is to create a number of attribute-value pairs for that element. In addition, relations among elements can be expressed at the attribute level or at the element level. As in the case of relations, XIML supports definitions and statements for attributes, and also predefines some typical attributes of interest for user interfaces.

7.2.4. VALIDATION EXERCISES

As mentioned in our introduction, it is necessary to conduct a number of validation exercises that enable us to assess the *feasibility* of XIML as a potential representation for industry. Whereas in the purely scientific world having an initial assessment of feasibility may not be in order, for our purposes, the project could not move into the next phase unless we could demonstrate such feasibility. We sought to answer two questions with these validation exercises:

1. Is the language expressive and flexible enough to represent simple user interfaces?
2. Can we build small demo applications based on XIML that enable functionality that is of general interest for industry, but that may not be easily implemented with current capabilities?

In order to proceed with the project, we needed "yes" answers for those questions. This section describes the various validation exercises undertaken.

7.2.4.1. Hand Coded Interface Specifications

It is useful with any new language schema to hand code a few real-world target samples. This allows language designers to ascertain the range of expressiveness of the language as well as its verbosity – the size and number of expressions that would be necessary to code one example. The first of these properties determines if the language is rich enough to cover common situations, the second one is useful in understanding potential implementation challenges to the language, such as computing resources needed for its storage and processing. It should be noted nevertheless that it is not expected that developers will write XIML directly, but rather that they will use tools that will read and write the language.

An XIML specification can contain as few as one of the standard components described in the previous section. Therefore, our hand coded specifications ranged from a single domain component describing the catalog items of a store, to a task model for a supply-chain management application, to presentation components for simple C++ interface controls for Windows, to entire interface definitions for a number of applications (a geographical data visualization application, a baseball box-score keeper, and a dictionary-based search tool, among others). In all of these examples, we found XIML to be sufficiently expressive to capture the relevant interaction data. We did find the language somewhat verbose but well under any threshold that could pose practical implementation problems.

7.2.4.2. Multi-Platform Interface Development

One of the important uses of XIML can be in the development of user interfaces that must be displayed in a variety of devices. XIML can be used to effectively display a single interface definition on any number of target devices. This is made possible by the strict separation that XIML makes between the definition of a user interface and the rendering of that interface – the actual display of the interface on a target device. In the XIML framework, the definition of the interface is the actual XIML specification and the rendering of the interface is left up to the target device to handle. In the past, many model-based interface development systems [Puerta 1997] and many user-interface management systems [Olsen 1992] have not had this separation established clearly, and therefore developers ended up mixing up interface logic with interface definition.

Figure 7.3 illustrates the XIML framework for multi-platform development for a couple of sample device targets. The language is not restricted to those two types of devices but can theoretically support many types of stationary and mobile devices. In the case shown, there is a single XIML interface specification for the data to be displayed, the navigation to be followed, and the user tasks to be supported. Then, by simply defining one presentation component per target device, the entire specification can support multiple platforms. Specifying presentation components simply means determining what widgets, interactors, and controls will be used to display each data item on each of the target devices. As far as the rendering of the interface is concerned, an XML-capable device is able to process an XIML specification directly. For the case when the target device is not XML-capable, a converter needs to be used to produce the target language. To support

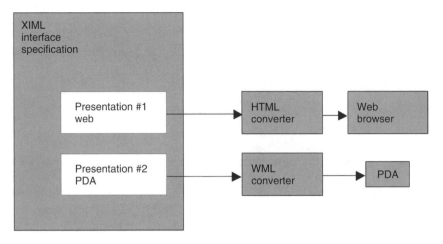

Figure 7.3. XIML provides a framework for developing user interfaces for multiple targets.

the XIML validation effort, we have developed converters for popular target languages including HTML and WML.

The multi-platform framework described above saves development time and helps ensure consistency. However, there is still the chore of creating a presentation component for each target device. To solve that problem, XIML enables additional capabilities that can provide a high-degree of automation to the multi-platform interface development process. Figure 7.4 illustrates this automation framework. Instead of creating and managing one presentation component per target device, developers would work with a single "intermediate" presentation component. XIML would then predefine, via relations, how the intermediate component maps to a specific widget or control on the target display device.

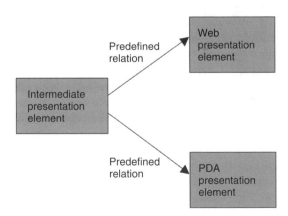

Figure 7.4. Automation framework for multi-platform interface development in XIML.

As an example, a designer could specify in XIML that a particular data type, say a geographical location, be presented with an intermediate presentation element called a 'map-location widget'. By using the established relations, XIML will then automatically map the map-location widget to an actual graphical map control for the Web and to a text-based data display on the PDA.

Clearly, specifying the relations between intermediate presentation objects and device presentation objects in a static manner would be too inflexible to be of practical use. There are many considerations that go into selecting an appropriate widget to use in a given instance. These considerations would include screen size, what other elements are on the screen at the same time, user preferences, contextual issues and so on. Therefore, it is expected that intelligent tools would be necessary to handle the task of creating and updating the relations between intermediate and device elements. For the purposes of our project, we first conducted a small-scale validation exercise for a map-annotation user interface to be displayed in two devices (desktop and cell phone). Then, we used this same problem area to build an XIML-based pilot application, which we report in detail in Section 7.3.

7.2.4.3. Intelligent Interaction Management

Another important application of XIML in industry may be as a resource for the management of a user interface at runtime. By centralizing in a single definition the interaction data of an interface, it is hoped that we can build tools that will similarly centralize a range of functions related to the operation of that interface. In order to validate the feasibility of using XIML for interaction-management functions, we explored three runtime functions: (a) dynamic presentation reorganization, (b) personalization, and (c) distributed interface management.

- *Dynamic presentation reorganization*. Figure 7.5 shows a sequence of views of a single web page that displays the system load of a server. The page displays different widgets or controls according to the screen area available for display. When that area is minimal, the page displays the most basic data item. As the area increases, additional text and then a graphical view are added. Finally, when the display area is maximized, the page displays the most sophisticated control available for that target data item. To implement this function, we built a simple application that read the XIML specification for the interface and dynamically adjusted the presentation component of the specification according to a set of thresholds on the value of the display area available. It is clear that a system that would support sophisticated dynamic reorganization of the presentation would probably need a good degree of sophistication itself. However, our goal at this point was not to build such a system, but rather to validate that this type of problem, can be represented and solved using XIML as an interaction data repository and in a very straightforward manner.
- *Personalization*. Figure 7.6 shows a simple example of a personalization feature. The widgets display a reading of a data source (in this case system load as in the previous example). The corresponding XIML specification indicates that there are a number of widgets that can be used to display that data source. In addition, the widgets can be

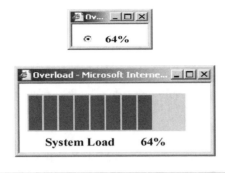

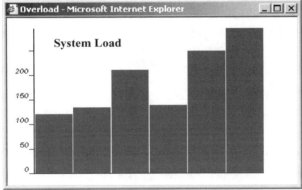

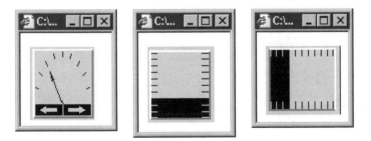

Figure 7.5. Dynamic presentation based on available display area.

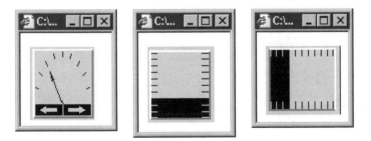

Figure 7.6. Widgets available for personalization in an XIML specification.

oriented in various manners on the screen. For this feature, we wrote a small application that selected the widget to display according to criteria based on the user component of the XIML specification. As in the previous example, the sophistication of the personalization issue was not the focus of the experiment. The focus was to ensure that XIML has capabilities to support personalization features and that, as the various examples are added together, the XIML framework can offer the value of a single repository of interaction data to support many user-interface management functions.

- *Distributed interface management.* One of the drawbacks of any client-based software application is that the update of the client software is problematic since each individual client needs to be updated. Server-side applications reduce that problem to a large extent but then have the tradeoff that a server update affects every user of the application at the same time whether these users desire the change or not. In either case, an update is not a trivial task and can be initiated solely by the software provider.

  XIML could be used to provide a mechanism for the distributed update of user interface components. Figure 7.7 illustrates this mechanism. As we saw in the previous two examples, the widget or control being displayed on a page can be changed easily via an XIML specification. That framework assumes that the widget to be displayed is available, but it does not confine it to be on a specific server or a client. It simply treats the widget as a black box that performs a function. We have taken advantage of that flexibility to allow the widget to simply be available somewhere on the network be it on a client, a peer, or a server machine. The XIML specification can be set to link to providers of the widget or it can rely on a search-and-supply application. In this manner, for example, a calendar widget on a travel-reservations page can be provided by any number of XIML-compliant calendar-widget suppliers. The choice of suppliers can be made dependant on any XIML-supported criteria such as user preferences.

7.2.4.4. Task Modelling

One of the critical requirements that we set for XIML was the ability to represent abstract concepts such as user tasks, domain objects, and user profiles (a process that can be referred to as task modelling). Our group has previously developed a number of model-based interface development tools. These tools included, among others, an informal user-task model specification tool called U-TEL [Tam *et al.* 1998], and a formal interface-model development environment called MOBI-D [Puerta 1997]. Both of these tools have advanced modelling facilities to represent interface models, including the contextual concepts of user tasks, domain objects, and user profiles. The tools have been used to model

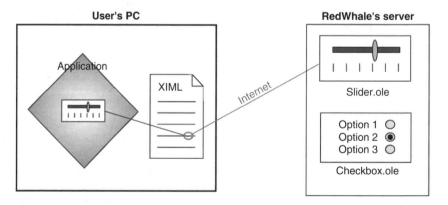

Figure 7.7. XIML mechanism for distributed interface management.

a wide variety of applications such as a military-logistics management tool, a medical-data visualization application, and a supply-chain management tool, among others. The interface modelling language used by U-TEL and MOBI-D is a frame-based language that shares some characteristics with XIML. To verify that the task-modelling capabilities of XIML were at least at the same level as those of MOBI-D, we successfully built a converter that can take any MOBI-D model specification and convert it into an XIML specification. We applied the converter successfully to all models previously built with MOBI-D.

7.2.4.5. Reverse Engineering

While the benefits of XIML are potentially many, practical reality indicates that a very substantial amount of code has been written in HTML. It would be ideal that in the same way that XIML can be converted into HTML, that HTML code could be reverse engineered into XIML. In this manner, the benefits of XIML could be brought to existing applications through some level of automated support. The reverse engineering of HTML into XIML has successfully been accomplished by a research group working independently from us [Bouillon and Vanderdonckt 2002]. The implementation is currently at the prototype level and it has been applied to simple examples such as converting the CHI conference online registration form to XIML.

7.2.4.6. Summary of Validation Exercises

The validation exercises performed for XIML allowed us to conclude that there is enough evidence to justify the engineering feasibility of XIML as a universal interface-definition language. The next step therefore is the development of a pilot application of XIML, which we report in Section 7.3.

7.3. AN XIML PILOT APPLICATION

In order to satisfy our exit criteria for the first phase of the development of XIML, we must (1) successfully complete the validation exercises and (2) be able to build a proof-of-concept prototype of interest. When we say 'of interest', we refer principally to two items:

- The prototype must implement a solution to a problem that the software industry is currently facing and for which there is no efficient solution under current technologies.
- The prototype must demonstrate a new set of features and/or methodologies that are valuable and that could be incorporated into mainstream industry frameworks within a short period of time (one or two years at most).

It is clear that the theme of this book – that of multiple user interfaces – fulfills the requirements of the first interest item. The ubiquity of internet-capable devices, both mobile and stationary, is now a reality. Users of these devices demand an integrated interactive experience across devices but the software industry lacks methods and tools to efficiently design and implement such an experience. For the second interest item, we rely on the widespread acceptance of XML to transition XIML into industry. In addition, we will demonstrate via the proof-of-concept prototype that there are a number of valuable middleware functions that can easily be based on the XIML framework.

In the remainder of this section, we first present the problem application domain and then we illustrate how we define the abstract components of the corresponding XIML representation for the problem domain. Next, we detail various middleware functions useful within the context of our sample problem.

7.3.1. MANNA: THE MAP ANNOTATION ASSISTANT

MANNA is a hypothetical software application that reveals many of the challenges posed by user-interface development for multiple user interfaces, including mobile ones. MANNA is a multimedia application that must run on several platforms and can be utilized collaboratively over the Internet. It is intended for use by geologists, engineers, and military personnel to create annotated maps of geographical areas. Annotations can include text, audio, video, or even virtual reality walk-through.

For this application domain, we created an interaction scenario that demanded various display devices and posed a number of presentation and dialog constraints. In our scenario, a geologist from the United States Geological Survey has been dispatched to a remote location in northern California to examine the effects of a recent earthquake. Using a desktop workstation, our geologist downloads existing maps and reports on the area to prepare for his/her visit (see Figure 7.8). The desktop workstation poses few limiting

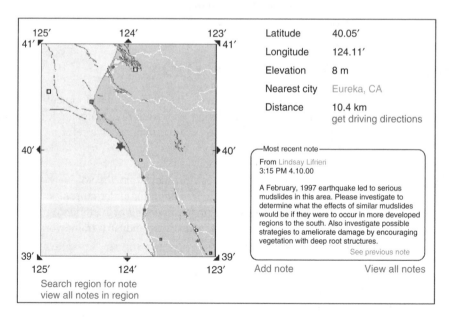

Figure 7.8. The desktop UI design for MANNA. Hyperlinks are in grey.

constraints to UI development, but obviously it is a stationary device. The documents are therefore downloaded to a laptop, and the geologist boards a plane for the site.

On the plane, the laptop is not networked, so commands that rely on a network connection are disabled. When the geologist examines video of the site, the UI switches to a black-and-white display and reduces the rate of frames per second. This helps to conserve battery power. In addition, because many users find laptop touch pads inconvenient, interactors that are keyboard-friendly are preferred, e.g., list boxes replace drop-lists.

After arriving at the airport, the geologist rents a car and drives to the site. She receives a message through the MANNA system to her cellular phone, alerting her to examine a particular location. Because a cellular phone offers extremely limited screen-space, the map of the region is not displayed (see Figure 7.9). Instead, the cell phone shows the geographical location, driving directions, and the geologist's current GPS position. A facility for responding to the message is also provided.

Finally arriving at the site, our geologist uses a palmtop computer to make notes on the region (see Figure 7.10). Since the palmtop relies on a touch pen for interaction, interactors

Figure 7.9. The cell phone UI design for MANNA.

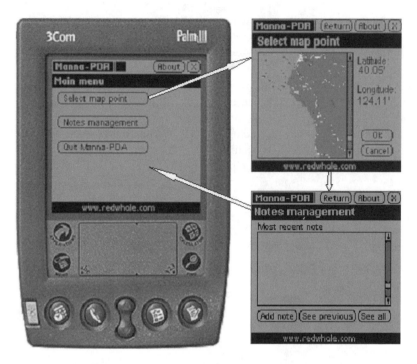

Figure 7.10. The PDA UI design for MANNA.

that require double-clicks and right-clicks are not permitted. Screen size is a concern here, so a more conservative layout is employed. Having completed the investigation, our geologist prepares a presentation in two formats. First, an annotated walk-through is presented on a heads-up display (HUD). Because of the HUD's limited capabilities for handling textual input, speech-based interactors are used instead. A more conventional presentation is prepared for a high-resolution large-screen display. Since this is a final presentation, the users will not wish to add information, and interactors that are intended for that purpose are removed. The layout adapts to accommodate the larger screen space, and important information is placed near the center and top, where everyone in the audience can see it.

Clearly in this scenario, we are faced with quite a number of presentation and dialog constraints. Obviously, we could build each interface separately, but that would duplicate work multiple times and would make it difficult to maintain consistency among interfaces. Ideally therefore, we can use XIML to define a single abstract interface specification (see Section 7.2.4.2), and then implement appropriate middleware that can correctly generate each interface, under each set of constraints, from that single specification.

7.3.2. THE MANNA ABSTRACT XIML COMPONENTS

The first step in building our proof-of-concept prototype is to define the abstract XIML components for the MANNA application. These components are: task, domain, and user.

With these abstract components in hand, we will be able to develop middleware that allows us to manage the creation of multiple target interfaces. For this task, we used a basic XML editor. In what follows, we present samples of the resulting MANNA abstract XIML components.

7.3.2.1. Task Component

The task component captures the business process and/or user tasks that the interface supports. The component defines a hierarchical decomposition of tasks and subtasks, and also defines the expected flow among those tasks and the attributes of those tasks. It should be noted that when referring to a business process that is captured by this component, we are referring to that part of the business process that requires interaction with a user. Therefore, this component is not aimed at capturing application logic. The granularity of tasks is not set by XIML so valid tasks can range from simple one-step actions (e.g., Enter Date, View Map) to complicated multi-step processes (e.g., Perform Contract Analysis). For MANNA, we created a task/subtask decomposition of the user tasks based on our interaction scenario. A portion of this task component is as follows:

```
<TASK_MODEL ID='tm1'>
    <TASK_ELEMENT ID='t1' name='Make annotation'>
        <TASK_ELEMENT ID='t1.1' name='Select location'/>
        <TASK_ELEMENT ID='t1.2' name='Enter note'/>
        <TASK_ELEMENT ID='t1.3' name='Confirm Annotation'/>
    </TASK_ELEMENT>
</TASK_MODEL>
```

7.3.2.2. Domain Component

The domain component is an organized collection of data objects and classes of objects. Objects to be included in this component are restricted to those that are viewed or manipulated by a user; internal program data structures are not represented here. Objects can be typed, using either simple types, e.g. integer, enumeration, or using more complex structures.

The domain model is typically structured into a 'contains' hierarchy. In this case, inheritance relations are specified using relation statements. However, as with all model components, the structure of the domain component can be set by the user [XIML 2003]. Here is a portion of the XIML domain component for MANNA:

```
<DOMAIN_MODEL ID='dm1'>
    <DOMAIN_ELEMENT ID='d1.1' name='map annotation'>
        <DOMAIN_ELEMENT ID='d1.1.1' name='location'/>
        <DOMAIN_ELEMENT ID='d1.1.2' name='note'/>
        DOMAIN_ELEMENT ID='d1.1.3' name='entered_by'/>
        <DOMAIN_ELEMENT ID='d1.1.4' name='timestamp'/>
    </DOMAIN_ELEMENT>
</DOMAIN_MODEL>
```

Note that the type of each domain element is not indicated here. In XIML, type is represented using attributes, a generic formalism for describing characteristics of elements.

Attributes are features or properties of elements that can be assigned a value. The value of an attribute can be one of a basic set of data types, or it can be an instance of another existing element. Multiple values are allowed, as well as enumerations and ranges. The basic mechanism in XIML to define the properties of an element is to create a number of attribute–value pairs for that element. XIML supports attribute definitions that specify the canonical form of an attribute, including its type, and the elements to which it may apply. Attribute statements specify actual instances of attributes.

7.3.2.3. User Component

The user component describes relevant characteristics of the various individuals or groups who will use the interface. User elements are organized in an inheritance hierarchy; more general user types are represented by high-level elements, while their children have more specific features. For example, a highest-level user type might be 'Medical Personnel', containing a child element 'Technician', which in turn contains a child element 'Nuclear Medicine Technician'. Attribute–value pairs define the characteristics of these users. As currently defined, the user component of XIML does not attempt to capture the mental model (or cognitive states) of users, but rather, data and features that are relevant in the functions of design, operation and evaluation. Relevant features include user preferences, permissions, and levels of expertise. The following is a portion of the user component for MANNA:

```
<USER_MODEL ID='umodel'>
        <USER_ELEMENT ID='u1.1' NAME='field researcher'>
                <USER_ELEMENT ID='u1.1.1' NAME='field supervisor'/>
                <USER_ELEMENT ID='u1.1.2' NAME='field geologist'/>
        </USER_ELEMENT>
        <USER_ELEMENT ID='u1.2' NAME='analyst'/>
</USER_MODEL>
```

7.3.2.4. Using XIML Relations

The three abstract components detailed above constitute a basic representation of the MANNA application domain from a user interface point of view. There is however, a further body of knowledge that is not captured as of yet by those components. Namely, how do we answer a basic question such as what domain objects are involved in completing a defined user task.

To answer those questions and more, XIML uses relations. In short, a relation links two or more XIML elements (e.g., tasks, users) with each other and may define a semantic meaning to the relation. We have identified a number of relations among XIML elements that are of interest [Puerta and Eisenstein 1999], such as relations between task elements and domain elements. The semantics of a relation are not mandated by XIML. Instead, the language allows the application handling of the relation to process it according to its own semantics. It is expected, however, that a set of basic XIML predefined relations will be applied with similar semantics across applications. The mechanism to manage and enforce such basic semantics is currently under conceptualization.

7.3.3. XIML-BASED MIDDLEWARE FOR MANNA

Once the abstract components of MANNA are defined in XIML, we can proceed to build the user interface for each of our target platforms (desktop, PDA, and cell phone). Basically, this entails defining complete XIML presentation and dialog components for each of the platforms. By defining those concrete components and relating them to the abstract components, we already have a fairly positive gain as our three target user interfaces are consistently linked to a single set of tasks, domain objects, and user types.

However, we are more interested in ways to automate – at least to some degree – the process of creating the three target presentation and dialog components. To that effect for the MANNA application, we defined three *middleware units*. These units are able to examine a partial XIML specification and augment and/or modify that specification based on their internal logic and available knowledge base. Of the various middleware units considered [Eisenstein 2001; Eisenstein and Rich 2002; Eisenstein *et al.* 2001], we focus here in particular on three units to support the following functions: (1) interactor selection, (2) presentation structure definition, and (3) contextual adaptation.

7.3.3.1. Interactor Selection

The presentation component of XIML is used to predefine the presentation elements and attributes of display devices. For each device, XIML captures all of the interactors (widgets) available in that device as well as any associated parameters and attributes (e.g., screen resolution). As part of the building process for a user interface, it is necessary to select an appropriate interactor for each user action and domain object to be displayed. Clearly for multiple user interfaces, this can turn out to be a challenging, if not tedious task. Therefore, our first middleware unit automates the selection of interactors for a wide range of situations.

First, the unit simplifies the selection of interactors by managing sets of Abstract Interaction Objects (AIO) and Concrete Interaction Objects (CIO) [Vanderdonckt and Bodart 1993]. An AIO is a platform-neutral XIML presentation element that symbolizes an action or a domain object display. A CIO is an actual widget available in a target platform. Figure 7.11 illustrates the AIO–CIO relationship.

The middleware unit manages the set of AIO–CIO relationships. It can set or cancel those relationships based on designer input, or by examining the XIML platform specifications for a device and using its knowledge base to determine the appropriate CIOs for a given AIO. In this manner, a user interface designer can create a single design based on AIO and the middleware unit completes the interaction selection process.

Secondly, the unit helps with dynamic aspects of interactor selection. There are a number of constraints that can affect the selection of an interactor at runtime. These include attributes such as screen resolution, available display area, current bandwidth, user preferences, and so on. For MANNA, we explored adaptation based on two constraints: available display area and screen resolution. Based on the CIO established by the unit as appropriate for a given presentation situation, the unit attempts to change the value of the CIO's XIML attribute values to meet the applicable constraint values. For example, this

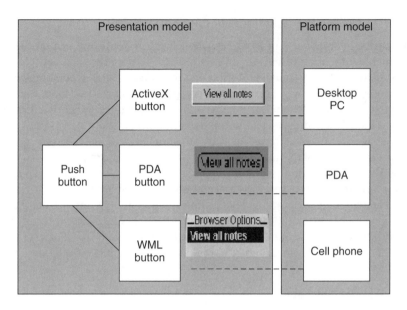

Figure 7.11. An abstract interaction object is subclassed into concrete interaction objects for specific platforms.

can take the form of shrinking a widget. Alternatively, it may replace one valid CIO with another based on the stated constraints (see Figure 7.5). In this manner, a developer of multiple user interfaces gains a considerable degree of freedom by not having to account explicitly for constraint-satisfaction issues in interactor selection [Eisenstein and Puerta 2000; Kawai *et al.* 1996; Vanderdonckt and Berquin 1999].

7.3.3.2. Presentation Structure Definition

Another key aspect of building a user interface is the definition of the presentation structure. For example, in a desktop device this would entail defining how many windows will be created, what user tasks will be completed in each window, and correspondingly, what interactors will be placed in each window. Our second middleware unit helps automate this task. The unit helps us deal with two interrelated issues: (1) how user tasks translate into a presentation structure and (2) how to choose an optimal solution for a device among various presentation structure options.

In XIML, we call the presentation structure the union of the language's two concrete components: the presentation component and the dialog component. The first one represents the hierarchy of presentation elements in the interface (e.g., windows, widgets) whereas the second one defines the actions between user and system (i.e., navigation, input and responses).

The middleware unit supports the translation of user tasks into presentation structures in two steps. First, it creates an initial XIML dialog component that is based on the abstract user task component. The following is an example of this initial dialog component (note the similarities with the task model in 3.2.1):

```
<DIALOG_MODEL ID='im1'>
    <DIALOG_ELEMENT ID='i1.1' NAME='Make annotation'>
        <DIALOG_ELEMENT ID='i1.2' NAME='Select location'>
        <DIALOG_ELEMENT ID='i1.2.1' NAME='Select map point'/>
        <DIALOG_ELEMENT ID='i1.2.2' NAME='Specify latitude'/>
        <DIALOG_ELEMENT ID='i1.2.3' NAME='Specify longitude'/>
        </DIALOG_ELEMENT>
        <DIALOG_ELEMENT ID='i1.3' NAME='Enter note'/>
        <DIALOG_ELEMENT ID='i1.4' NAME='Confirm annotation'/>
    </DIALOG_ELEMENT>
</DIALOG_MODEL>
```

Furthermore, it is necessary to link the dialog elements to the interactors selected as in Section 7.3.3.1. The XIML sample below exemplifies a dialog component after interactors have been linked. Note that in Section 7.3.3.1 we linked interactors to domain objects, which in turn are linked to user tasks. By now linking tasks with dialog components, we can complete the full circle and link the dialog elements with the presentation elements (interactors). In the sample XIML dialog component, the <ALLOWED_CLASSES> tag indicates which types of elements are allowed to participate in the dialog-interactor relation; in this case, any child of the dialog model component i1 can appear on the left side, and any child of the presentation model component p2 can appear on the right side. Note the presence of an attribute definition inside the relation definition; this indicates that the relation itself can be parameterized. In this case, the parameter indicates the specific interaction technique that is required to perform the command indicated by the dialog element on the left side of the relation. The relation definition is applied to the dialog element 'Select map point', stating that to perform this command, the user must double click on the map.

```
<DIALOG_MODEL ID='im1'>
  <DEFINITIONS>
      <RELATION_DEFINITION NAME='is_performed_using'>
          <ALLOWED_CLASSES>
              <CLASS REFERENCE='im1' INHERITED='true' SLOT='left'/>
              <CLASS REFERENCE='pm2' INHERITED='true' SLOT='right'/>
          </ALLOWED_CLASSES>
        <ATTRIBUTE_DEFINITION NAME='interaction_technique'>
        <TYPE='enumeration'>
        <DEFAULT>onClick</DEFAULT>
        <ALLOWED_VALUES>
          <VALUE>onClick</VALUE>
          <VALUE>onScroll</VALUE>
          <VALUE>onChange</VALUE>
          <VALUE>onDoubleClick</VALUE>
        </ALLOWED_VALUES>
      </ATTRIBUTE_DEFINITION>
    </RELATION_DEFINITION>
  </DEFINITIONS>
```

```
<DIALOG_ELEMENT ID='i1.1' NAME='Make annotation'>
  <DIALOG_ELEMENT ID='i1.2' NAME='Select location'>
   <DIALOG_ELEMENT ID='i1.2.1' NAME='Select map point'>
    <FEATURES>
     <RELATION_STATEMENT DEFINITION='is_performed_by' REFERENCE='2.1.1'>
      <ATTRIBUTE_STATEMENT DEFINITION='interaction_technique'>
      onDoubleClick
      </ATTRIBUTE_STATEMENT>
     </RELATION_STATEMENT>
    </FEATURES>
    <DIALOG_ELEMENT ID='i1.2.2' NAME='Specify latitude'/>
    <DIALOG_ELEMENT ID='i1.2.3' NAME='Specify longitude'/>
   </DIALOG_ELEMENT>
   <DIALOG_ELEMENT ID='i1.3' NAME='Enter note'/>
   <DIALOG_ELEMENT ID='i1.4' NAME='Confirm annotation'/>
  </DIALOG_ELEMENT>
</DIALOG_MODEL>
```

Note that to this point, we have linked many of the elements of the XIML components, but do not have yet a definition for how, for example, we would distribute the user tasks and selected interactors among windows in a desktop device. This is a design problem that must take into consideration various factors, such as target device and its constraints, interactors selected, and distribution strategy (e.g., many windows vs single window). The middleware unit can give support to this function via a mediator agent [Arens and Hovy 1995]. As Figure 7.12 shows, a mediator can examine the XIML specification, including the device characteristics. It can also offer a user task distribution based on an appropriate strategy for the device in question.

After the process shown in this section, we have a comprehensive XIML specification for a user interface (concrete and abstract components). The specification is fully integrated and can now be rendered into the appropriate device. The middleware unit significantly simplifies the development work associated with completing the specification.

7.3.3.3. Contextual Adaptation

The context in which interaction occurs has an obvious impact on what user tasks may or may not be performed at any given point in time. For example, in the scenario in Section 7.3.1, we know that the cellular phone is especially suited for finding driving directions. If the user were not driving, she could be using the PDA. The desktop workstation cannot be brought out into the field, so it is unlikely that it will be used to enter new annotations about a geographical area; rather, it will be used for viewing annotations. Conversely, the highly mobile PDA is the ideal device for entering new annotations.

A middleware unit can take advantage of this knowledge and optimize the user interface for each device. The designer creates mappings between platforms (or classes of platforms) and tasks (or sets of tasks) at the abstract level. Additional mappings are then created between task elements and presentation layouts that are optimized for a given set of tasks. We can assume these mappings are transitive; as a result, the appropriate presentation model is associated with each platform, based on mappings through the task model. The procedure is depicted in Figure 7.13. In this figure, the task model is shown to be a simple

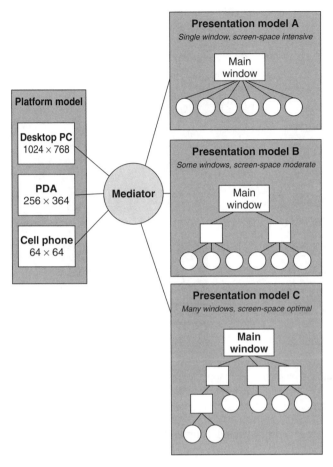

Figure 7.12. A mediating agent dynamically selects the appropriate presentation model for each device.

collection of tasks. This is for simplicity's sake; in reality, the task model is likely to be a highly structured graph where tasks are decomposed into subtasks at a much finer level than shown here.

There are several ways in which a presentation model can be optimized for the performance of a specific subset of tasks. Tasks that are thought to be particularly important are represented by AIOs that are easily accessible. For example, on a PDA, clicking on a spot on the map of our MANNA application allows the user to enter a new note immediately. However, on the desktop workstation, clicking on a spot on the map brings up a rich set of geographical and meteorological information describing the selected region, while showing previously entered notes (see Figure 7.8). On the cellular phone, driving directions are immediately presented when any location is selected. On the other devices, an additional click is required to get the driving directions (see Figure 7.9). The 'down

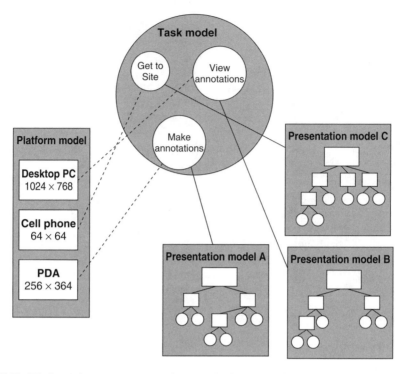

Figure 7.13. Platform elements are mapped onto task elements which are mapped onto presentation models.

arrow' button of the cellular phone enables the user to select other options by scrolling between them.

This middleware unit therefore benefits designers and developers by managing the localized contextual changes that apply across devices for a given task. The treatment by this unit of optimizations for the task structure of each device is similar to our treatment of screen-space constraints: global and structural modifications to the presentation model are often necessary, and adaptive interactor selection alone will not suffice.

7.4. DISCUSSION

We conclude by offering a perspective on the ongoing development of the XIML framework, examining related work, and highlighting the main claims of this chapter.

7.4.1. THE XIML ROADMAP

We consider that our exit criteria for the initial phases of the development of the XIML framework have been satisfied. Therefore, we plan to continue this effort. We have devised a number of stages that we plan to follow to build and refine XIML into an industry resource. Each of the stages constitutes a development and evaluation period. The stages are as follows:

1. *Definition.* This phase includes the elicitation of requirements and the definition of language constructs, which we have completed.
2. *Validation.* Experiments are conducted on the language to assess its expressiveness and the feasibility of its use. This phase is being carried out.
3. *Dissemination.* The language is made available to interested parties in academia and industry for research purposes (www.ximl.org). Additional applications, tests, and language refinements are created.
4. *Adoption.* The language is used by industry in commercial products.
5. *Standardization.* The language is adopted by a standards body under a controlled evolution process.

There is no single measure of success in this process. The language may prove to be very useful and successful at certain levels but not at others. In addition, the analysis of the functional and theoretical aspects of XIML is just one of several considerations that must be made in order to develop a universal language for user interfaces. It should be noted first that the meaning of the word 'universal' in this context is a language that has broad applicability and scope. The term should not be considered to mean a language that is used by every developer and every application. We do consider, however, that the evidence produced so far seems to indicate that further efforts are warranted.

7.4.2. RELATED WORK

The work on XIML draws principally from previous work in three areas: model-based interface development [Puerta 1997; Szekely 1996; Szekely *et al.* 1993], user-interface management systems [Olsen 1992], and knowledge representation for domain ontologies [Neches *et al.* 1991]. XIML shares some of the goals of these fields but is not directly comparable to them. For example, the main focus of model-based interface development systems has usually been the design and construction of the user interface. For XIML, this is just one aspect, but the goal is to have a language that can support runtime operations as well. In this point, it mirrors the aims of user-interface management systems. However, those systems have targeted different computing models and their underlying definition languages do not have the scope and expressiveness of XIML.

There are also some related efforts in the area of creating XML-based user interface specification languages. UIML [Abrams *et al.* 1999] is a language geared towards multiplatform interface development. UIML provides a highly detailed level of representation for presentation and dialog data, but provides no support for abstract components and their relation to the concrete user interface design. Consequently, while UIML is well suited for describing a user interface design, it is not capable of describing the design rationale of a user interface. Ali *et al.* [2002] have recently begun to explore the integration of external task modelling languages with UIML (see Chapter 6). However, at present XIML remains the only language to provide integrated support for both abstract and concrete components of the user interface.

There are several existing or developing standards that overlap with one or more of XIML's model components. XUL [Cheng 1999], Netscape's XML-based User Interface Language, provides a thin layer of abstraction above HTML for describing the presentation components of a web page. Just as XUL overlaps with XIML's presentation component, CC/PP [Butler 2001; W3C] and UAProf [Butler 2001] provide similar functionality to XIML's platform component. ConcurTaskTrees provides an XML representation of a UI's task structure [Paterno *et al.* 1997] (see Chapter 11). DSML [OASIS 2003], the Directory Services Markup Language, offers an adequate representation for the domain structure, at least in the case of e-commerce applications.

Of course, none of these languages provides support for relational modelling between components, which is the essence of XIML's representation framework. Since all of these languages are based in XML, it is straightforward to translate them into XIML. Consequently, we view the existence of these languages as an advantage. In the future, we hope to show how XIML can exploit existing documents in each of these languages.

7.4.3. SUMMARY OF FINDINGS

In this chapter, we have reported on the following results:

- XIML serves as a central repository of interaction data that captures and interrelates the abstract and concrete elements of a user interface.
- We have established a roadmap for building XIML into an industry resource. The roadmap balances the requirements of a research effort with the realities of industry.
- We have performed a comprehensive number of validation exercises that established the feasibility of XIML as a universal interface-specification language.
- We have completed a proof-of-concept prototype that demonstrates the usefulness of XIML for multiple user interfaces.
- We have successfully satisfied our exit criteria for the first phases of the XIML framework development and are proceeding with the established roadmap.

ACKNOWLEDGEMENTS

Jean Vanderdonckt made significant contributions to the development of the MANNA prototype and to the XIML validation exercises. We also thank the following individuals for their contribution to the XIML effort: Hung-Yut Chen, Eric Cheng, Ian Chiu, Fred Hong, Yicheng Huang, James Kim, Simon Lai, Anthony Liu, Tunhow Ou, Justin Tan, and Mark Tong.

REFERENCES

Abrams, M., Phanouriou, C., Batongbacal, A., *et al.* (1999) UIML: An appliance-independent XML user interface language. *Computer Networks*, 31, 1695–1708.

Ali, M., Perez-Quinonez, M., Abrams, M., *et al.* (2002) Building multi-platform user interfaces using UIML, in *Computer Aided Design of User Interfaces 2002* (eds C. Kolski and J. Vanderdonckt). Springer-Verlag.

Arens, Y., and Hovy, E. (1995) The design of a model-based multimedia interaction manager. *Artificial Intelligence Review*, 9, 167–88.

Bouillon, L., and Vanderdonckt, J. (2002) *Retargeting Web Pages to Other Computing Platforms*. *Proceedings of the IEEE 9th Working Conference on Reverse Engineering (WCRE '2002)*, 339–48, IEEE Computer Society Press.

Butler, M. (2001) Implementing Content Negotiation Using CC/PP and WAP UAPROF. Technical Report HPL-2001-190, Hewlett Packard Laboratories.

Cheng, T. (1999) *XUL: Creating Localizable XML GUI*. Proceedings of the Fifteenth Unicode Conference.

Eisenstein, J. (2001) *Modeling Preference for Abstract User Interfaces*. Proceedings First International Conference on Universal Access in Human-Computer Interaction. Lawrence Erlbaum Associates.

Eisenstein, J., and Puerta, A. (2000) *Adaptation, in Automated User-Interface Design*. Proceedings of the 5th International Conference on Intelligent User Interfaces (IUI '00), 74–81. ACM Press.

Eisenstein, J., and Rich, C. (2002) *Agents and GUIs from Task Models*. Proceedings of the 7th International Conference on Intelligent User Interfaces (IUI '02), 47–54. ACM Press.

Eisenstein, J., Vanderdonckt, J., and Puerta, A. (2001) *Applying Model-Based Techniques to the Development of UIs for Mobile Computers*. Proceedings of the 6th International Conference on Intelligent User Interfaces (IUI '01), 69–76. ACM Press.

Kawai, S., Aida, H., and Saito, T. (1996) *Designing Interface Toolkit with Dynamic Selectable Modality*. Proceedings of Second International Conference on Assistive Technologies (ASSETS '96), 72–9. ACM Press.

Neches, R., Fikes, R., Finin, T. *et al.* (1991) Enabling technology for knowledge sharing. *AI Magazine*, Winter 1991, 36–56.

Olsen, D. (1992) *User Interface Management Systems: Models and Algorithms*. Morgan Kaufmann, San Mateo.

Organization for the Advancement of Structured Information Systems (2003). http://www.oasis-open.org.

Paterno, F., Mancini, C., and Meniconi, S. (1997) *Concurtasktrees: A Diagrammatic Notation for Specifying Task Models*. Proceedings of IFIP International Conference on Human-Computer Interaction (Interact '97), 362–9. Chapman and Hall.

Puerta, A., (1997) A model-based interface development environment. *IEEE Software*, 14 (4) (July/August), 41–7.

Puerta, A., and Eisenstein, J. (1999) Towards a general computational framework for model-based interface development systems. *Knowledge-Based Systems*, 12, 433–442.

Szekely, P. (1996) Retrospective and challenges for model-based interface development, in *Computer Aided Design of User Interfaces (CADUI '96)*, (eds F. Bodart and J. Vanderdonckt), Springer-Verlag. 1–27.

Szekely, P., Luo, P., and Neches, R. (1993) *Beyond Interface Builders: Model-Based Interface Tools*. Proceedings of 1993 Conference on Human Factors in Computing Systems (InterCHI '93), 383–390. ACM Press.

Szekely, P., Sukaviriya, P., Castells, P., *et al.* (1995) Declarative interface models for user interface construction tools: the mastermind approach. In *Engineering for Human-Computer Interaction* (eds L.J. Bass and C. Unger), 120–50. London: Chapman and Hall.

Tam, C., Maulsby, D., and Puerta, A. (1998) *U-TEL: A Tool for Eliciting User Task Models from Domain Experts*. Proceedings of the 3rd International Conference on Intelligent User Interfaces (IUI '98), 77–80. ACM Press.

Vanderdonckt, J., and Berquin, P. (1999) *Towards a Very Large Model-Based Approach for User Interface Development*. Proceedings of First International Workshop on User Interfaces to Data Intensive Systems (UIDIS '99), 76–85. IEEE Computer Society Press.

Vanderdonckt, J., and Bodart, F. (1993) *Encapsulating Knowledge for Intelligent Automatic Interaction Objects Selection*. *Proceedings of 1993 Conference on Human Factors in Computing Systems (InterCHI '93)*, 424–9. ACM Press.
VoiceXML Forum. http://www.voicexml.org.
World Wide Web Consortium. http://www.w3c.org.
XIML Forum. http://www.ximl.org.

AUIT: Adaptable User Interface Technology, with Extended Java Server Pages

John Grundy[1,2] and Wenjing Zou[2]

[1] Department of Electrical and Electronic Engineering and
[2] Department of Computer Science,
University of Auckland, New Zealand

8.1. INTRODUCTION

Many web-based information systems require degrees of adaptation of the system's user interfaces to different client devices, users and user tasks [Vanderdonckt et al. 2001; Petrovski and Grundy 2001]. This includes providing interfaces that will run on conventional web browsers, using Hyper-Text Mark-up Language (HTML), as well as wireless PDAs, mobile phones and pagers using Wireless Mark-up Language (WML) [Marsic 2001a; Han et al. 2000; Zarikas et al. 2001]. In addition, adapting to different users and user tasks is required [Eisenstein and Puerta 2000; Grunst et al. 1996; Wing and Colomb 1996]. For example, hiding 'Update' and 'Delete' buttons if the user is a customer or if the user is

Multiple User Interfaces. Edited by A. Seffah and H. Javahery
© 2004 John Wiley & Sons, Ltd ISBN: 0-470-85444-8

a staff member doing an information retrieval-only task. Building such interfaces using current web-based systems implementation technologies is difficult, time-consuming and results in hard-to-maintain solutions.

Developers can use proxies that automatically convert e.g. HTML content to WML content for wireless devices [Marsic 2001a; Han et al. 2000; Vanderdonckt et al. 2001]. These include web clipping services and portals with multi-device detection and adaptation features [Oracle 1999; Palm 2001; IBM 2001]. Typically these either produce poor interfaces, as the conversion is difficult for all but simple web interfaces, or require considerable per-device interface development work. Some systems take XML-described interface content and transform it into different HTML or WML formats depending on the requesting device information [Marsic 2001a; Vanderdonckt et al. 2001]. The degree of adaptation supported is generally limited, however, and each interface type requires often complex, hard-to-maintain XSLT-based scripting. Intelligent and component-based user interfaces often support adaptation to different users and/or user tasks [Stephanidis 2001; Grundy and Hosking 2001]. Most existing approaches only provide thick-client interfaces (i.e. interfaces that run on the client device, not the server), and most provide no device adaptation capabilities. Some recent proposals for multi-device user interfaces [Vanderdonckt et al. 2001; Han et al. 2000; Marsic 2001b] use generic, device-independent user interface descriptions. Most of these do not typically support user and task adaptation, however, and many are application-specific rather than general approaches. A number of approaches to model-driven web site engineering have been developed [Ceri et al. 2000; Bonifati et al. 2000; Fraternali and Paolini 2002]. Currently these do not support user task adaptation and their support for automated multi-device layout and navigation control is limited. These approaches typically fully-automate web site generation, and while valuable they replace rather than augment current development approaches.

We describe the Adaptable User Interface Technology (AUIT) architecture, a new approach to building adaptable, thin-client user interface solutions that aims to provide developers with a generic screen design language that augments current JSP (or ASP) web server implementations. Developers code an interface description using a set of device-independent XML tags to describe screen elements (labels, edit fields, radio buttons, check boxes, images, etc.), interactors (buttons, menus, links, etc), and form layout (lines, tables and groups). These tags are device mark-up language independent i.e. not HTML or WML nor specific to a particular device screen size, colour support, network bandwidth etc. Tags can be annotated with information about the user or user task they are relevant to, and an action to take if not relevant (e.g. hide, disable or highlight). We have implemented AUIT using Java Server Pages, and our mark-up tags may be interspersed with dynamic Java content. At run-time these tags are transformed into HTML or WML mark-up and form composition, interactors and layout determined depending on the device, user and user task context.

The following section gives a motivating example for this work, a web-based collaborative job management system, and reviews current approaches used to build adaptable, web-based information system user interfaces. We then describe the architecture of our AUIT solution along with the key aspects of its design and implementation. We give examples of using it to build parts of the job management system's user interfaces, including examples of device, user and user task adaptations manifested by these AUIT-implemented user interfaces. We discuss our development experiences with AUIT using it to build the user interfaces for three variants of commercial web-based systems and report results of two empirical studies of AUIT. We conclude with a summary of future research directions and the contributions of this research.

8.2. CASE STUDY: A COLLABORATIVE JOB MANAGEMENT SYSTEM

Many organisations want to leverage the increasingly wide-spread access of their staff (and customers) to thin-client user interfaces on desktop, laptop and mobile (PDA, phone, pager etc.) devices [Amoroso and Brancheau 2000; Varshney *et al.* 2000]. Consider an organisation building a job management system to co-ordinate staff work. This needs to provide a variety of functions allowing staff to create, assign, track and manage jobs within an organisation. Users of the system include employees, managers and office management. Key employee tasks include login, job creation, status checking and assignment. In addition, managers and office management maintain department, position and employee data. Some of the key job management screens include creating jobs, viewing job details, viewing summaries of assigned jobs and assigning jobs to others. These interactions are outlined in the use case diagram in Figure 8.1.

All of these user interfaces need to be accessed over an intranet using multi-device, thin-client interfaces i.e. web-based and mobile user interfaces. This approach makes the system platform- and location-independent and enables staff to effectively co-ordinate their work no matter where they are.

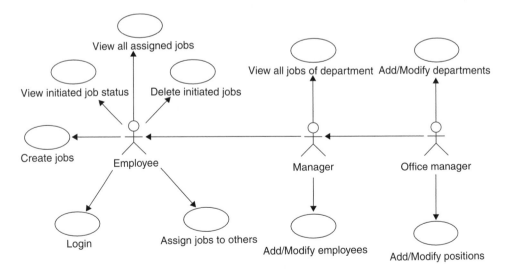

Figure 8.1. Use cases in the job management system.

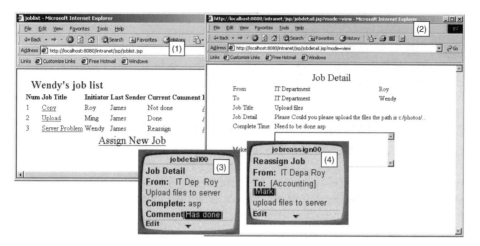

Figure 8.2. Adaptive job management information system screens.

Some of the thin-client, web-based user interfaces our job management information system needs to provide are illustrated in Figure 8.2. Many of these interfaces need to 'adapt' to different users, user tasks and input/output web browser devices. For example, the job listing screens (1) for job managers and other employees are very similar, but management has additional buttons and information fields. Sometimes the job details screen (2) has buttons for modifying a job (when the owning user is doing job maintenance) but at other times not (when the owning user is doing job searches or analysis, or the user is not the job owner). Sometimes interfaces are accessed via desktop PC web browsers (1 and 2) and at other times the same interface is accessed via a WAP mobile phone, pager or wireless PDA browser (3 and 4), if the employee wants to access job information when away from their desktop or unable to use their laptop.

8.3. RELATED WORK

To build user interfaces like the ones illustrated in Figure 8.2, we can use a number of approaches. We can build dedicated server-side web pages for each different combination of user, user task and device, using Java Server Pages, Active Server Pages, Servlets, PhPs, CGIs, ColdFusion and other technologies and tools [Marsic 2001a; Fields and Kolb 2000; Evans and Rogers 1997; Petrovski and Grundy 2001]. This is currently the 'standard' approach. It has the major problem of requiring a large number of interfaces to be developed and then maintained – for M different information system screens and N different user, user task and device combinations, we have to build and then maintain M*N screens. We can improve on this a little by adding conditional constructs to the screens for user and to some degree user task adaptations, reducing the total number of screens to build somewhat. However, for even small numbers of different users and user tasks, this approach makes screen implementation logic very complex and hard to maintain [Vanderdonckt et al. 2001; Grundy and Hosking 2001]. Each different device that may use the same screen still needs a dedicated server-side implementation [Fox *et al.* 1998; Marsic 2001a] due to different device mark-up language, screen size, availability of fonts and colour and so on [Vanderdonckt *et al.* 2001].

Various approaches have been proposed or developed to allow different display devices to access a single server-side screen implementation. A specialised gateway can provide automatic translation of HTML content to WML content for WAP devices [Fox *et al.* 1998; Palm 2001]. This allows developers to ignore the device an interface will be rendered on, but has the major problem of producing many poor user interfaces due to the fully-automated nature of the gateway. Often, as with Palm's Web Clipping approach, the translation cuts out much of the content of the HTML document to produce a simplified WML version, not always what the user requires. The W3C consortium has also been looking at various ways of characterising different display devices, such as Composite Capabilities/Preferences Profile (CC/PP) descriptions using RDF [W3C 2002a], and more generally at Device Independence Activity (DIA) [W3C 2002b], aiming to support seamless web interface descriptions and authoring. These approaches aim to capture different device characteristics to support accurate and appropriate dynamic user interface adaptation, and to allow write-once-display-anywhere style web page descriptions and design.

Another common approach is to use an XML encoding of screen content and a set of transformation scripts to convert the XML encoded data into HTML and WML suitable for different devices [Han *et al.* 2000; Vanderdonckt *et al.* 2001; Marsic 2001b]. For example, Oracle's Portal-to-go™ approach [Oracle 1999] allows device-specific transformations to be applied to XML-encoded data to produce device-tailored mark-up for rendering. Such approaches work reasonably well, but don't support user and task adaptation well and require complex transformation scripts that have limited ability to produce good user interfaces across all possible rendering devices. IBM's Transcoding [IBM 2002] provides for a set of transformations that can be applied to web content to support device and user preference adaptations. However a different transformation must be implemented for each device/user adaptation and it is unclear how well user task adaptation could be supported.

Some researchers have investigated the use of a database of screen descriptions to convert, at run-time, this information into a suitable mark-up for the rendering device, possibly including suitable adaptations for the user and their current task [Fox *et al.* 1998; Zarikas *et al.* 2001]. An alternative approach is the use of conceptual or model-based web specification languages and tools, such as HDM, WebML, UIML and Autoweb [Abrams *et al.* 1998; Ceri *et al.* 2000; Bonifati *et al.* 2000; Fraternali and Paolini 2002; Phanouriou 2000]. These all provide device-independent specification techniques and typically generate multiple user interface implementations from these, one for each device and user. WebML describes thin-client user interfaces and can be translated into various concrete mark-up languages by a server. UIML provides an even more general description of user interfaces in an XML-based format, again for translation into concrete mark-up or other user interface implementation technologies. The W3C work on device descriptors and device independence for web page descriptions are extending such research work. All of these approaches require sophisticated tool support to populate the database or generate web site implementations. They are very different to most current server-side implementation technologies like JSPs, Servlets, ASPs and so on. Usually such systems must fully generate web site server-side infrastructure, making it difficult for developers to reuse existing development components and approaches with these technologies.

Various approaches to building adaptive user interfaces have been used [Dewan and Sharma 1999; Rossel 1999; Eisenstein and Puerta 2000; Grundy and Hosking 2001]. To date, most of these efforts have assumed the use of thick-client applications where client-side components perform adaptation to users and tasks, but not to different display devices and networks. The need to support user interface adaptation across different users, user tasks, display devices, and networks (local area, high reliability and bandwidth vs wide-area, low bandwidth and reliability [Rodden *et al.* 1998]) means a unified approach to supporting such adaptivity is desired by developers [Vanderdonckt *et al.* 2001; Marsic 2001b; Zarikas *et al.* 2001; Han *et al.* 2000].

8.4. OUR APPROACH

We have developed an approach to building adaptive, multi-device thin-client user interfaces for web-based applications that aims to augment rather than replace current server-side specification technologies like Java Server Pages and Active Server Pages. User interfaces are specified using a device-independent mark-up language describing screen elements and layout, along with any required dynamic content (currently using embedded Java code, or 'scriptlets' [Fields and Kolb 2000]). Screen element descriptions may include annotations indicating which user(s) and user task(s) to which the elements are relevant. We call this Adaptive User Interface Technology as (AUIT). Our web-based applications adopt a four-tier software architecture, as illustrated in Figure 8.3.

Clients can be desktop or laptop PCs running a standard web-browser, mobile PDAs running an HTML-based browser or WML-based browser, or mobile devices like pagers and WAP phones, providing very small screen WML-based displays. All of these devices connect to one or more web servers (the wireless ones via a wireless gateway) accessing

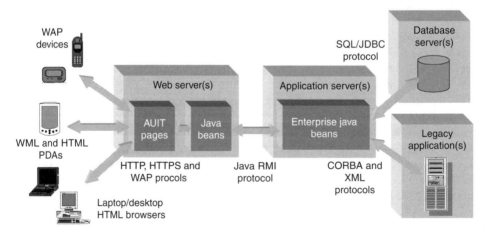

Figure 8.3. The four-tier web-based information system architecture.

a set of AUIT-implemented screens (i.e. web pages). The AUIT pages detect the client device type, remember the user associated with the web session, and track the user's current task (typically by which page(s) the current page has been accessed from). This information is used by the AUIT system to generate an appropriately adapted thin-client user interface for the user, their current task context, and their display device characteristics. AUIT pages contain Java code scriptlets that can access JavaBeans holding data and performing form-processing logic [Fields and Kolb 2000]. These web server-hosted JavaBeans communicate with Enterprise JavaBeans which encapsulate business logic and data processing [Vogal 1998]. The Enterprise JavaBeans make use of databases and provide an interface to legacy systems via CORBA and XML.

The AUIT pages are implemented by Java Server Pages (JSPs) that contain a special mark-up language independent of device-specific rendering mark-up languages, that are like HTML and WML but which contain descriptions of screen elements, layout and user/task relevance, unlike typical data XML encodings. Developers implement their thin-client web screens using AUIT's special mark-up language, specifying in a device, user and task-independent way each screen for their application i.e. only M screens, despite the N combinations of user, user task and display device combinations possible for each screen. While AUIT interface descriptions share some commonalities with model-based web specification languages [Ceri *et al.* 2000; Fraternali and Paolini 2002], they specify in one place a screen's elements, layout, and user and task relevance.

An AUIT screen description encodes a layout grid (rows and columns) that contains screen elements or other layout grids. The layout grids are similar to Java AWT's GridBagLayout manager, where the screen is comprised of (possibly) different-sized rows and columns that contain screen elements (labels, text fields, radio buttons, check boxes, links, submit buttons, graphics, lines, and so on), as illustrated in Figure 8.4. Groups and screen elements can specify the priorities, user roles and user tasks to which they are relevant. This allows AUIT to automatically organise the interface for different-sized display devices into pages to fit the device and any user preferences. The structure of the screen description is thus a logical grouping of screen elements that is used to

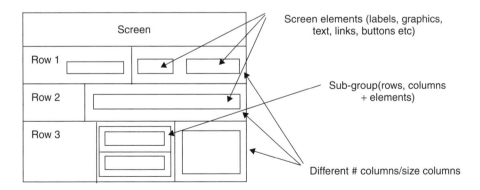

Figure 8.4. Basic AUIT screen description logical structure.

generate a physical mark-up language for different device, user role and user task combinations.

Unlike generic web mark-up languages and XML-encoded screen descriptions, AUIT screen descriptions include embedded server-side dynamic code. Embedded Java scriptlets currently provide this dynamic content for AUIT web pages and conventional JSP JavaBeans are used to provide data representation, form processing and application server access. When a user accesses an AUIT-implemented JSP page, the AUIT screen description is interpreted with appropriate user, user task and display device adaptations being made to produce a suitable thin-client interface. Using embedded dynamic content allows AUIT tags to make use of data as a device-specific screen description is generated. It also allows developers to use their existing server-side web components easily with AUIT screen mark-up to build adaptive web interfaces.

8.5. DESIGN AND IMPLEMENTATION

Java Server Pages are the Java 2 Enterprise Edition (J2EE) solution for building thin-client web applications, typically used for HTML-based interfaces but also usable for building WML-based interfaces for mobile display devices [Fields and Kolb 2000; Vogal 1998]. To implement AUIT we have developed a set of device-independent screen element tags for use within JSPs that allow developers to specify their screens independent of user, task and display device. Note that we could implement AUIT in various ways, for example we could populate an AUIT-encoded screen description with data then transform it from its XML format into a device-specific mark-up language, or extract AUIT screen descriptions from a database at run-time, generating device-specific mark-up from these. AUIT screen descriptions are typically lower-level than those of conceptual web specification languages e.g. WebML and HDM [Ceri *et al.* 2000; Bonifati *et al.* 2000] but we don't attempt to generate full web-side functionality from AUIT, rather we interpret the custom AUIT tags and embedded Java scriptlets. AUIT descriptions have some similarities to some XML-based web screen encoding approaches, but again our focus is on providing JSP (and ASP) developers a device, user and user task adaptable mark-up language rather than requiring them to generate an XML encoding which is subsequently transformed for an output device.

Some of the AUIT tags and their properties are shown in Table 8.1, along with some typical mappings to HTML and WML mark-up tags. Some AUIT tag properties are not used by HTML or WML e.g. graphic, alternate short text, colour, font size, user role and task information, and so on. Reasonably complex HTML and WML interfaces can be generated from AUIT screen descriptions. This includes basic client-side scripting with variables and formulae – currently we generate JavaScript for HTML and WMLScript for WML display devices. AUIT tags generally control layout (screen, group, table, row, paragraph etc), page content (edit field, label, line, image etc), or inter-page navigation (submit, link). Each AUIT tag has many properties the developer can specify, some mandatory and some optional. All tags have user and user task properties that list the users and user tasks to which the tag is relevant. Screen tags have a specific task property allowing the screen to specify the user's task context (this is passed onto linked pages

Table 8.1. Some AUIT tags and properties.

AUIT Tag/Properties	Description	HTML	WML
<auit:screen> • title, alternate • width, height • template • colour, bgcolour • font, lcolour	Encloses contents of whole screen. Title and short title alternative are specified. Can specify max width/height and AUIT appearance template to use. Default colours, fonts and link appearance can be specified.	<html>	<wml>, <card>
<auit:form> • action • method	Indicates an input form to process (POSTable). Specify processing action URL and processing method.	<form>	<do type = accept>
<auit:group> • width, height • rows, columns • priority • user, task	Groups related elements of screen. Group can have m rows, with each row 1 to n columns (may be different number). Number of rows can be dynamic i.e. determined from data iteration.	–	–
<auit:table> • border, width • colour, bgcolour • rows, columns	Table (grid) with fixed number rows and columns. Can specify border width and colour, 3D or shaded border, fixed table rows/columns (if known).	<table>	<table>
<auit:row> • width, height • columns • user, task	Group or table row information. Can specify # columns, width and height row encloses. Can also restrict relevance of enclosed elements to specified user/task.	<tr>	<tr>

(continued overleaf)

Table 8.1 (*continued*)

AUIT Tag/Properties	Description	HTML	WML
<auit:column> • width, height	Group or table column information.	<td>	<td>
<auit:iterator> • bean, variable	Iterates over data structure elements. Uses JavaBean collection data structure.	–	–
<auit:paragraph>	Paragraph separator.	<p>	<p>
<auit:line> • height, colour	Line break. Optional height (produces horizontal line).	 , <hr>	 , <hr>
<auit:heading> • level, colour, font • user, task	Heading level and text.	<h1>, <h2> etc	Plain text
<auit:label> • colour, font • alternate, image • user, task	Label on form. Can have short form, image.	Plain text	Plain text
<auit:textbox> • colour, font • user, task • script	Edit field description. Can define colour, fonts.	<input type = text>	<input type = text>
<auit:radio>	Radio button.	<input type = radio>	<input type = radio>
<auit:select>	Popup menu item list.	<select . . . >	<select . . . >
<auit:image> • source, alternate	Image placeholder, has alternate text (short and long forms).	<img src = . . . >	<img src = . . . >
<auit:link> • url, image • user, task	Hypertext link, has label or image.	<a href = . . . >	<go href = . . . >
<auit:submit> • user, task • colour, image • url, script	Submit button/action (for form POSTing).	<input type = submit	<do><go href = . . . >
<% . . . %>	Embedded Java scriptlet code.	–	–

by auit:link tags to set the linked form's user task). Grouped tags, as well as table rows and columns have a 'priority' indicating which grouped elements can be moved to linked screens for small-screen display devices and which must be shown. Table, row and column tags have minimum and maximum size properties, used when auto-laying out AUIT elements enclosed in table cells. Edit box, radio button, check boxes, list boxes and pop-up menus have a name and value, obtained from JavaBeans when displayed and that set Java-Bean properties when the form is POSTed to the web server. Images have alternate text values and ranges of source files (typically, .gif, .jpg and wireless bit map .wbm formats). Links and submit tags specify pages to go to and actions for the target JSP to perform.

Users, as well as user task relevance and priority can be associated with any AUIT tag (if with a group, then this applies to all elements of the group). We use a hierarchical role-based model to characterise users: a set of roles is defined for a system with some roles being generalisations of others. Specific users of the system are assigned one or more roles. User tasks are hierarchical and sequential i.e. a task can be broken down into sub-tasks, and tasks may be related in sequence, defining a basic task context network model. Any AUIT tag may be denoted as relevant or not relevant to one or more user roles, sub-roles, tasks or sub-tasks, and to a task that follows one or more other tasks. In addition, elements can be 'prioritised' on a per-role basis i.e. which elements must be shown first, which can be moved to a sub-screen, which must always be shown to the user.

A developer writes an AUIT-encoded screen specification, which makes use of JavaBeans (basically Java classes) to process form input data and to access Enterprise JavaBeans, databases and legacy systems. At run-time the AUIT tags are processed by the JSPs using custom tag library classes we have written. When the JSP encounters an AUIT tag, it looks for a corresponding custom tag library class which it invokes with tag properties. This custom tag class performs suitable adaptations and generates appropriate output text to be sent to the user's display device. Link and submit tags produce HTML or WML markups directing the display device to other pages or to perform a POST of input data to the web server as appropriate. Figure 8.5 outlines the way AUIT tags are processed. Note that dynamic content Java scriptlet code can be interspersed with AUIT tags.

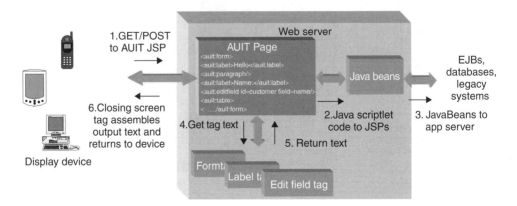

Figure 8.5. JSP custom tag-implemented AUIT screen descriptions.

One of the more challenging things to support across multiple devices and in the presence of user and task adaptations (typically inappropriate screen elements being hidden) is providing a suitable screen layout for the display device. HTML browsers on desktop machines have rich layout support (using multi-layered tables), colour, a range of fonts and rich images. PDAs have similar layout capability, a narrower font range, and some have no colour and limited image display support. Mobile devices like pagers and WAP phones have very small screen size, one font size, typically no colour, and need low-bandwidth images. Hypertext links and form submissions are quite different, using buttons, clickable text or images, or selectable text actions.

AUIT screen specifications enable users to use flexible groups to indicate a wide variety of logical screen element relationships and to imply desired physical display layouts. These allow our AUIT custom tags to perform automatic 'splitting' of a single AUIT logical screen description into multiple physical screens when all items in a screen can not be sensibly displayed in one go on the display device. Group rows and columns are processed to generate physical mark-up for a device, and then are assembled to form a full physical screen. If some rows and/or columns will not fit the device screen, AUIT assembles a 'first screen' (either top, left rows and columns that fit the device screen, or top-priority elements if these are specified). Remaining screen elements are grouped by the rows and columns and are formed into one or more 'sub-screens', accessible from the main screen (and each other) via hypertext links. This re-organisation minimises the user needing to scroll horizontally and vertically on the device, producing an interface that is easier to use across all devices. It also provides a physical interface with prioritised screen elements displayed. AUIT has a database of device characteristics (screen size, colour support and default font sizes etc.), that are used by our screen splitting algorithm. Users can specify their own preferences for these different display device characteristics, allowing for some user-specific adaptation support.

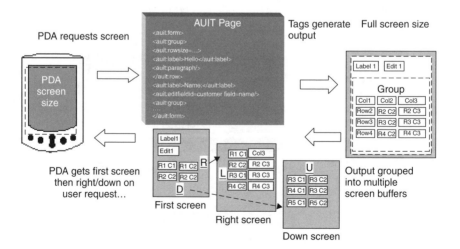

Figure 8.6. Screen splitting adaptation of form layout.

When processing an AUIT screen description, as each AUIT tag is processed, it generates physical mark-up output text that is cached in a buffer. When group, row or column text will over-fill the display device screen, text to the right and bottom over-filling the screen is moved to separate screens, linked by hypertext links and organised using the specified row/column groupings. An example of this process is outlined in Figure 8.6. Here a PDA requests a screen too big for it to display, so the AUIT tag output is grouped into multiple screens. The PDA gets the first screen to display and the user can click on the _R_ight and _D_own links to get the other data. Some fields can be repeated in following screens (e.g. the first column of the table in this example) if the user needs to see them on each screen.

8.6. JOB MANAGEMENT SYSTEM EXAMPLES

We illustrate the use of our AUIT system in building some adaptable, web-based job maintenance system interfaces as outlined in Section 8.2. Figure 8.7 shows examples of the job listing screen being displayed for the same user in a desktop web browser (1), mobile PDA device (2 and 3) and mobile WAP phone (4–6). The web browser can show all jobs (rows) and job details (columns). It can also use colour to highlight information and hypertext links. The PDA device can not show all job detail columns, and so additional job details are split across a set of horizontal screens. The user accesses these additional details by using the hypertext links added to the sides of the screen. The WAP phone similarly can't display all columns and rows, and links are added to access these. In addition, the WAP phone doesn't provide the degree of mark-up the PDA and

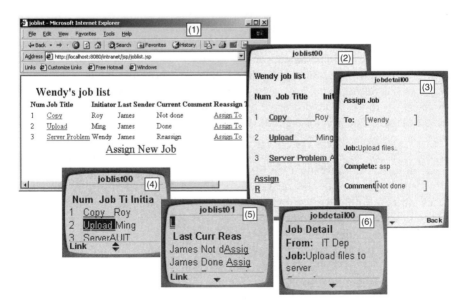

Figure 8.7. Job listing screen on multiple devices.

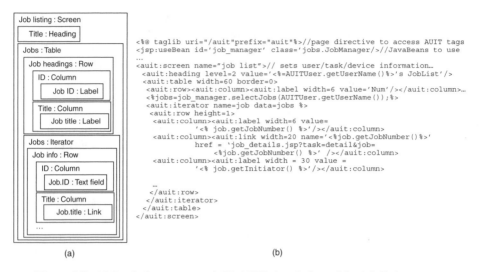

(a)   (b)

Figure 8.8. (a) Logical structure and (b) AUIT description of the job listing screen.

web browser can, so buttons and links and colour are not used. The user instead accesses other pages via a text-based menu listing.

Figure 8.8(a) shows the logical structure of the job listing screen using the AUIT tags introduced in Section 8.4. The screen is comprised of a heading and list of jobs. The first table row displays column headings, the subsequent rows are generated by iterating over a list of job objects returned by a Java Bean. Figure 8.8(b) shows part of the AUIT Java Server Page that specifies this interface. The first lines in the JSP indicate the custom tag library (AUIT) and 'JavaBean' components accessible by the page e.g. the job manager class provides access to the database of jobs. The screen tag sets up the current user, user task and device information obtained from the device and server session context for which the page is being run. The heading tag shows the user whose job list is being displayed. The table tag indicates a table, and in this example, one with a specified maximum width (in characters per row) and no displayed border. The first row shows headings per column, displayed as labels. The iterator tag loops, displaying each set of enclosed tags (each row) for every job assigned to the page user. The job list is obtained via the embedded Java code in the $<\% \ldots \%>$ tags. The column values for each row include labels (number, initiator, comment etc) and links (Job title, Assign To).

Figure 8.9(a) shows examples of the job details screen in use for different users and user tasks. In (1), a non-owning employee has asked to view the job details. They are not able to modify it or assign other employees to this job. The same screen is presented to the job manager if they go to the job details screen in a 'viewing' (as opposed to job maintenance) task context. In (2), the job manager has gone to the job details screen in a 'job re-assignment' task context from the job listing screen in Figure 8.7. They can assign employees to the job but not modify other job details. In (3) the job manager has gone to the screen to update a job's details and the job detail fields are now editable (a similar interface is presented when creating a new job). If an assigned employee accesses

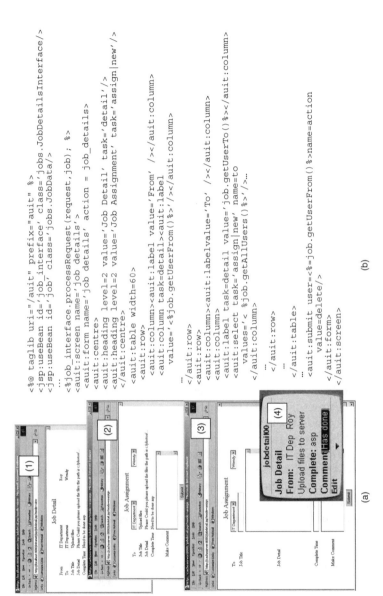

Figure 8.9. (a) Adapted job detail screens and (b) part of its AUIT description.

```
<%@ taglib uri="/auit" prefix="auit" %>
<jsp:useBean id='job_interface' class='jobs.JobDetailsInterface/>
<jsp:useBean id='job' class='jobs.JobData/>
...
<%job_interface.processRequest(request,job); %>
<auit:screen name='job details' >
<auit:form name='job details' action = job_details>
<auit:centre>
<auit:heading level=2 value='Job Detail' task='detail'/>
<auit:heading level=2 value='Job Assignment' task='assign|new'/>
</auit:centre>
<auit:table width=60>
<auit:row>
  <auit:column><auit:label value='From' /></auit:column>
  <auit:column task=detail><auit:label
    value='<%job.getUserFrom()%>'/></auit:column>
  ...
</auit:row>
<auit:row>
  <auit:column><auit:labelvalue='To' /></auit:column>
  <auit:column>
  <auit:label task=detail value='job.getUserTo()%></auit:column>
  <auit:select task='assign|new' name=to
    values='< %job.getAllUsers()%>'/>...
</auit:column>
  ...
</auit:row>
...
</auit:table>
...
<auit:submit user=<%=job.getUserFrom()%>name=action
    value=delete/>
</auit:form>
</auit:screen>
```

(b)

the job details in a job maintenance task context, only the comment and complete fields are editable, as shown in the WAP phone-displayed job details screen in (4).

Part of the AUIT specification of the job details screen is shown in Figure 8.9(b). The screen encloses a form which is posted to the web server for processing (done by the job_interface JavaBean component) when the user fills out values for it. The heading is task-dependent – if the user is viewing job details, the heading is not the same as when they are assigning or adding a job. The 'task' attribute of the two heading tags is used to determine which heading is shown. A table is used to achieve the layout. Some columns are common to all screens e.g. the left-hand side labels. Some rows are not shown for some screens e.g. the 'From' row is not shown if the user task is *assign* or *new*. Sometimes a different kind of form element is used e.g. if viewing a job, labels are used but when adding or assigning a job, some fields for the job have editable elements (text box, pop-up menu etc). If the user of the screen is not the job owner, then the delete button is not shown.

8.7. EXPERIENCES

To date, we have built three substantial web-based applications with AUIT technology: the job management system, an on-line car sales E-commerce site, and an on-line collaborative travel planning system. Each of these systems has over two dozen AUIT screens, JavaBean and EJB application server components, and database tables. We have also built, for other projects, 'hard-coded' versions of these systems using conventional JSP technology – a commercial version of the on-line car retailer system, a commercial, in-house company job management system, and a large prototype collaborative travel planning system. Each of these systems has JSPs specifically built for different users, user tasks and display devices. The AUIT systems all have less than a third of the screen specifications of these hard-coded systems. These AUIT screen specifications are easier to extend and maintain as new data and functions are added to the systems, as only a single specification needs modifying rather than up to half a dozen for some screens in some of these hard-coded systems. Our AUIT technology allows developers to use all the usual JSP and Servlet functionality in conjunction with the AUIT adaptable tags. This means the expressive power for building dynamic web applications with AUIT-based JSPs is preserved, unlike when using XML-based translation approaches.

We have run two empirical evaluations of our AUIT-based systems, with a dozen end users comparing the AUIT and hard-coded system interfaces, and with half a dozen experienced, industry web-based information system developers comparing the use of AUIT technology to conventional JSP technology. End users in our studies using the job management and car site systems found the AUIT-implemented user interfaces to be as good from a usability perspective as the hard-coded ones. In fact, some found them better since they could change their device preferences and have the AUIT interfaces change to suit them; this was not possible with the hard-coded interface implementations. We had the software developers extend an existing JSP-implemented system by using AUIT to build two interfaces, both needing device, user and user task adaptations. They built two AUIT screen implementations and up to half a dozen conventional JSP implementations (generating HTML and WML respectively). These developers found AUIT to be straight-forward to use and much more powerful and easier than the conventional JSP technology for building adaptable user interfaces. They found the range of functionality supported by AUIT tags to be large enough for most of their web-based user implementation needs.

We have tried to provide developers with a superset of screen specification tags and facilities enabling a wide range of both HTML- and WML-coded interfaces to be generated. However AUIT trades off the power of developers to tailor a web-based user interface to a specific device, user role and user task with ease of specification and maintenance of adaptable user interfaces. One consequence of specifying logical screen groups, elements and relevancies is that quite different physical screen layouts and interactions may result depending on device, user and task accessing the interface. The interface the user sees is thus variable and may not be optimal i.e. usability is less for end users than hard-coded interfaces. An interesting and unintended consequence we found with AUIT user preferences is that users can specify dynamically their own preferred screen size limits, colour and image usage for different devices. AUIT then provides user-tailored interfaces using these preferences, an unintended facility, but one that our end users found useful. We have found AUIT limited for highly image-oriented interfaces and interfaces using large numbers of embedded tables to provide very fine-tuned screen layout. AUIT is quite scalable in terms of number of users, imposing quite low overheads on the hosting server.

Our evaluation of AUIT has identified some areas for further research. The design of AUIT-based systems is radically different to user interface design of conventional web-based application interfaces. Designers need to work with logical structures like that illustrated in Figure 8.8(b), rather than with fixed-format layout as in conventional web user interface design. AUIT groups provide a reasonably flexible facility to layout screens, which work well for WML and moderately complex HTML interfaces. Very complex, fine-tuned HTML layout for desktop browsers is difficult to achieve with the current AUIT grouping components. Estimating the amount of room that rendered screen elements will take up, used by the screen splitting algorithm to move some information to linked screens, is difficult, as users may configure their device browsers with different default fonts and font sizes.

We are currently developing a new design method for adaptable web application user interfaces, along with a GUI specification tool that will generate AUIT implementations from these graphical designs. This will make it easier for developers to specify such systems interactively. We are continuing to enhance the layout control in AUIT grouping constructs to give developers more control over complex screen layout across display devices. Current task adaptation support is limited and we are extending this to allow developers to use more workflow-like information to support such adaptations. Extending user preference control and device characteristics, like network bandwidth, will allow further detailed specification of interface adaptation. AUIT could be applied to specifying thick-client adaptable user interfaces i.e. client-side user interface objects. Our aim to date however has been to focus on augmenting JSP server-side, thin-client adaptable user interface implementation, due to the limited client-side support in many small-screen devices.

8.8. SUMMARY

We have developed a new approach for the development of adaptable, web-based information system user interfaces. This provides developers with a set of device-independent mark-up tags used to specify thin-client screen elements, element groupings, and user and user task annotations. We have implemented this with Java Server Page custom tag libraries, making our system fully compatible with current J2EE-based information system architectures. We have developed a novel automated approach for splitting too-large screens into parts for different display devices. We have developed several systems with our technology, all evaluated by end users and commercially deployable. Developers report that they find our technology easier to use and more powerful for building and maintaining adaptable web-based user interfaces than other current approaches. End users report they find the adaptive interfaces suitable for their application tasks, and in some instances prefer them to hard-coded, device- and user-tailored implementations.

REFERENCES

Abrams, M., Phanouriou, C. Batongbacal, A.L., Williams, S., and Shuster, J.E. (1998) UIML: An Appliance-Independent XML User Interface Language, In *Proceedings of the 8th World Wide Web Conference*, Toronto, Canada.

Amoroso, D.L., and Brancheau, J. (2001) Moving the Organization to Convergent Technologies e-Business and Wireless, in *Proceedings of the 34th Annual Hawaii International Conference on System Sciences*, Maui, Hawaii, Jan 3–6 2001. IEEE CS Press.

Bonifati, A., Ceri, S., Fraternali, P., and Maurino, A. (2000) Building multi-device, content-centric applications using WebML and the W3I3 Tool Suite, in Proceedings of. Conceptual Modelling for E-Business and the Web, LNCS 1921, pp. 64–75.

Ceri, S., Fraternali, P., and Bongio, A. (2000) Web Modelling Language (WebML): a modelling language for designing web sites. *Computer Networks*, 33 (1–6).

Dewan, P., and Sharma, A. (1999) An experiment in inter-operating, heterogeneous collaborative systems, in *Proceedings of the 1999 European Conference on Computer-Supported Co-operative Work*, 371–390. Kluwer.

Eisenstein, J., and Puerta, A. (2000) Adaptation in automated user-interface design, in *Proceedings of the 2000 Conference on Intelligent User Interfaces*, New Orleans, 9–12 January 2000, 74–81. ACM Press.

Evans, E., and Rogers, D. (1997) Using Java Applets and CORBA for multi-user distributed applications. *Internet Computing*, 1 (3). IEEE CS Press.

Fields, D., and Kolb, M. (2000) *Web Development with Java Server Pages*. Manning.

Fox, A., Gribble, S., Chawathe, Y., and Brewer, E. (1998) Adapting to Network and Client Variation using Infrastructural Proxies: lessons and perspectives. *IEEE Personal Communications*, 5 (4), 10–19.

Fraternali, P., and Paolini, P. (2002) Model-Driven Development of Web Applications: the Autoweb system, to appear in *ACM Transactions on Office Information Systems*.

Grundy, J.C., and Hosking, J.G. (2001) Developing Adaptable User Interfaces for Component-based Systems. *Interacting with Computers*. Elsevier.

Grunst, G., Oppermann, R., and Thomas, C.G. (1996) Adaptive and adaptable systems, in *Computers as Assistants: A New Generation of Support Systems* (ed. P. Hoschka), 29–46. Hillsdale: Lawrence Erlbaum Associates.

Han, R., Perret, V., and Naghshineh, M. (2000) WebSplitter: A unified XML framework for multi-device collaborative web browsing, in *Proceedings of CSCW 2000*, Philadelphia, Dec 2–6 2000. ACM Press.

IBM Corp (2001) IBM Transcoding™ White Paper. http//www.research.ibm.com/networked_data_ systems/transcoding/transcodef.pdf.

Marsic, I. (2001a) Adaptive Collaboration for Wired and Wireless Platforms. *IEEE Internet Computing* July/August 2001, 26–35.

Marsic, I. (2001b) An architecture for heterogeneous groupware applications, in *Proceedings of the International Conference on Software Engineering*, May 2001, 475–84. IEEE CS Press.

Oracle Corp (1999) Oracle Portal-to-go™ White Paper. http//www.alentus.com/library/oracle/wp_ portal.pdf.

Palm Corp (2001) Web Clipping services. www.palm.com.

Petrovski, A., and Grundy, J.C. (2001) Web-enabling an Integrated Health Informatics System, in Proceedings of the 7th International Conference on Object-oriented Information Systems, Calgary, Canada, August 26–29 2001. Lecture Notes in Computer Science.

Phanouriou, C. (2000) UIML: A Device-Independent User Interface Markup Language. Ph.D. dissertation, Virginia Tech, September 2000.

Rodden, T., Chervest, K., Davies, N., and Dix, A. (1998) Exploiting context in HIC Design for Mobile Systems, in Proceedings of the first Workshop on Human Computer Interaction with Mobile Devices.

Rossel, M. (1999) Adaptive support: the Intelligent Tour Guide. *International Conference on Intelligent User Interfaces*. New York: ACM.

Stephanidis, C. (2001) Concept of Unified User Interfaces, in *User Interfaces for All: Concepts, Methods and Tools*, 371–88. Lawrence Erlbaum.

Vanderdonckt, J., Limbourg, Q., Florins, M., Oger, F., and Macq, B. (2001) Synchronised, model-based design of multiple user interfaces, in *Proceedings of the 2001 Workshop on Multiple User Interfaces over the Internet*.

Varshney, U., Vetter, R.J., and Kalakota, R. (2000) Mobile Commerce a New Frontier. *Computer*, 33 (10). IEEE Press.

Vogal, A. (1998) CORBA and Enterprise Java Beans-based Electronic Commerce. *International Workshop on Component-based Electronic Commerce*. Fisher Center for Management & Information Technology, UC Berkeley.

W3C (2002a) *Composite Capabilities/Preferences Profile*. World Wide Web Consortium, http://www.w3.org/Mobile/CCPP/.

W3C (2002b) *Device Independence Activity: Working towards seamless Web access and authoring.* http://www.w3.org/2001/di/.

Wing, H., and Colomb, R.M. (1996) Behaviour sharing in adaptable user interfaces. *Proceedings of the Sixth Australian Conference on Computer-Human Interaction*, Los Alamitos, CA, USA, 197–204. IEEE CS Press.

Zarikas, V., Papatzanis, G., and Stephanidis, C. (2001) An architecture for a self-adapting information system for tourists, in *Proceedings of the 2001 Workshop on Multiple User Interfaces over the Internet*.

Part IV

Model-Based Development

Adaptive Task Modelling: From Formal Models to XML Representations

Peter Forbrig, Anke Dittmar, and Andreas Müller

University of Rostock, Department of Computer Science, Germany

9.1. INTRODUCTION

What does it mean for software designers to be more aware of the context of use? First, they have to analyze thoroughly the situations under which the interactive application will be executed. This analysis should consider the variety of tasks users want to perform with the aid of the interactive system, as well as the variety of environments under which tasks must be performed. Questions such as "Which platform is available in a certain task situation?" must be considered. Secondly, designers must ensure that the designed system shows a reasonable behavior in specific situations of use on different platforms. This chapter shows that research results of model-based approaches can serve as a useful basis for the development of context-aware applications.

We start with a short introduction to model-based development of interactive systems and to some important ideas about task modelling. Then, we consider new kinds of applications and the requirements that emerge with them. A design process which can better

Multiple User Interfaces. Edited by A. Seffah and H. Javahery
© 2004 John Wiley & Sons, Ltd   ISBN: 0-470-85444-8

cope with changing environments is introduced, and general modelling ideas are supported by presenting two specification techniques in more detail. The first technique, which can be applied to task models, allows a separate description of the general temporal relations within a task model and the specific temporal constraints which are imposed by concrete situations. The second technique uses XML technology to distinguish between an abstract interaction model of an interactive system and its specific representations. Specific features of devices are given by XML-descriptions. This approach allows a smooth integration of newly developed devices into already existing software. Finally, an example illustrates the integration of the proposed techniques into a general modelling approach.

9.2. MODEL-BASED SOFTWARE DEVELOPMENT

9.2.1. MODELS USED IN THE DESIGN PROCESS

In most model-based approaches, several models are created and combined to characterize a domain of interest from different perspectives. Among the many models used for the design of interactive systems, 'the task model has today gained much attention and many acceptances in the scientific community to be the model from which a development should be initiated' [Forbrig and Dittmar 2001]. This model is based on the idea that interactive systems should reflect the tasks that users have to perform in certain areas of application. This position, which is known as a *task-based* approach, emphasizes the fact that the design and the introduction of new software systems is not only a technical problem, but also affects the whole labor situation, including the domain model and the user model.

Figure 9.1 shows the most important models in use. It distinguishes between the existing and the envisioned labor situation as situations before and after the introduction of the designed interactive system [van der Veer *et al.* 1996; Lim and Long 1994]. A precise analysis of the existing situation is necessary if one is to make reasonable suggestions for an envisioned one. The interactive system can be conceptualized as consisting of two main parts (as seen in Figure 9.1), which can be divided into further sub-models. Whereas the application model describes the functionality of the system, the interaction model specifies the user interface in an abstract way. The requirements for the interactive system should be derived from the models of the existing and the envisioned situations and can further influence the envisioned sub-models.

9.2.2. TASK MODELLING

Task theories such as TKS [Johnson and Wilson 1993] form the basis for modelling the conceptual knowledge a user requires to fulfill a task. The underlying assumption of task-based approaches is that models representing this knowledge can support the designers in constructing adequate systems and user interfaces.

There is a general agreement that people develop hierarchical representations of tasks [Sebilotte 1988]. Our limited working memory means that we are not able to survey more complex tasks. Consequently, we form intermediate tasks from subtasks [Hacker 1995]. However, besides the hierarchical representations we also employ behavioral descriptions of tasks. Whereas a hierarchy says something about the subtasks which have to be executed

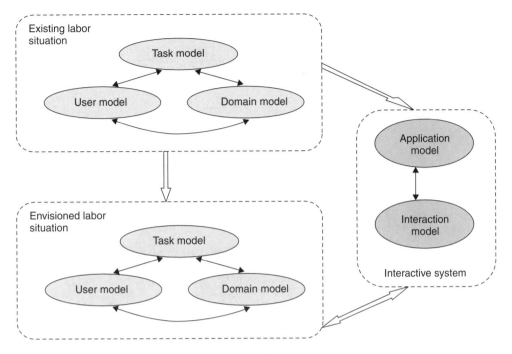

Figure 9.1. Models used in the design process.

to fulfill the whole task (*what* to do), a behavioral representation takes the conditions for performing a subtask into consideration (*when* to do things) [Shepherd 1989]. Following [Hacker 1995], we argue that a task has both a hierarchical and a sequential character.

The simple example in Figure 9.2 illustrates the modelling ideas mentioned above. Basically, the model describes the task of E-shopping. Figure 9.2a shows the hierarchical decomposition. The task *E-shopping* consists of the subtasks *Looking for the product, Checking the offer*, and *Ordering the product,* which have to be decomposed themselves until the level of atomic tasks is achieved. Each atomic task is linked with an operation which causes a change in the task environment.

The sequential description of the task can be found in Figure 9.2b. It specifies the sequential order that must be applied to subtasks, and to their components, so that they are executed successfully. This sequential description is developed by using temporal operators known from process algebras like CSP [Hoare 1985] or CCS [Milner 1989]. Most of the task modelling approaches proceed in this manner [Johnson and Wilson 1993; Paterno *et al*. 1997; Scapin and Pierret-Golbreich 1990].

For instance: Let T_1 and T_2 be subtasks, then the expression $T_1 \mid T_2$ means that the subtasks T_1 and T_2 can be performed in any order. If we want to indicate that subtask T_2 has to be executed after T_1 then the sequential operator has to be applied (T_1; T_2).

To describe the impact a task performance has on the environment, we need a domain model, which includes a description of goals, preconditions, and effects of the task.

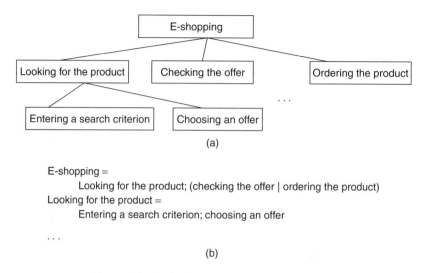

(a)

E-shopping =
        Looking for the product; (checking the offer | ordering the product)
Looking for the product =
        Entering a search criterion; choosing an offer

. . .

(b)

Figure 9.2. A simple task model of E-shopping.

Whereas a goal describes an intended or desired state which is supported by appropriate tasks, the precondition of a task is a state necessary for its execution. The effect of a task is the state which exists after performing it. Ideally, the goal to which a task is assigned and its effect describe the same state. Many approaches use object-oriented techniques to specify the task environment [Biere *et al.* 1999; Shepherd 1989; Tauber 1990]. Figure 9.3 illustrates the preconditions and effects of the E-shopping subtask 'choosing an offer,' from Figure 9.2.

The goal of *Choosing an offer* is to select an offer, which is indeed one part of the effect of this subtask. But other consequences – that the list of offers and the means of

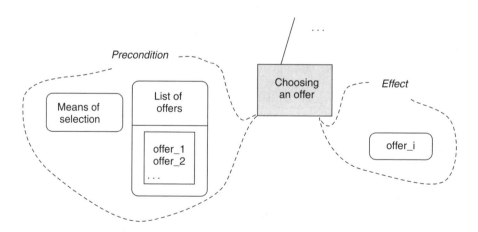

Figure 9.3. Precondition and effect of the subtask *Choosing an offer*.

selection do not exist after the execution – are certainly not part of the goal. Of course, one can imagine a more precise description of the task environment. The level of precision that must be applied in specifying the elements of the preconditions and effect depend on the goals of the modelling procedure.

9.2.3. NEW CHALLENGES FOR MODELLING

Devices such as the mobile phone and the PDA allow a user to perform subtasks by using an interactive system in previously inaccessible environments. For example, consider an engineer who is able now to input measured data directly at the place of measuring (by using a PDA) or later sitting in his office (by using a desktop). Both situations can be characterized not only by the different applied input devices, but also by the different subtasks the engineer has also to perform there.

There are different ways to develop software that runs on different devices. Of course, a separate software design for each device is possible, but then designers must find a way to combine all subsystems into a complete and consistent system which can be considered 'task-oriented.'

Furthermore, most of the functionality has to be designed several times. Instead, we suggest a design process utilizing task-based approaches, so that the interactive system is developed from a task model. The physical devices which a user requires to interact with the system can be considered as part of the specific situation of use.

However, by even following the second design approach, we have a large design space [Dey and Abowd 1999]. As Dey points out, 'we could require users explicitly to express all information relevant to a given situation'; alternatively, and preferably, we can '. . . let the application designer decide what information is relevant and how to deal with it' [Dey and Abowd 1999]. The consequences for designers are that 'the application designer should have spent considerably more time analyzing the situations under which the application will be executed and can more appropriately determine what information is relevant and how to react to it' [Dey and Abowd 1999].

We consider an interactive system to be context-aware if it is able to adapt its behavior to a specific situation of use. Task analysis processes can supply models specifying the user's tasks and their intentions. However, task models are not an appropriate means of describing a context which can be changed dynamically [Pribeanu *et al.* 2001].

To illustrate this point, let us look back to the example in Figure 9.2. Basically, the model allows a user to *Check an offer* and to *Order the product* in any order (the only constraint is that the subtask *Looking for the product* has to be completed). Consequently, an appropriate interactive system has to present at the same time the opportunity to perform both subtasks. However, if a mobile phone is used, for example, the screen could be too small to show all necessary information. In other words, the model neglects specific constraints in the actual environment as well as individual preferences or abilities.

It is widely accepted that the use of different models during software development leads to more mature systems because they can accommodate different aspects of the design. Undoubtedly, the use of task models and other sub-models like domain models and user models is particularly important in the design process of context-aware systems in order to develop an appropriate design model of the interactive system [Forbrig 1999].

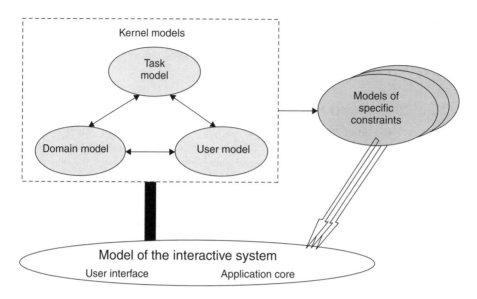

Figure 9.4. A more flexible task-based approach.

Although model-based approaches contribute to a flexible specification of software, the sub-models are often too rigid to describe different user needs. Consequently, we are often confronted with specifications which are either too restrictive for single users or give them too much freedom within the application context. New modularization principles on different modelling levels are needed to describe specific constraints.

A flexible task-based approach proposes to separate a sub-model into a fixed kernel part and additional specific parts to allow more flexible specifications. Whereas the kernel is stable or static (as far as you can consider a model as stable), the addition of a specific part depends on the specific situation of use. The general idea is illustrated in Figure 9.4. The model of the interactive system is derived from the kernel sub-models as usual. However, only the general constraints given by the field of application are specified. Models of specific constraints can adapt a kernel model to a specific situation of use and can change the model of the interactive system dynamically. To support this process, the interaction and the application models should be described on different levels of abstraction.

9.3. ADAPTIVE SPECIFICATION TECHNIQUES

To implement the general modelling approach of context-aware interactive systems introduced above, we need improved specification techniques. In this section, two proposals are made. In the first subsection, an adaptation mechanism for task models is demonstrated. Models of specific temporal constraints adapt a kernel model by restriction without changing its nature completely. The second subsection shows how we can transform the interaction model into an abstract sub-model and various specific sub-models. Moreover, this modelling approach supports the use of different devices.

9.3.1. ADAPTED TASK MODELS

A 'classical' or simple task model consists of two parts, as mentioned in Section 9.2. In the abstract example given in Figure 9.5, the *hierarchical description* H decomposes the task T into subtasks down to the level of atomic tasks T_{11}, T_{12}, T_{21}, and T_{22}. For each atomic task T_i an operation op_i is assigned, which must then be executed in order to fulfill T_i. The task tree, the most stable part of the task model, summarizes what has to be done. The *sequential description* S specifies the order in which the operations can be performed to fulfill the whole task. Therefore in S, the task T is considered as a process of an algebra common to task-based approaches e.g., [Paterno 2000].

S is composed of two types of equations. The first type describes temporal constraints between those subtasks which have a common parent node in the task tree (equ_T, equ_{T1}, and equ_{T2} in Figure 9.5). Equations of the second type define the mapping between atomic tasks and operations of the set OP (equ_{T11}, equ_{T12}, equ_{T21}, and equ_{T22} in the example). For brevity, the following examples omit equations of the second type, as well as operations. Temporal dependencies between subtasks can be formulated by temporal operators [Johnson and Wilson 1993], and the following operators are often used (possibly in different notation):

$\mid$	(parallel) composition	;	sequence (enabling)
$+$	choice	[]	option
$*$	iteration	[>	deactivation
$\mid$>	interruption		

In Figure 9.5 the temporal operators for parallel ($\mid$) and sequential execution (;) of subtasks are applied. Thus, T can be fulfilled by executing one of the following six sequences of atomic tasks.

$$\langle T_{11}, T_{12}, T_{21}, T_{22}\rangle, \langle T_{11}, T_{21}, T_{12}, T_{22}\rangle, \langle T_{11}, T_{21}, T_{22}, T_{12}\rangle,$$

$$\langle T_{21}, T_{11}, T_{12}, T_{22}\rangle, \langle T_{21}, T_{11}, T_{22}, T_{12}\rangle, \langle T_{21}, T_{22}, T_{11}, T_{12}\rangle$$

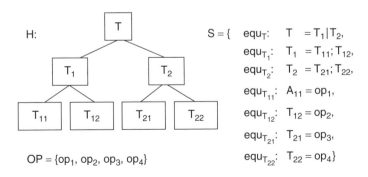

Figure 9.5. A simple model of the abstract task T.

In this chapter, a simple task model is used as a *kernel task model*. The task tree determines which subtasks can be performed in general. However, the sequential description contains only those temporal relations which come from the problem domain. In Figure 9.5, T_1 and T_2 could be interpreted, for example, as tasks decomposed into further subtasks concerning the input of some data to a software system (T_{11} and T_{21}) and the calculation of the appropriate results (T_{12} and T_{22}). Furthermore, T_{12} and T_{22} require a long time to execute (even with the best processor). Obviously, T_{11} has to be performed before T_{12}, just as T_{21} must be performed before T_{22}. However, there are no temporal constraints between T_1 and T_2. A user interface derived from this task model might consist of two windows W_1 and W_2 displayed in parallel, W_1 reflecting T_1 and W_2 reflecting T_2. The calculation is disabled in both windows until all input data are given. With open user communities, constraints change with the specific situations of use. In our example, a user could work with a small device which can display only one of the windows W_1 and W_2 at a time; that window is blocked until a calculation is finished. The task model could be adapted to this situation, for example, by saying that T_1 has to be executed to enable T_2 because T_1 has a higher priority.

In our approach, a kernel task model is adapted to a specific situation of use by additional temporal constraints expressed as equations similar to the equations in the sequential description of the kernel model. Thus, the task hierarchy itself is considered stable, but the set of possible execution sequences can be restricted. Disabling some of the optional subtasks is also possible. Figure 9.6 demonstrates an adaptation of the kernel model presented in Figure 9.5. The situation described above is reflected in the specific constraint equ_C which restricts the six possible execution sequences to only one:

$$\langle T_{11}, T_{12}, T_{21}, T_{22} \rangle.$$

At this point we should mention another temporal operator, denoted by $\langle \langle \rangle \rangle$ as it can be very useful in describing additional temporal constraints. If a subtask T_i is enclosed by $\langle \langle \rangle \rangle$, then the execution of T_i cannot be disturbed by any other subtask. An example is given in Figure 9.7. The kernel model does not impose temporal constraints on subtasks, but with the additional equation equ_c we specify that T_{12} has to be performed *immediately after* T_{22}.

Figure 9.8 illustrates an adapted model for the introductory E-shopping example in Figure 9.2; this model could represent a mobile phone being used for E-shopping. It

$$
\begin{aligned}
S \ = \{ \ &equ_T: && T = T_1 | T_2, \\
&equ_{T_1}: && T_1 = T_{11}; T_{12}, \\
&equ_{T_2}: && T_2 = T_{21}; T_{22} \ \}
\end{aligned}
$$
} Part of the kernel model

$$
\begin{aligned}
equ_c: \quad C = T_1; T_2
\end{aligned}
$$
} Specific constraints

Figure 9.6. An adaptation of Figure 9.5.

$$S = \{ \text{equ}_T: \quad T = T_1 | T_2,$$
$$\text{equ}_{T_1}: \quad T_1 = T_{11} | T_{12} | T_{13},$$
$$\text{equ}_{T_2}: \quad T_2 = T_{21} | T_{22} \}$$

$$\text{equ}_c: \quad C = <<T_{22}; T_{12}>> \} \quad \text{Specific constraints}$$

Figure 9.7. Another adapted model.

E-shopping =
       Looking for the product; (checking the offer | ordering the product)
Looking for the product =
       Entering a search criterion ; choosing an offer

. . .

equ_c: C = <<Looking for the product ; ordering the product>>

Figure 9.8. An adaptation of the specification from Figure 9.2.

includes the constraint that the completion of the subtask *Looking for the product* has to be immediately followed by executing *Ordering the product*. In other words, *Checking the offer* is not possible in this environment.

9.3.2. SPECIFICATION OF DEVICE FEATURES BY XML

Different devices with different capabilities are used by modern information and communication systems to present application logic. For instance, modern banking software should provide an HTML-based user interface for a web-based usage of the service as well as a WAP-based mobile communication service. A customer should also be able to use the software on automatic teller machines.

This subsection shows how the conceptual parts of design models can be reused. XML technology [Goldfarb and Prescod 2001] is used for representation and transformation features. Using XML technology allows a flexible development process from an abstract interaction model to a specific representation. The specific features of devices are given by XML-descriptions as well, and the transformation process is based on this information. Thus, we can facilitate a smooth integration of newly developed devices into already existing software and a better access to information resources by multiple users.

There are already some approaches that enable us to specify user interfaces with mark-up languages. An example is UIML [Abrams *et al.* 1999] – an XML-based language for describing user interfaces. It comes with different rendering programs allowing the generation of user interfaces on different platforms. However, a special renderer for each target-device is needed, limiting the flexibility for selecting target-device features. A separate chapter of this book describes the UIML approach and its further development in more detail (Chapter 6).

XUL (www.xulplanet.com) focuses on window-based graphical user interfaces specified by XML. It is not applicable to interfaces of small mobile devices because no abstraction of interaction functionality is available. It was developed for the Mozilla project rather than for device-independent specification of user interfaces.

XIML (www.xulplanet.com) is a more general approach and may eventually provide a way to standardize the specification of user interfaces and all necessary models. This approach is also described in a separate chapter of this book (Chapter 7). The advantage of XIML is its ability to specify a whole set of models. While it enables one to express many different ideas, it can also be used as a communication language for different model-based tools. In the future an XIML notation will be available for our model-based approach, but at the moment XIML lacks tool support, especially renderers for different devices.

In our approach, a user interface is considered to be a simplified version of the MVC model (Model View Control) and is separated into a model component and a presentation component. The model component describes the features of the user interface on an abstract level. It is also called an abstract interaction model. The user-interface objects and their representations are specified in the presentation component. During the development process, a mapping that shows the transformation of the abstract interaction model to the specific interaction model is necessary.

The tool TADEUS [Dittmar and Forbrig 1999] represents both abstract interaction and specific interaction models by terms. The transformation process is described by attribute grammars. For the user interface, these grammars play the role of a renderer. However, different platforms require different grammars. The development of distinct grammars for different mobile devices can be time consuming, and the integration of the specific features of different devices can be especially difficult. Tool support, such as with XML technology, can facilitate the transformation process.

XML technology enables the description of the abstract interaction model, the description of specific characteristics of different devices and the specification of the transformation process, as will be shown later. If there were only one concrete user interface object for each abstract one, the transformation process could be executed automatically. However, often there are more than one, meaning that the appropriate object has to be selected. This selection can be done interactively or according to stored profiles.

Later in the development process, some concrete user interface object attributes must be set, according to certain design rules, and then the specific user interface can be generated. Figure 9.9 illustrates the process of transforming an abstract interaction model into a specific user interface.

The process is best described by an example, so Section 9.4 describes a specific way of using models and XML in the proposed manner. It starts with the specification of task models for an electronic shop and discusses tool support. The example in this next section will be used to illustrate the transformations based on XML technology, so that the reader can follow the development from abstract specification to concrete user interfaces step by step as shown in Figure 9.9.

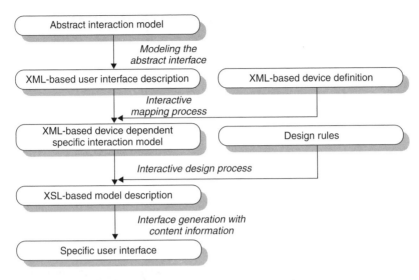

Figure 9.9. Creation of a user interface.

9.4. EXAMPLE OF AN ELECTRONIC SHOP

In this section, we extend the E-shopping example by applying both of the specification techniques which were proposed in the last section in order to improve the adaptability of interactive systems. We start with a kernel task model and an adaptation to a specific situation. Then, we present the process of generating two specific user interfaces.

9.4.1. THE TASK MODEL OF E-SHOPPING

Let us assume that the task of E-shopping can be decomposed into three subtasks. First, you have to look for the product, then check the offer, and finally, order the product. These subtasks are further refined as shown in Figure 9.10. The hierarchical (left side of the window) and sequential (right side of the window) descriptions of the task are presented using a prototyping tool.

Even at the level of task modelling, checks of structural or behavioral characteristics are helpful. The tool allows the animation of a task specification. Figure 9.11 shows a screenshot of the animation of the example in Figure 9.10. It visualizes the state of the animation after performing the subtasks Looking_for_the_product (operations *enter* and *choose*) and Short_checking the offer (operation *short_read*). The next possible atomic tasks are Reading_detailed_information (operation *long_read*), *Ordering* (operation *order*) or *Discarding* the offer (operation *discard*). The appropriate nodes within the task tree are highlighted. The user of the prototype can double-click on one of the highlighted nodes and the animation goes ahead. An animation allows a first impression of the behavior of the prospective system, making possible some preliminary usability tests.

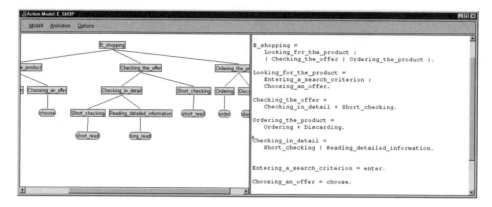

Figure 9.10. A refined task model of E_shopping.

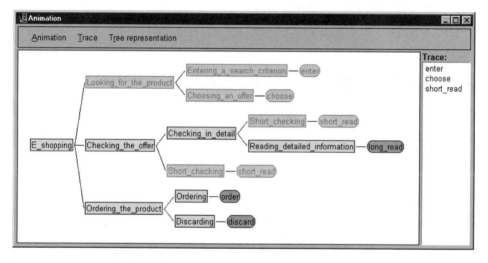

Figure 9.11. Task animation of E_shopping.

We consider the simple task model (introduced so far) as a part of the kernel model as illustrated in Figure 9.4. It is a general description of the task of E-shopping. For example, it specifies that the check of the offer can be done briefly or in more detail. This task model makes sense if a user works with a personal computer. However, if he uses a mobile phone, we can assume that he is not interested in a detailed check of the offer. Rather, he already knows what he wants to order. Furthermore, a mobile phone restricts concurrent sub-dialogues with the software system.

Figure 9.12 shows the full kernel task model (as a sequential description in the form of process equations) and an additional equation specifying the constraints given by the situation where a person uses a mobile phone. The additional constraint determines that

```
Kernel Model:
E_shopping =
    Looking_for_the_product;
    (Checking_the_offer|ordering_the_product.

Looking_for_the_product =
    Entering_a_search_criterion;
    Choosing_an_offer.

Checking_the_offer =
    Checking_in_detail + Short_checking.

Ordering_the_product =
    Ordering + Discarding.

Checking_in_detail =
    Short_check|Reading_detailed_information.

/* Atomic tasks */
Entering_a_search_criterion = enter.
Choosing_an_offer = choose.
Short_checking = short_read.
Short_check = short_read.
Reading_detailed_information = long_read.
Ordering = order.
Discarding = discard.

Additional temporal constraint for the situation of using a mobile phone:
C = Short_checking ; Ordering_the_product.
```

Figure 9.12. The sequential description of E_shopping.

the subtasks Short_checking and Ordering_the_product have to be executed in sequential order. There is no room for Reading_detailed_information.

The following subsection demonstrates the development of the user interface of two devices. This development is based on an abstract interaction model, which itself is based on a device-specific task model. The transformation from a task model to an interaction model has to be done by hand at the moment, but tool support is possible and some tools are under development.

9.4.2. THE GENERATION OF SPECIFIC USER INTERFACES

9.4.2.1. XML-Based Abstract Interaction Model

The abstract interaction model is transformed into a notation which is based on a language of the XML family. The basic document type definition (DTD) was introduced by Müller *et al.* [2001]. It has the following structure:

Example 1. DTD for an abstract user interface description.

```
<?xml version ="1.0" encoding ="ISO-8859-1"?>
<!-- version 0.5   14.02.2001 -->
<!--MARK 1 -->
```

```
<!ELEMENT ui, uio+, context*)>
    <!ATTLIST ui ui_name CDATA #REQUIRED>
<!ELEMENT uio, uio |, (output+, context*)* ,, input+, context*)*))>
    <!ATTLIST uio name CDATA #REQUIRED>
<!--MARK 2 -->
<!ELEMENT context, context_value+)>
<!ELEMENT context_value, #PCDATA)>

<!ELEMENT output, (output_string | output_1-n | output_m-n | output_table),
    context*, optional?)>
    <!ATTLIST output name CDATA #REQUIRED>
. . . .
<!ELEMENT output_table, #PCDATA)>
    <!ATTLIST output_table
        num_x CDATA #IMPLIED
        num_y CDATA #IMPLIED>
<!ELEMENT input, (input_1-n | input_m-n | input_trigger | input_string |
    input_table), context*, optional?)>
    <!ATTLIST input name CDATA #REQUIRED>
<!--MARK 3 -->
<!ELEMENT input_1-n, #PCDATA)>
    <!ATTLIST input_1-n
        range_min CDATA #IMPLIED
        range_max CDATA #IMPLIED
        range_interval CDATA #IMPLIED
        list_n CDATA #IMPLIED>
<!--MARK 4 -->
. . . .
<!ELEMENT input_table, #PCDATA)>
    <!ATTLIST input_table
        num_x CDATA #IMPLIED
        num_y CDATA #IMPLIED>

<!ELEMENT optional, optional_value+)>
. . . .
```

At this stage, the user interface is composed of one or more user interface objects (UIOs). A UIO can be composed of other UIOs or one or more input/output values (Mark 1–2). These values are in an abstract form and describe interaction features of the user interface in an abstract way. Later these abstract UIOs will be mapped to concrete ones. Different types of these abstract UIOs can be seen between Marks 2 and 3. These UIOs have different attributes to specify their behavior. One example is the input_1-n value (Mark 3–4). The range_min attribute specifies the minimum value of input possibilities, while the range_max attribute specifies the maximum value. The range_interval determines the interval within the range. A list of input values can also be specified.

According to the analyzed task model (Figure 9.12), some abstract interaction objects such as title and search are identified. Due to a lack of space, only a few objects and attributes are listed here. The interested reader may look ahead to the final generated user interfaces (Figures 9.15, and 9.16).

9.4.2.2. XML-Based Device Definition

A universal description language for describing properties of various target devices (and comparable matter) is necessary to support the transformation process from an abstract interaction model to a device-dependent abstract interaction model. Such a language has been developed, based on [Mundt 2000]. XML documents written in this language describe the specific features of devices. Consider the corresponding DTD shown in Example 2.

Example 2. DTD for Device Definition.

```
<?xml version ="1.0" encoding ="ISO-8859-1"?>
<!-- DevDef-File  Version 0.2  Date: 01.02.2001 -->
<!ELEMENT device, device*, service*, vendor, version)>
    <!ATTLIST device
        id ID #REQUIRED>
<!ELEMENT service, name+, feature+, service*)>
    <!ATTLIST service
        kind CDATA #REQUIRED>
<!ELEMENT name, #PCDATA)>
    <!ATTLIST name
        language CDATA #REQUIRED>
<!ELEMENT feature, description)>
    <!ATTLIST feature
        id ID #IMPLIED>
<!ELEMENT description, text+, value, property)>
<!ELEMENT property, #PCDATA)>
<!ELEMENT value, #PCDATA)>
<!ELEMENT vendor, url*, text)>
<!ELEMENT url EMPTY>
    <!ATTLIST url
        ref NMTOKEN #REQUIRED>
<!ELEMENT version, #PCDATA)>
<!ELEMENT text, #PCDATA)>
```

Such device specifications are not only necessary for the described transformations but also influence the specification of formatting rules. The following example shows a very short fragment of a device definition of java.AWT:

Example 3. Device definition for java.AWT.

```
<?xml version ="1.0" encoding ="ISO-8859-1"?>
<?xml-stylesheet type ="text/xsl" href ="?.xsl"?>
<!DOCTYPE device SYSTEM "devdef-1.dtd">
<device id =''Advanced Windowing Toolkit''>
```

```
<device id =''java.awt.Button''>
    <service id =''?'' kind =''trigger''>
        <name>Location</name>
        <feature>java.awt.Button.setLocation</feature>
    </service>
</device>
<vendor>SUN Microsystems</vendor>
<version>2.0</vendor>
</device>
```

9.4.2.3. XML-Based Device Dependent Specific Interaction Model

This model, while still on an abstract level, already fulfills some of the constraints of device specification. It uses available features and omits unavailable services. The result of the mapping process is a file in which all abstract UIOs are mapped to concrete UIOs of a specific representation (Figure 9.13).

This file is specific to a target device and includes all typical design properties of concrete UIOs, such as color, position, size, and so on. It consists of a collection of option–value pairs. The content of the values is specified later on in the design process and describes the 'skeleton' of a specific user interface. Designers develop the user interface on the basis of this skeleton. A portion (i.e., one UIO) of our simple example of an E-shop system mapped to a HTML-representation is shown in the following example:

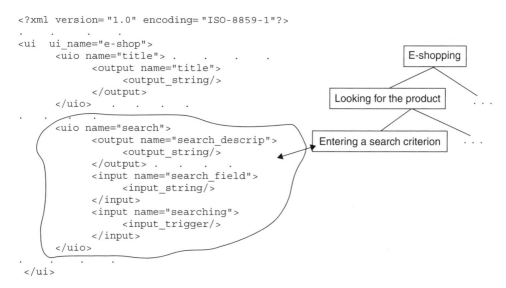

```
<?xml version= "1.0" encoding= "ISO-8859-1"?>
   .     .     .     .
<ui   ui_name="e-shop">
     <uio name="title"> .     .     .     .
          <output name="title">
               <output_string/>
          </output>
     </uio>   .     .     .     .
      .     .     .     .
     <uio name="search">
          <output name="search_descrip">
               <output_string/>
          </output> .     .     .     .
          <input name="search_field">
               <input_string/>
          </input>
          <input name="searching">
               <input_trigger/>
          </input>
     </uio>
 .     .     .     .
</ui>
```

Figure 9.13. Part of the description of a simple user interface of an E-shop system.

Example 4. Device dependent (HTML) interaction model.

```
            . . .
    <uio name ="search">
        <input name ="searching">
            <input_trigger/>
            <device>
                <type>BUTTON</type>
                <parameter>
                    <option>TYPE</option>
                    <value>submit</value>
                </parameter> ....
            </device> ....
        </uio> ....
    </ui_map>
```

Tool support for this specific part of the process is demonstrated in Figure 9.14. Necessary features of user interfaces (Window 1) are mapped to the available services of a device (Window 2). If this mapping process is not uniquely defined, an interactive decision has to be made. The tool shows device services which fit to the current features of the user interface (Window 3). In some cases, a selection has to be made.

The resulting specification is the basis for the development of the final user interface. There could be a separate abstract interaction model for other devices like Java or WML. However, we do not present the entire specification. A fragment of the specification of the same user interface mapped to java.AWT is shown below.

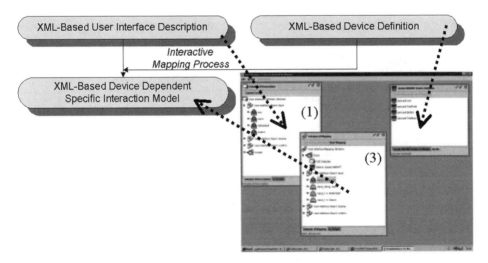

Figure 9.14. Tool support for the XML mapping process.

Example 5. Abstract device dependent (java.AWT) interaction model.

```
                    . . .
    <uio name ="search">
       <input name ="searching">
          <input_trigger/>
          <device>
              <type>java.awt.Button</type>
              <parameter>
                  <option>java.awt.Button.setLocation</option>
                  <value></value>
              </parameter> ...
          </device> ...
       </uio> ...
    </ui_map>
```

In Example 5, the value of a parameter (setLocation) is still undefined. This value can be set during the XSL transformation. The WML-example is omitted here because of its similarity to the HTML document. Figure 9.16 presents the final user interface for WML.

9.4.2.4. XSL-Based Model Description

The creation of the XSL-based model description is based on the knowledge of available UIOs for specific representations. It is necessary to know which property values of a UIO are available in the given context. The XML-based device-dependent abstract interaction model (skeleton) and available values of properties are used to create a XSL-based model description specifying a complete user interface.

Example 6. XSL-file for generation of a HTML user interface of the E-shop.

```
<?xml version ="1.0"?>
<xsl:stylesheet xmlns:xsl ="http://www.w3.org/XSL/Transform/1.0">
<xsl:output method = "html"
     xml-declaration ="yes"/>
     <html> ...
         <h3> E-Shop-System </h3>
         <br/>
         <SELECT size =1>
             <option>"content-input" </option>
             <option>"content-input" </option> ...
         </SELECT>
         <br/>  ...
         <BUTTON type =submit>Ok</BUTTON>
         <br/> ...
       <TABLE>
           <CAPTION>purchase list</CAPTION>
           <TR>
               <TD>"content-input"</TD>
               <TD>"content-input"</TD>
```

```
                </TR> ...
            </TABLE>
          </html>
      </xsl:stylesheet>
```

The wildcard 'content-input' refers to content from applications or databases at a later step of development.

9.4.2.5. Specific User Interface

A file describing a specific user interface will be generated by XSL transformation. Some examples for java.AWT and HTML are given by [Müller *et al.* 2001]. The XSL transformation process consists of two sub-processes. First, one creates a specific file representing the user interface (Java.AWT, Swing, WML, VoiceXML, HTML,...). Then, one integrates content (database, application) into the user interface.

The generated model is already a specification that will run on the target platform. There are two different cases. In the first case, the user interface is generated once (e.g. java.AWT). Therefore, there is no need for a new generation of the user interface if the contents change. The content handling is included within the generated description file. Figure 9.15 shows the final user interface for HTML on a personal computer for the E-shop.

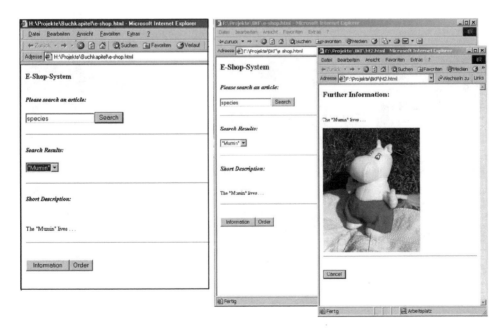

Figure 9.15. Generated user interface for HTML.

In the second case, the user interface has to be generated dynamically several times (e.g. WML). The content will be handled by instructions within the XSL-based model description. Each modification of the contents results in a new generation of the description file. Figure 9.16 demonstrates the final result of the user interface for a mobile device with restricted capabilities. The consequences of the additional constraints can be seen.

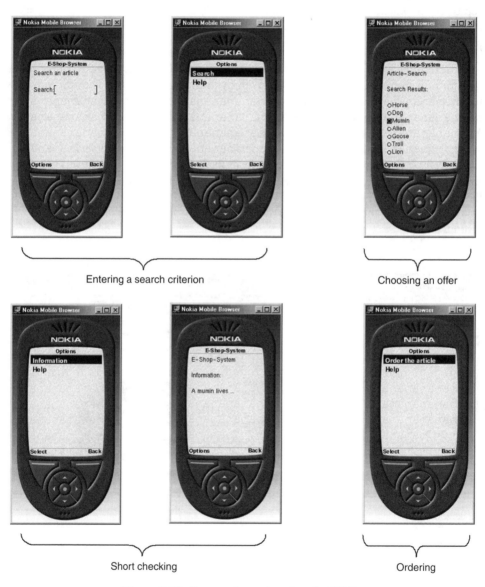

Figure 9.16. Generated user interface for WML.

9.5. CONCLUSIONS

This chapter demonstrates how the model-based approach can be used to develop optimal interactive systems by considering the tasks which have to be fulfilled, the devices which can be used, and other aspects concerning the context of use. Task models can be used to specify the problem domain. Based on the theory of process algebra, it is possible to modularize task specifications into a stable kernel part and additional parts specifying situational constraints. This technique is illustrated by an example which also shows that experiments and usability tests can be performed at a very early stage of the software development.

XML technology can be used in the process of developing user interfaces for mobile devices. The current work presents an XML-based language for describing user interfaces and specifies an XML language which allows the description of the process of mapping from abstract to concrete user interface objects. This concept is illustrated by several examples. A tool supporting the mapping and the design process is currently under development, and already delivers promising results. So far, however, it has not been fully integrated in previous phases of the design process.

Our experiments show that XML is a promising technology for the platform-independent generation of user interfaces. The ability to specify features of devices and platforms separately seems to be important and could be incorporated into other approaches as well. Further studies will show whether dialogue sequences of already-developed applications can be integrated into the design process, as such patterns could enhance the development process. Seffah and Forbrig [2002] discuss this problem in more detail.

Applications of ubiquitous computing must demonstrate how the module concept of task models can be applied to such problems. An interpreter of task models can run on a server or perhaps on the mobile device itself. In both cases, the application is controlled by the task model. Some problems can be solved by interpreting task models or by using XML technology. The future will reveal how detailed task models should be specified and on which level XML technology is most promising.

REFERENCES

Abrams, M., Phanouriou, C., Batongbacal, A.L., Williams, S.M., and Shuster, J.E.(1999) *UIML: An appliance-independent XML user interface language.* Proceedings of WWW8.http://www8.org/w8-papers/5b-hypertext-media/uiml/uiml.html

Biere, M., Bomsdorf, B., Szwillus, G. (1999) The Visual Task Model Builder. *Proceedings of the CADUI'99*, Louvain-la-Neuve, 245–56. Kluwer Academic Publishers.

Dey, A., and Abowd, G. (1999) *International Symposium on Handheld and Ubiquitous Computing – HUC'99*, Karlsruhe, Germany.

Dittmar, A. (2000) More Precise Descriptions of Temporal Relations within Task Models. In [Palanque and Paternó 2000] 151–168.

Dittmar, A., and Forbrig, P. (1999) Methodological and Tool Support for a Task-oriented Development of Interactive Systems. *Proceedings of the CADUI'99*, Louvain-la-Neuve. Kluwer Academic Publishers.

Forbrig, P. (1999) *Task- and Object-Oriented Development of Interactive Systems: How many models are necessary?* DSVIS'99, Braga, Portugal.

Forbrig, P., and Dittmar, A. (2001) Software Development and Open User Communities. *Proceedings of the HCI*, New Orleans, August 2001, 175–9.

Forbrig, P., Müller, A., and Cap, C. (2001) Appliance Independent Specification of User Interfaces by XML. *Proceedings of the HCI*, New Orleans, August 2001, 170–4.

Forbrig, P., Limbourg, Q., Urban, B., Vanderdonckt, J., (Eds) (2002) *Proceedings of Interactive Systems: Design, Specification, and Verification*. 9th International Workshop, DSV-IS 2002, Rostock Germany, June 12–14, 2002, LNCS Vol. 2545. Springer Verlag.

Hacker, W. (1995) *Arbeitstätigkeitsanalyse: Analyse und Bewertung psychischer Arbeitsanforderungen*. Heidelberg: Roland Asanger Verlag.

Hoare, C.A.R. (1985) *Communicating Sequential Processes*. Prentice Hall.

Goldfarb, C.F., and Prescod, F. (2001) *Goldfarb's XML Handbook* Fourth Edition.

Johnson, P., and Wilson, S. (1993) A framework for task-based design, in *Proceedings of VAMMS 93*, second Czech-British Symposium, Prague, March, 1993. Ellis Horwood.

Lim, K.Y., and Long, J. (1994) *The MUSE Method for Usability Engineering*. Cambridge University Press.

Milner, R. (1989) *Communicating and Concurrency*. Prentice Hall.

Müller, A., Forbrig, P., and Cap, C. (2001) Model-Based User Interface Design Using Markup Concepts. *Proceedings of DSVIS 2001*, Glasgow.

Mundt, T. (2000) *DEVDEF*. http://wwwtec.informatik.uni-rostock.de/IuK/gk/thm/devdef/.

Palanque, P., and Paternò, F. (Eds) (2000) Proceedings of 7th Int. *Workshop on Design, Specification, and Verification of Interactive Systems DSV-IS'2000*. Lecture Notes in Computer Science, 1946. Berlin: Springer-Verlag.

Paterno, F., Mancini, C., and Meniconi, S. (1997) ConcurTaskTrees: A Diagrammatic Notation for Specifying Task Models. *Proceedings of Interact'97*, 362–9, Sydney: Chapman and Hall.

Paterno, F. (2000) *Model-based Design and Evaluation of Interactive Applications*. Springer Verlag.

Pribeanu, C., Limbourg, Q., and Vanderdonckt, J. (2001) Task Modeling for Context-Sensitive User Interfaces. *Proceedings of DSVIS 2001*. Glasgow.

Scapin, D.L., and Pierret-Golbreich, C. (1990) *Towards a Method for Task Description: MAD in work with display units 89*. Elsevier Science Publishers, North-Holland.

Sebilotte, S. (1988) Hierarchical Planning as a Method for Task Analysis: The example of office task analysis. *Behaviour and Information Technology*, 7, 275–93.

Seffah, A., and Forbrig, P. (2002) Multiple User Interfaces: Towards a task-driven and patterns-oriented design model. In [Forbrig, *et al.* 2002].

Shepherd, A. (1989) Analysis and training in information technology tasks. In *Task Analysis for Human-Computer Interaction* (ed. D. Diaper). John Wiley & Sons, New York, Chichester.

Tauber, M.J. (1990) ETAG: Extended task action grammar. In Human-computer interaction – Interact'90 (ed. D. Diaper). 163–8. Amsterdam: Elsevier.

Vanderdonckt, J., and Puerta, A. (1999) Introduction to Computer-Aided Design of User Interfaces. *Proceedings of the CADUI'99*, Louvain-la-Neuve. Kluwer Academic Publishers.

van der Veer, G.C., Lenting, B.F., and Bergevoet, B.A.J. (1996) GTA: Groupware Task Analysis – Modeling Complexity. *Acta Psychologica*, 91, 297–322.

Multi-Model and Multi-Level Development of User Interfaces

Jean Vanderdonckt,[1] Elizabeth Furtado,[2] João José Vasco Furtado,[2] Quentin Limbourg,[1] Wilker Bezerra Silva,[2] Daniel William Tavares Rodrigues,[2] and Leandro da Silva Taddeo[2]

[1] *Université catholique de Louvain ISYS/BCHI Belgium*
[2] *Universidade de Fortaleza NATI-Célula EAD Brazil*

10.1. INTRODUCTION

In universal design [Savidis *et al.* 2001], user interfaces (UIs) of interactive applications are developed for a wide population of users in different contexts of use by taking into account factors such as preferences, cognitive style, language, culture, habits and system experience. Universal design of single or multiple UIs (MUIs) poses some difficulties due to the consideration of these multiple parameters. In particular, the multiplicity of parameters dramatically increases the complexity of the design phase by adding a large number of design options. The number and scope of these design options increase the variety and complexity of the design. In addition, methods for developing UIs have difficulties with this variety of parameters because the factors are not necessarily identified and manipulated in a structured way nor truly considered in the standard design process.

Multiple User Interfaces. Edited by A. Seffah and H. Javahery
© 2004 John Wiley & Sons, Ltd ISBN: 0-470-85444-8

The goal of this chapter is to present a structured method addressing certain parameters required for universal design. The method is supported by a suite of tools based on two components: (i) an ontology of the domain of discourse and (ii) models that capture instantiations of concepts identified in this ontology in order to produce multiple UIs for multiple contexts of use. These different UIs exhibit different presentation styles, dialogue genres and UI structures.

The remainder of this chapter is structured as follows:

- Section 10.2 provides a discussion of the state of the art of methods for developing UIs with a focus on universal design.
- Section 10.3 defines a model in this development method, along with desirable properties. As this chapter adopts a conceptual view of the problem, the modelling activity will be organized into a layered architecture manipulating several models.
- The three levels are then described respectively in the three next sections: conceptual in Section 10.4, logical in Section 10.5, and physical in Section 10.6.
- The application of this method is demonstrated through the example of a UI for patient admissions at a hospital.
- The last section summarizes the main points of the chapter.

10.2. RELATED WORK

The Authors' Interactive Dialogue Environment (AIDE) [Gimnich *et al.* 1991] is an integrated set of interactive tools enabling developers to implement UIs by directly manipulating and defining UI objects, rather than by the traditional method of writing source code. AIDE provides developers with a more structured way of developing UIs as compared to traditional "rush-to-code" approaches where unclear steps can result in a UI with low usability.

The User-Centered Development Environment (UCDE) [Butler 1995] is an object-oriented UI development method for integrating business process improvements into software development. Business-oriented components are software objects that model business rules, processes and data from the end-user's perspective. The method maps data items onto UI objects that are compatible with the parameters of the data item (e.g., data type, cardinality). The advantage of UCDE is that it provides a well-integrated process from high-level abstraction to final UI.

Another methodological framework for UI development described by Hartson and Hix [Hartson and Hix 1989; Hix 1989] integrates usability into the software development process from the beginning. The focal point of this approach is a psychologically-based formal task description, which serves as the central reference for evaluating the usability of the user interface under development. This framework emphasizes the need for a task model as a starting point for ensuring UI usability, whereas UCDE emphasizes the need for a domain model.

The MUSE method [Lim and Long 1994] uses structured notations to specify other elements of the context of use, such as organizational hierarchies, conceptual tasks and domain semantics. In addition, structured graphical notations are provided to better communicate the UI design to users.

The above approaches illustrate the importance of using a structured method to capture, store, and manipulate multiple elements of the context of use, such as task, domain and user. Although the above methods partially consider this information, they do not consider the design of multiple UIs where task [Card *et al.* 1983; Gaines 1994], domain and user parameters vary, sometimes simultaneously. The Unified User Interface design method [Savidis *et al.* 2001] was the first method to suggest deriving multiple variations of a task model so as to take into account individual differences between users. The different variations of a task model are expressed by alternative branches showing what action to perform depending on particular interaction styles.

Thevenin and Coutaz [Thevenin and Coutaz 1999; Thevenin 2001] go one step further by introducing the concept of decorating a task, which is a process that introduces graphical refinements in order to express contextualization (see Chapter 3). The task is first modelled independently of any context of use, and thus independently of any type of user. Depending on the variations of the context of use to be supported, including variations of users, the initial task model is refined into several decorated task models that are specific to those contexts of use.

Paternò and Santoro [2002] show the feasibility of deriving multiple UIs from a single model by decomposing the task differently based on different contexts of use. For each context of use, different organizations of presentation elements are selected for each task (see Chapter 11).

In this paper, we consider how a single task model can represent the same task across different user profiles. In the following sections we address the following questions: Do we need to create a single task model where all differences between users are factored out? In this case, do we start by creating a different task model for each user stereotype and then create a unified model that describes only the commonalities? Or do we start with a single task model that includes both commonalities and differences between users? And if so, how do we isolate commonalities from differences? To address these questions, we first set up the foundations for our modelling approach.

10.3. DEFINITION OF MODEL

Several computer science methodologies decompose a modelling activity into a multi-level architecture where models are manipulated explicitly or implicitly: model engineering (e.g., Object-Modelling Technique [Rumbaugh *et al.* 1990], UML [Booch *et al.* 1998]), database engineering and certain information system development methodologies (e.g., SADT [Marca and McGowan 1988]). Similarly, the method proposed here structures the UI development process into three levels of abstraction (Figure 10.1):

1. The *conceptual level* allows a domain expert to define the ontology of concepts, relationships, and attributes involved in the production of multiple UIs.
2. The *logical level* allows designers to capture requirements for a specific UI design case by instantiating concepts, relationships, and attributes with a graphical editor. Each set of instantiations results in a set of models for each design case (*n* designs in Figure 10.1).

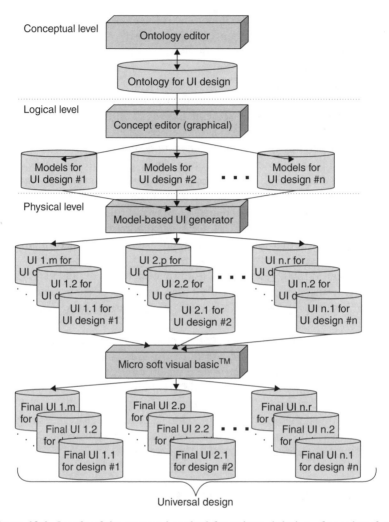

Figure 10.1. Levels of the proposed method for universal design of user interfaces.

3. The *physical level* helps developers derive multiple UIs from each set of models with a model-based UI generator: in Figure 10.1, *m* possible UIs are obtained for UI design #1, *p* for UI design #2,..., *r* for UI design #*n*. The generated UI is then exported to a traditional development environment for manual editing. Although the editing can be performed in any development environment, the tools discussed here support code generation for Microsoft Visual Basic V6.0.

A *UI model* is a set of concepts, a representation structure and a series of primitives and terms that can be used to explicitly capture knowledge about the UI and its related interactive application using appropriate abstractions. A model is assumed to abstract aspects of the real world. Any concept of the real world can therefore lead to multiple

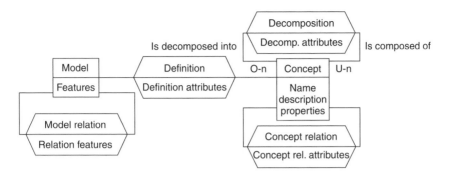

Figure 10.2. Definition of the user interface model.

possibilities of abstraction depending on how we want to develop UIs. Ideally, a model should be declarative rather than imperative or procedural. It should also be editable, preferably through tools, and finally it should be analysable, so as to allow some degree of automation.

A model consists of a number of features (Figure 10.2). It is typically built as a hierarchical decomposition of abstract concepts into more refined sub-concepts. Any concept can then be characterized by a name, a description and properties of interest. A model should also encompass relationships between these concepts with roles. These relationships apply both within models (called *intra-model relationships*) and between models (called *inter-model relationships*). Any of these relationships (i.e., the definition, the decomposition, the intra- or inter-model relationships) can possess a number of attributes.

How many models do we need? A single UI model is probably too complex to handle because it combines all static and dynamic relationships in the same model. It is also preferable to avoid using a large number of models, because this requires establishing and maintaining a large number of relationships between the models. *Model separability* is desirable in this case. Model separability adheres to the *Principle of Separation of Concerns*, which states that each concept should be clearly separated from the others and classified in only one category. Therefore, the quality of separability depends on the desired results and the human capacity to properly identify and classify concepts. Table 10.1 summarizes a list of desirable model properties.

With respect to these properties, some research proposes that *model integrability* (where all abstractions are concentrated into one single model or, perhaps, a few of them) can avoid model proliferation and can better achieve modelling goals than model separability (where the focus is only on one particular aspect of the real world at a time to be represented and emphasized). On one hand, integrability promotes integration of concepts and relationships, thus reducing the need to introduce artificial relationships to maintain consistency. Integrability also improves access to the concepts. On the other hand, integrability demands a careful consideration of the concepts to be integrated and may become complex to manipulate. In contrast, with *separability*, each concept is unequivocally classified into one and only one model. However, this may increase the number of models and relationships needed to express all dependencies between the

Table 10.1. Desirable properties of a model.

Property	Definition
Completeness	Ability of a model to abstract all real world aspects of interest via appropriate concepts and relationships
Graphical completeness	Ability of a model to represent all real world aspects of interest via appropriate graphical representations of the concepts and relationships
Consistency	Ability of a model to produce an abstraction in a way that reproduces the behaviour of the real world aspect of interest in the same way throughout the model and that preserves this behaviour throughout any manipulation of the model
Correctness	Ability of a model to produce an abstraction in a way that correctly reproduces the behaviour of the real world aspect of interest
Expressiveness	Ability of a model to express any real world aspect of interest via an abstraction
Conciseness	Ability of a model to produce compact abstractions of real world aspects of interest
Separability	Ability of a model to classify any abstraction of a real world aspect of interest into one single model (based on the *Principle of Separation of Concerns*)
Correlability	Ability of two or more models to establish relationships between themselves so as to represent a real world aspect of interest
Integrability	Ability of a model to bring together abstractions of real world aspects of interest into a single model or a small number of models

models. In the following sections, different types of models will be defined for different levels of abstraction. We begin with the conceptual level and continue with the subsequent levels.

10.4. CONCEPTUAL LEVEL

10.4.1. DEFINITION

Each method for developing UIs possesses its own set of concepts, relationships and attributes, along with possible values and ways to incorporate them into the method. However, this set is often hidden or made implicit in the method and its supporting tool, thus making the method insufficiently flexible to consider multiple parameters for universal design. When the set of concepts, relationships and attributes is hidden, we risk manipulating fuzzy and unstructured pieces of information. The conceptual level is therefore intended to enable domain experts to identify common concepts, relationships and attributes of the models involved in universal design. The identification of these concepts, relationships and attributes will govern how they will be used in future models when manipulated by the method.

An *ontology* explicitly defines any set of concepts, relationships, and attributes that need to be manipulated in a particular situation, including universal design [Gaines 1994;

Savidis *et al.* 2001]. The concept of ontology [Guarino 1995] comes from Artificial Intelligence where it is identified as the set of formal terms with which one represents knowledge, since the representation completely determines what exists for the system. We hereby define a *context of use* as the global environment in which a user population, perhaps with different profiles, skills and preferences, carries out a series of interactive tasks on one or multiple semantic domains [Pribeanu *et al.* 2001]. In universal design, it is useful to consider many types of information (e.g., different user profiles, different skills, different user preferences) in varying contexts of use. This information can be captured in different models [Paternò 1999; Puerta 1997].

A *model* is a set of postulates, data and inferences presented as a declarative description of a UI facet. Many facets exist that are classified into one of the following models: task, domain, user, interaction device, computing platform, application, presentation, dialogue, help, guidance, tutorial or organizational environment. A model is typically built as a hierarchical decomposition of abstract concepts into several refined sub-levels. Relationships between these concepts should be defined with roles, both within and between models.

To avoid incompatible models, a *meta-model* defines the language with which any model can be specified. One of the most frequently used meta-models, but not the only one, is the UML meta-model. The concepts and relationships of interest at this level are *meta-concepts* and *meta-relationships* belonging to the *meta-modelling* level. Figure 10.3 exemplifies how these fundamental concepts can be defined in an ontology editor. In Figure 10.3, the core entity is the concept, characterized by one or many attributes, each having a data type (e.g., string, real, integer, Boolean, or symbol). Concepts can be related to each other. Relationships include inheritance (i.e., 'is'), aggregation (i.e., 'composed of') and characterization (i.e., 'has'). At the meta-modelling stage, we do not yet know what type of concepts, relationships, and attributes will be manipulated. Therefore, any definition of a UI model, as represented in Figure 10.2, can be expressed in terms of the basic entities as specified in Figure 10.3.

10.4.2. CASE STUDY

The context of use can theoretically incorporate any real world aspect of interest, such as the user, the software/hardware environment, the physical and ambient environment, the socio-organizational environment, etc. For the simplicity of this paper, the context of use focuses on three models:

1. A *domain model* defines the data objects that a user can view, access, and manipulate through a UI [Puerta 1997]. These data objects belong to the domain of discourse. A domain model can be represented as a decomposition of information items, and any item may be iteratively refined into sub-items. Each such item can be described by one or many parameters such as data type and length. Each parameter possesses its own domain of possible values.

2. A *task model* is a hierarchical decomposition of a task into sub-tasks and then into actions, which are not decomposed [Paternò and Santoro 2002; Pribeanu *et al.* 2001; Top and Akkermans 1994]. The model can then be augmented with temporal relationships stating when, how and why these sub-tasks and actions are carried out.

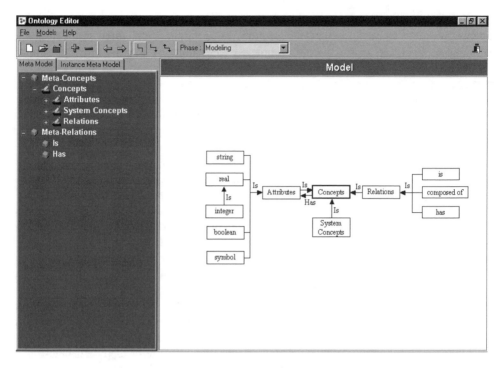

Figure 10.3. The ontology editor at the meta-modelling level.

Similarly to the domain model, a task model may have a series of parameters with domains of possible values – for instance, task importance (low/medium/high), task structure (low/medium/high decomposition), task critical aspects (little/some/many), and required experience (low/moderate/high).

3. A *user model* consists of a hierarchical decomposition of the user population into stereotypes [Puerta 1997; Vanderdonckt and Bodart 1993]. Each stereotype brings together people sharing the same value for a given set of parameters. Each stereotype can be further decomposed into sub-stereotypes. For instance, population diversity can be reflected by many user parameters such as language, culture, preference (e.g. manual input vs selection), level of task experience (elementary, medium, or complex), level of system experience (elementary, medium, or complex), level of motivation (low, medium, high), and level of experience of a complex interaction medium (elementary/medium/complex).

Other characterizations of these models in terms of their parameters, or even other model definitions, can be incorporated depending on the modelling activity and the desired level of granularity.

Figure 10.4 graphically depicts how the ontology editor can be used at the *modelling stage* to input, define and structure concepts, relationships and attributes of models with respect to a context of use. Here, the three models are represented and they all share a

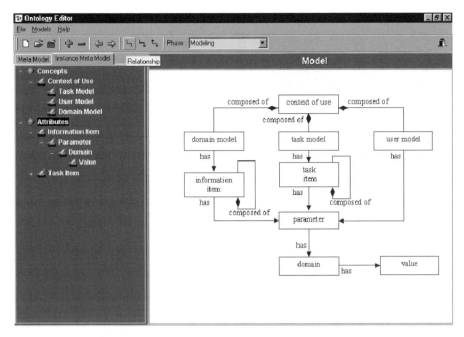

Figure 10.4. The ontology editor at the modelling level.

description through parameters. Each parameter has a domain; each domain has a set of values, possibly enumerated. The 'composed-of' relationship denotes aggregation, while 'has' denotes properties.

The definition of an ontology encourages structured UI design based on explicit concepts, relationships, and attributes. This structured approach contrasts with eclectic or extreme programming where the code is produced directly; it also contrasts with design methods that are not open to incorporating new or custom information as required by universal design. A UI ontology facilitates multi-disciplinary work that people from different backgrounds need to gather for collaborative or participatory design. The advantage of this level is that the ontology can be defined once and used as many times as desired. When universal design requires the consideration of more information within models or more models, the ontology can be updated accordingly, thereby updating the method for universal design of UIs. This does not mean that the subsequent levels will change automatically, but their definition will be subsequently constrained and governed so as to preserve consistency with the ontology. Once an ontology has been defined, it is possible to define the types of models that can be manipulated in the method, which is the goal of the logical level.

10.5. LOGICAL LEVEL

10.5.1. DEFINITION
Each model defined at the conceptual level is now represented with its own information parameters. For example, in the context of universal UIs, a user model is created

because different users might require different UIs. Multiple user stereotypes, stored as user models, allow designs for different user types in the same case study. Any type of user modelling can be performed since there is no predefined or fixed set of parameters. This illustrates the generality of the method proposed here to support universal design. The set of concepts and attributes defined in the ontology are instantiated for each context of use of a domain. This means each model, which composes a context of use, is instantiated by defining its parameters with domains of possible values.

10.5.2. CASE STUDY

In this example we use the ontology editor to instantiate the context of use, the relationships and attributes of models for the *Medical Attendance* domain involved in patient admission. Figure 10.5 graphically depicts the *Emergency Admission* context of use and the attributes of models of task, user and domain. Two tasks are instantiated: *Admit patient* and *Show patient data*. The first one is activated by a *Secretary* and uses *Patient information* during its execution. For the user model of the secretary, the following parameters are considered: the user's experience level, input preference and information density with the values *low* or *high*. The data elements describing a patient are the following: date, first name, last name, birth date, address, phone number, gender and civil status. Variables for insurance affiliation and medical regime can be described similarly. The variable parameters of a domain model depend on the UI design process. For instance, parameters and values of an information item used to generate UIs in [Vanderdonckt and Bodart 1993; Vanderdonckt and Berquin 1999] are: data type (date, Boolean, graphic, integer,

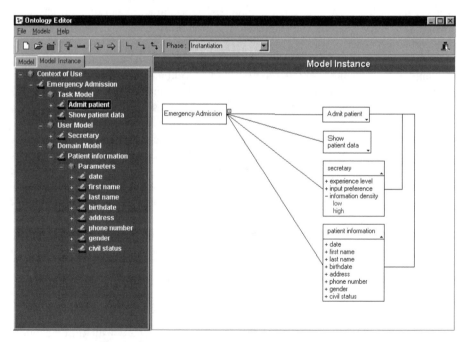

Figure 10.5. The ontology editor at the instance level.

(a)

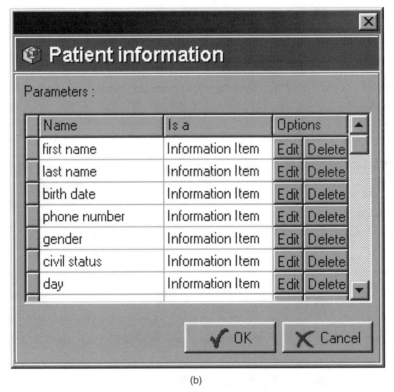

(b)

Figure 10.6. Some definitions at the modelling level.

real, or alphanumeric), length ($n > 1$), domain definition (known, unknown, or mixed), interaction direction (input, output, or input/output), orientation (horizontal, vertical, circular, or undefined), number of possible values ($n > 1$), number of values to choose ($n > 1$), and preciseness (low or high).

Figure 10.6 shows parameters of the model that was previously introduced. At this stage the parameters are listed but not instantiated. The parameters will be instantiated at the instance level. All of this information can then be stored in a model definition file that can be exported for future use.

Models defined and input at the logical level are all consistently based on the same ontology. The advantage is that when the ontology changes, all associated models change accordingly since the ontology is used as the input to the graphical editor. The graphical nature of the editor improves the legibility and the communicability of information, while information that cannot be represented graphically is maintained in text properties. The models are used for both requirements documentation and UI production in the next level.

As we have mentioned, it is possible to use any method for modelling tasks and users since there is no predefined or fixed set of parameters. Since there are many publications describing the rules for combining these parameters in order to deduce UI characteristics, we developed a rule module linked to the ontology editor. The rules can manipulate only entities in the logical level. The meta-definition at the conceptual level can be changed by domain experts (for example by adding/deleting/modifying any attribute, model or relationship), but not by designers. Once the meta-definition is provided to designers, they can only build models that are compatible with the meta-definition.

Figure 10.7 shows the editing of a rule for optimizing user interaction style based on several parameters [Vanderdonckt 2000]. The rule depicted in Figure 10.7 suggests natural language as the interaction style when the following conditions are met: the task

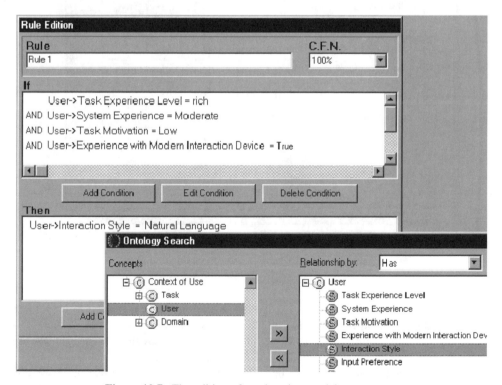

Figure 10.7. The editing of a rule using model parameters.

experience level (attribute coming from the user model) is rich, the system experience of the user is moderate, task motivation is low, and the user is experienced with modern interactive devices (e.g. touch screen, track point, trackball). When natural language is selected, appropriate design choices can be inferred. The advantage of this approach is that it is possible to easily define new rules when a new parameter is added to the models. Any rule can be produced to derive new design elements from user, task and system characteristics in a systematic way.

10.6. PHYSICAL LEVEL

10.6.1. DEFINITION

The main goal of the physical level lies in its ability to exploit instantiations captured in individual models to produce multiple UIs for different computing platforms, development environments and programming languages. This level is the only one that is dependent on the target hardware/software configuration intended to support the UI. Instantiations of the previously defined models, along with the values of their parameters, are stored in the logical level in specification files. Each specification file consists of a hierarchical decomposition of the UI models into models, parameters, values, etc. maintained in an ASCII file. Each instance of each concept is identified by an ID. All relationships and attributes are written in plain text in this file in a declarative way (e.g., Define Presentation Model Main;...; EndDef;). This file can in turn be imported into various UI editors as needed.

Here, the SEGUIA [Vanderdonckt and Berquin 1999] tool is used (Figure 10.8): it consists of a model-based interface development tool that is capable of automatically generating code for an executable UI from a file containing the specifications defined in the previous step. Of course, any other tool that complies with the model format, or that can import the specification file, can be used to produce an executable UI for other design situations, contexts of use, user models, or computing platforms. SEGUIA is able to automatically generate several UI presentations to obtain multiple UIs. These different presentations are obtained either:

(a) In an automated manner, where the developer launches the UI generation process by selecting which layout algorithm to use (e.g. two-column format or right/bottom strategy)

(b) In a computer-aided manner, where the developer can see the results at each step, can work in collaboration with the system, and can control the process.

In Figure 10.8, the left-hand column contains the information hierarchy that will be used to generate multiple presentation styles for MUIs. The right-hand column displays individual parameters for each highlighted data item in the hierarchy.

10.6.2. CASE STUDY

In our case study, information items and their values introduced at the modelling stage (Figures 10.5 and 10.6) are imported into an Import list box of data items (left-hand side of Figure 10.8). Each of these items is specified separately in the Current

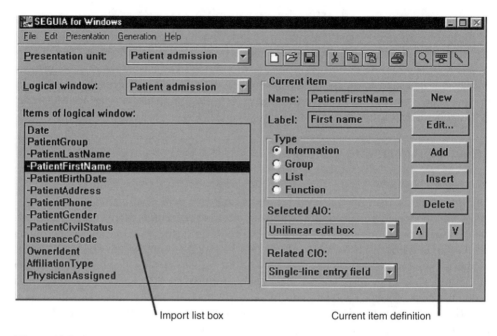

Import list box   Current item definition

Figure 10.8. Seguia environment displaying the case study specifications before UI generation.

item definition for viewing or editing (right-hand side of Figure 10.8). By selecting Generation in the menu bar, the developer launches the UI generation process and produces different user interfaces depending on the rules used for this purpose.

Selection rules automatically select concrete interaction objects (or *widgets*) by exploiting the values of parameters for each data item. For instance, the PatientFirstName information item is mapped onto a unilinear (single line) edit box (as Abstract Interaction Object) that can be further transformed into a Single-line entry field (as Concrete Interaction Object belonging to the MS Windows computing platform). Selection rules are gathered in different selection strategies. Once concrete widgets are defined, they can be automatically laid out.

Figure 10.9 shows the results of the code generation for the Patient Admission as defined in this case study. This layout strategy places widgets in two balanced columns by placing widgets one after another. This procedure can result in unused screen space, thus leading to a sub-optimal layout (Figure 10.9).

To avoid unused screen space and to improve aesthetics, a right/bottom layout strategy has been developed and implemented [Vanderdonckt and Bodart 1993; Vanderdonckt and Berquin 1999]. Figure 10.10 shows how this strategy can significantly improve the layout. Note the optimization of the upper-right space to insert push buttons, the right alignment of labels (e.g. between the date and the "Patient" group box) and the left alignment of edit fields. The method is very flexible at this stage, allowing the designer to experiment with different design options such as selection of widgets and layout strategies. For instance, some strategies apply rules for selecting purely textual input/output widgets (e.g. an edit

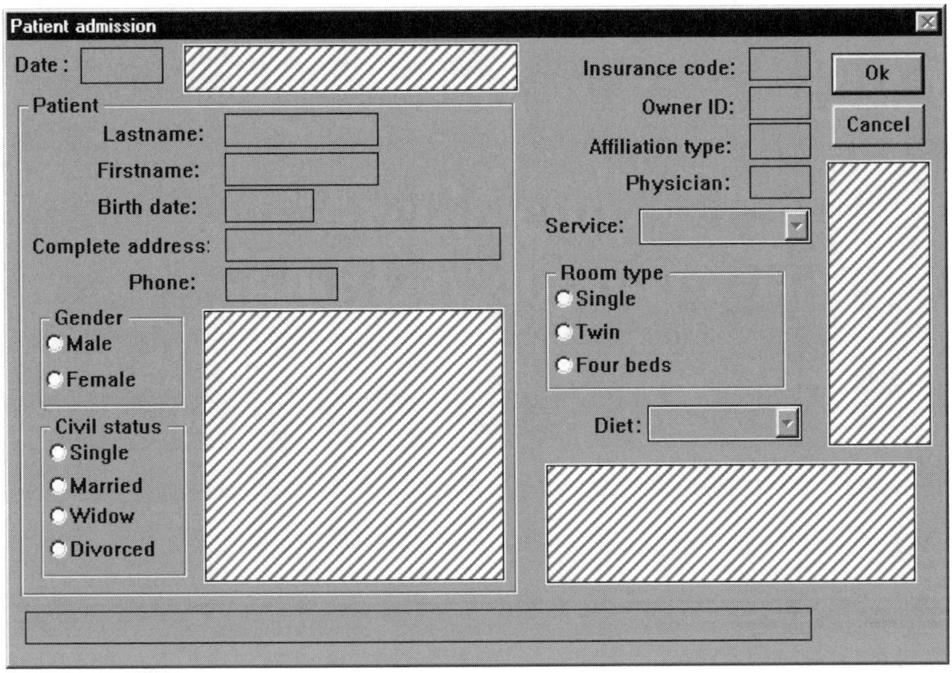

Figure 10.9. Patient admission window statically generated using the two-column strategy (with unused screen spaces).

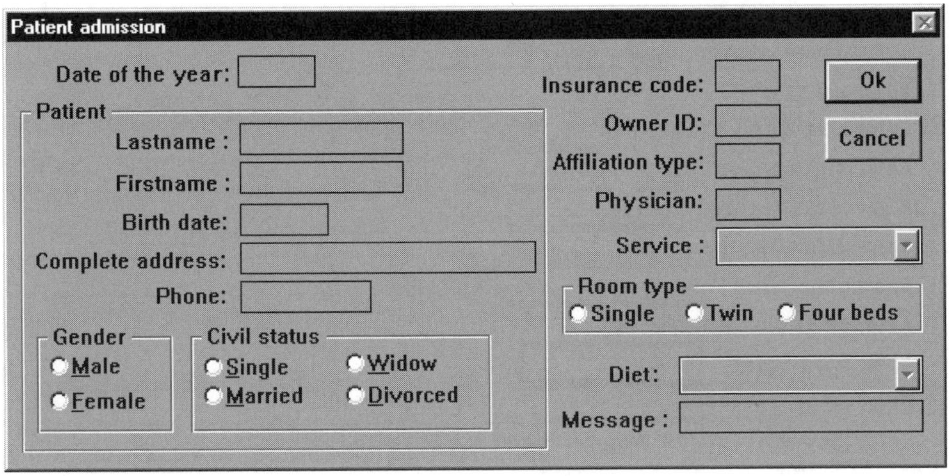

Figure 10.10. Patient admission window interactively generated using the right/bottom strategy.

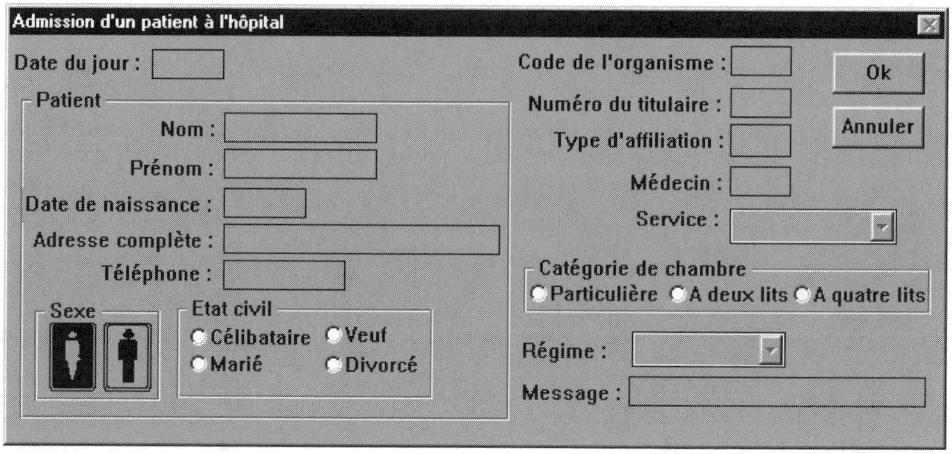

Figure 10.11. Patient admission window re-generated for a graphical icon and a different user language.

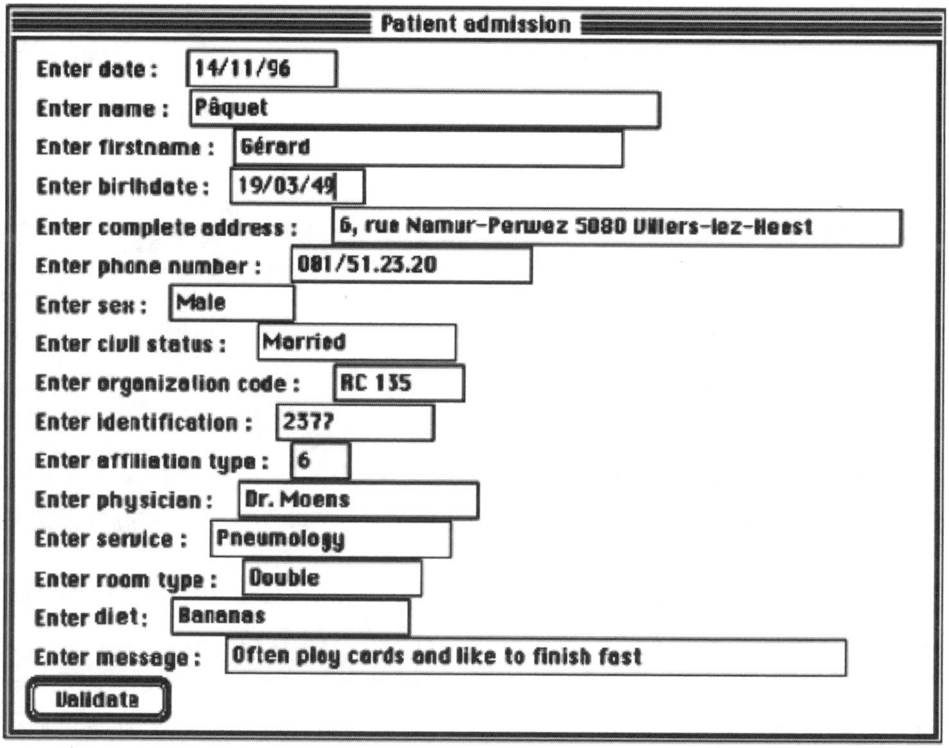

Figure 10.12. Patient admission re-generated for a different computing platform.

box), while others prefer widgets displaying graphical representations (e.g. a drawn button or an icon). The layout can also change according to the native language of the end-user.

Figure 10.11 shows the `Patient Admission` window re-generated after switching the `Gender` group box from a textual modality to a graphical modality (here, a radio icon) and switching the end user native language from English to French. The previous generation process can be reused for the logical positions of widgets. Thus by changing the value of just one parameter in the existing design options, it is possible to create a new UI in another language with an alternative design. Note in Figure 10.11 that some alignments changed with respect to Figure 10.10 due to the length of labels and new widgets introduced in the layout. The layout itself can be governed by different strategies ranging from the simplest (e.g., a vertical arrangement of widgets without alignment as in Figure 10.12 but for the Apple Macintosh computing platform) to a more elaborate one (e.g. layout with spatial optimization as in Figure 10.11).

At this stage it is possible to share or reuse previously defined models for several UI designs, which is particularly useful when working in the same domain as another UI. The approach also encourages users to work at a higher level of abstraction than merely the code and to explore multiple UI alternatives for the same UI design case. This flexibility can produce UIs with unforeseen, unexpected or under-explored features. The **advantage**

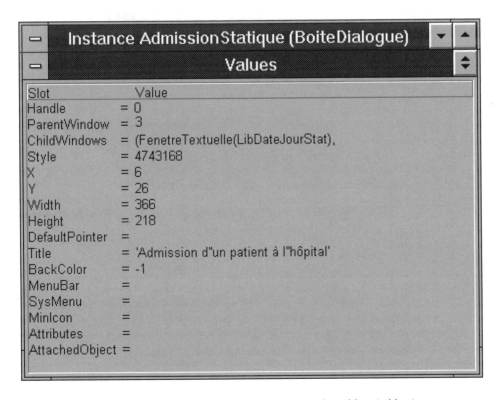

Figure 10.13. Parameters of a concrete interaction object (widget).

of this method is that when the set of models change, all UIs that were created from this set can change accordingly on demand.

The *design space* is often referred to as the set of all possible UIs that can be created from an initial set of models for one UI design. When a UI is generated for a target context of use, for example a target computing platform, it can be edited not only in SEGUIA, but also in the supported development environment for this computing platform. For example, Figure 10.13 represents the dialogue box of widget parameters generated in SEGUIA for a particular UI. The developer can of course edit these parameters, but should be aware that any change at this step can reduce the quality of what was previously generated. Changing the values of parameters in an inappropriate way should be avoided. Once imported into MS Visual Basic, for instance, these parameters can be edited by changing any value.

10.7. SUMMARY OF THE DEVELOPMENT PROCESS

The three levels of the development method are represented in Figure 10.14. This figure shows that each level, except the highest one, is governed by concepts defined at the next higher level. The meta-model level is assumed to remain stable over time, unless new high-level objects need to be introduced. Figure 10.2 includes only the main objects of the model definition. Possible models and their constituents are then defined at the model level as concepts, relationships, and attributes of the meta-model level. Again, the model level should remain stable over time, unless new models need to be introduced (e.g., a platform model, an organization model) or existing models need to be modified (e.g., the user model needs to include a psychological and cognitive profile).

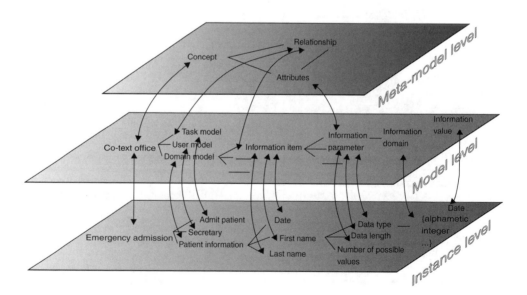

Figure 10.14. The different levels of the proposed method.

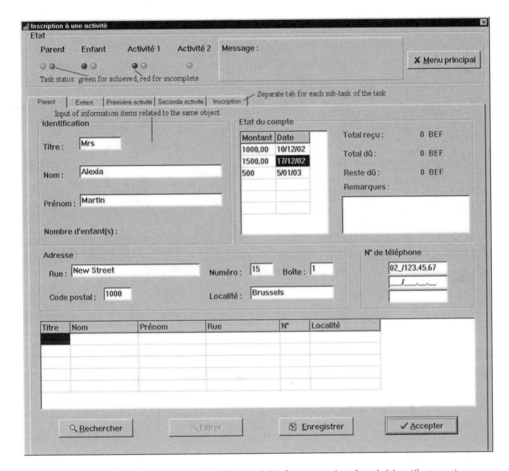

Figure 10.15. Case study: registering a child for recreational activities (first part).

For each case study, a new instance level will be produced for each model level. For instance, figures 10.15 and 10.16 exemplify UIs generated by the method with the SEGUIA tool for another context of use. The task consists of registering a child for recreational activities in a children's club. In the domain model, each child has specific information (such as first name, birth date), is related to one or many parents (described by their first name, last name, address, phone number, phone number in case of need) and at most two recreational activities (described by a name, a monitor, beginning and ending dates and times). The user model is restricted to only one profile: a secretary working in a traditional office environment. The upper left part of the window reproduced in Figure 10.15 illustrates the task status based on another rule defined at the model level: for the task to be completed, all information related to the child, at least one corresponding parent or relative, and two recreation activities are required. Therefore, a green or red flag is highlighted according to the current status of the task: here, the parent has been

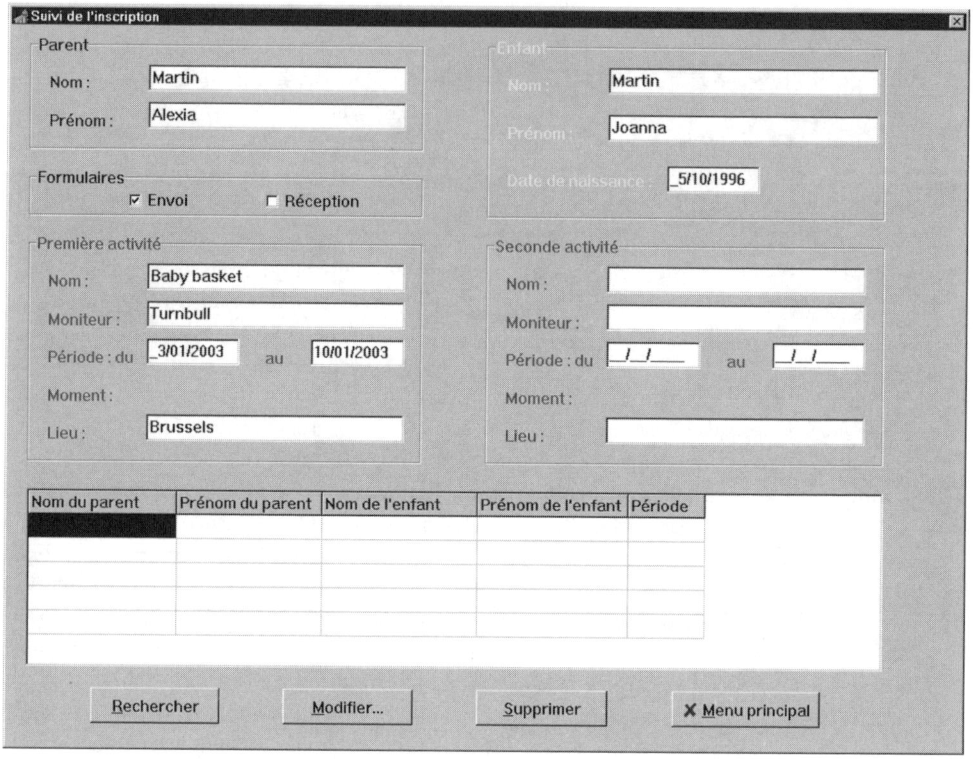

Figure 10.16. Case study: registering a child for recreational activities (second part).

completely and properly input, but not yet the child and the activities. The second activity is still greyed out as nothing has been entered yet. The presentation is governed by the following rule: each object should be presented in the same window (to avoid confusing users), but in different tabs of the window (here, 'Parent', 'Enfant', 'Activité 1', and 'Activité 2').

Figure 10.16 presents another rule working this time on the presentation model. Each group of information items related to the same entity (i.e., a child, a parent, an activity) is assembled in a group box; this box is laid out using the optimized algorithm presented above and emphasized in a distinctive colour scheme. The group boxes are therefore presented in blue (parent), yellow (child), red (first activity), and green (second activity). The colours are chosen from the standard palette of colours. Such rules cannot be defined in a logical way in traditional development environments. But they can of course be defined manually, which becomes particularly tedious for producing MUIs.

For the same case study, Figure 10.17 presents an alternative UI for a slightly different profile requiring more guidance to achieve the task step by step. Rather than using a graphical overview in the upper part of the window (as in Figure 10.15) and multiple tabs, the UI progressively proceeds from one task to the next by adding tabs each time a

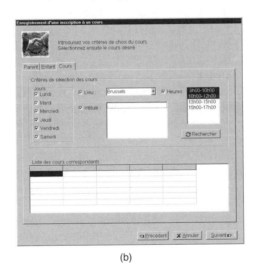

(a)

(b)

Figure 10.17. Alternate layout for the task: registering a child for recreational activities.

sub-task is finished. Therefore, the rule coded here is: present each object in a separate tab; once a sub-task is carried out, add a new tab with the next object for the next sub-task to be carried out and add buttons to move forward and backward as in a set-up wizard.

10.8. CONCLUSION

The UI design method can be explicitly structured into three levels (i.e., conceptual, logical, and physical), as frequently found in other disciplines such as databases, software engineering and telecommunications. Each level can be considered a level of abstraction of the physical level as represented in Figure 10.14. The physical level is the level where instances of the case study are analysed. The logical level is the model level where these instances are mapped onto relevant abstractions. The conceptual level is the meta-model level where abstractions manipulated in the lower levels can be combined to identify the concepts, relationships, and attributes used in a particular method.

The three levels make it possible to apply the *Principle of Separation of Concerns* as follows: (i) a definition of useful concepts by someone who is aware of UI techniques such as user-centred design, task analysis, and human factors; (ii) a model definition where, for each UI design, multiple sets of models can be defined on the same basis with no redefinition of previously defined concepts; and (iii) multiple UI creation: for each set of UI models, several UIs can be created by manipulating parameters supported by the UI generator and manual editing is allowed when needed.

The advantage of the method is that changes at any level are instantly propagated to other levels: when the ontology changes, all possible models change accordingly; when a model changes, all possible specifications change accordingly as well as the set of all possible UIs that can be created (the UI design space).

The use of an ontology editor not only allows model definition at the highest level of abstraction (i.e. the meta-model) but also prevents any designer from defining:

- Models that are not compatible with the model definition;
- Models that are error-prone and that do not satisfy desirable properties (Table 10.1); and
- Models that are not supported by code generation tools (e.g. when a design option is not supported).

Model-based approaches for designing UIs have been extensively researched for more than fifteen years. These different approaches have worked in similar ways at the model level as indicated in Figure 10.14. For the first time, a complete method and supporting environment has the ability to increase the level of abstraction of these approaches one step further. Today, we are not aware of any similar approach that offers the designer the ability to work at the meta-model level while maintaining the ability to propagate changes at the highest level to the underlying levels. This capability can be applied to a single UI or to multiple UIs, especially when multiple contexts of use need to be considered. In this way, any work done at the meta-model level for one particular context of use can be reused directly at the *same level of abstraction* for another context of use.

ACKNOWLEDGEMENTS

This work is a joint effort of the CADI-CADINET research project under the auspices of University of Fortaleza, Brazil (http://ead.unifor.br/) and the Seguia tool developed by J. Vanderdonckt, BCHI, Université catholique de Louvain, Belgium (http://www.isys.ucl.ac.be/bchi/research/seguia.htm), a component of the Trident research project, funded by Institut d'Informatique, Facultés Universitaires Notre-Dame de la Paix, Belgium (http://www.info.fundp.ac.be/~emb/Trident.html). We gratefully acknowledge the support of the above organizations for funding the respective projects. The authors would also like to thank the anonymous reviewers for their constructive comments on an earlier version of this chapter.

REFERENCES

Booch, G., Jacobson, I., and Rumbaugh, J. (1998) *The Unified Modelling Language User Guide.* Addison-Wesley, Reading.

Butler, K.A. (1995) Designing Deeper: Towards a User-Centered Development Environment Design in Context. *Proceedings of ACM Symposium on Designing Interactive Systems: Processes, Practices, Methods, & Techniques DIS'95* (Ann Arbor, 23–25 August 1995). ACM Press, New York. 131–142.

Card, S., Moran, T.P., and Newel, A. (1983) *The Psychology of Human-Computer Interaction.* Lawrence Erlbaum Associates, Hillsdale.

Gaines, B. (1994) A Situated Classification Solution of a Resource Allocation Task Represented in a Visual Language, *Special Issue on Models of Problem Solving, International Journal of Human-Computer Studies*, 40(2), 243–71.

Gimnich, R., Kunkel, K., and Reichert, L. (1991) A Usability Engineering Approach to the Development of Graphical User Interfaces, *Proceedings of the 4th International Conference on Human-Computer Interaction HCI International'91* (Stuttgart, 1–6 September 1991), 1, Elsevier Science, Amsterdam. 673–7.

Guarino, N. (1995) Formal Ontology, Conceptual Analysis and Knowledge Representation: The role of formal ontology in information technology. *International Journal of Human-Computer Studies*, 43(5–6), 625–40.

Hartson, H.R., and Hix, D. (1989) Human-Computer Interface Development: Concepts and Systems for its Management. *ACM Computing Surveys*, 21(1), 241–7.

Hix, D. (1989) Developing and Evaluating an Interactive System for Producing Human-Computer Interfaces. *Behaviour and Information Technology*, 8(4), 285–99.

Lim, K.Y., and Long, J. (1994) Structured Notations to Support Human Factors Specification of Interactive Systems Notations and Tools for Design. *Proceedings of the BCS Conference on People and Computers IX HCI'94* (Glasgow, 23–26 August 1994). Cambridge University Press, Cambridge. 313–26.

Marca, D.A., and McGowan, C.L. (1988) *SADT: Structured Analysis and Design Techniques.* McGraw-Hill, Software Engineering Series.

Paternò, F. (1999) *Model-based Design and Evaluation of Interactive Applications.* Springer-Verlag, Berlin.

Paternò, F., and Santoro, C. (2002) One Model, Many Interfaces, in C. Kolski and J. Vanderdonckt (Eds), *Computer-Aided Design of User Interfaces III, Proceedings of 4th International Conference on Computer-Aided Design of User Interfaces CADUI'2002* (Valenciennes, 15–17 May 2002) 143–54. Kluwer Academics, Dordrecht.

Pribeanu, C., Vanderdonckt, J., and Limbourg, A. (2001) Task Modelling for Context Sensitive User Interfaces, in C. Johnson (Ed.), *Proceedings of 8th International Workshop on Design, Specification, Verification of Interactive Systems DSV-IS'2001* (Glasgow, 13–15 June 2001), Lecture Notes in Computer Science, 2220, 49–68. Springer-Verlag, Berlin.

Puerta, A.R. (1997) A Model-Based Interface Development Environment. *IEEE Software*, 14(4), July/August 1997, 41–7.

Rumbaugh, J., Blaha, M., Premerlani, W., Eddy, F., and Lorenson, W. (1990) *Object-Oriented Modelling and Design.* Prentice Hall, New Jersey.

Savidis, A., Akoumianakis, D., and Stephanidis, C. (2001) The Unified User Interface Design Method, Chapter 21 in C. Stephanidis (Ed.), *User Interfaces for All: Concepts, Methods, and Tools*, 417–40. Lawrence Erlbaum Associates, Mahwah.

Top, J., and Akkermans, H. (1994) Tasks and Ontologies in Engineering Modelling, *International Journal of Human-Computer Studies*, 41(4), 585–617.

Vanderdonckt, J. (2000) A Small Knowledge-Based System for Selecting Interaction Styles, *Proceedings of International Workshop on Tools for Working with Guidelines TFWWG'2000* (Biarritz, 7–8 October 2000), 247–62 Springer-Verlag, London.

Vanderdonckt, J., and Bodart, F. (1993) Encapsulating Knowledge for Intelligent Automatic Interaction Objects Selection, in S. Ashlund, K. Mullet, A. Henderson, E. Hollnagel, and T. White (Eds), *Proceedings of the ACM Conf. on Human Factors in Computing Systems INTERCHI'93* (Amsterdam, 24–29 April 1993), 424–9. ACM Press, New York.

Vanderdonckt, J., and Berquin, P. (1999) Towards a Very Large Model-based Approach for User Interface Development, in N.W. Paton and T. Griffiths (Eds), *Proceedings of 1st International Workshop on User Interfaces to Data Intensive Systems UIDIS'99* (Edinburgh, 5–6 September 1999), 76–85 IEEE Computer Society Press, Los Alamitos.

Supporting Interactions with Multiple Platforms Through User and Task Models

Luisa Marucci, Fabio Paternò, and Carmen Santoro

ISTI-CNR, Italy

11.1. INTRODUCTION

In recent years, interest in model-based approaches has been increasing. The basic idea of such approaches is to identify useful abstractions highlighting the main aspects that should be considered when designing effective interactive applications. UML [Booch *et al.* 1999], the most common model-based approach in software engineering, has paid very little attention to supporting the design of the interactive component of software. Therefore, specific approaches have been developed for interactive system design. Of the relevant models, task models play a particularly important role because they indicate the logical activities that an application should support. A task is an activity that should be performed in order to reach a goal. A goal is either a desired modification of state or an inquiry about the current state.

For the generation of multiple user interfaces, task models play a key role in the adaptation to different contexts and platforms. The basic idea is to capture all the relevant

Multiple User Interfaces. Edited by A. Seffah and H. Javahery
© 2004 John Wiley & Sons, Ltd ISBN: 0-470-85444-8

requirements at the task level and then use this information to generate effective user interfaces tailored for different types of platforms. Information about the design of the final user interface can be derived from analysing task models. For example, the logical decomposition of a task can provide guidance for the generation of the corresponding concrete user interface. The task structure is reflected in the graphical presentation by grouping together interaction techniques and objects associated with the same sub-task. We have identified a number of possibilities for how tasks should be considered in the generation of multi-platform user interfaces, for example:

- When the same task can be performed on multiple platforms in the same manner;
- When the same task is performed on multiple platforms but with different user interface objects or domain objects;
- When tasks are meaningful only on specific platforms.

In addition to adapting interfaces at the design phase, it is possible to adapt them at run-time by considering users' preferences and environment (location, surrounding, etc.). Preference and environment information is used to adapt the navigation, presentation and content of the user interface to different interaction platforms (see Figure 11.1).

Figure 11.1. User model support for multiple platforms.

User modelling [Brusilovsky 1996] supports adaptive interfaces that change according to user interaction. It can also be helpful in designing for multiple interaction platforms. User models represent aspects such as knowledge level, preferences, background, goals, physical position, etc. This information provides user interfaces with adaptability, which is the ability to dynamically change their presentation, content and navigation in order to better support users' navigation and learning according to the current context of use. Several types of user models have been used. For example, some models use information about the level of users' knowledge for the current goals; other models employ user stereotypes and evaluate the probability of their relevance to the current user.

Various aspects of user interfaces can be adapted through user models. Text presentation can be adapted through techniques such as conditional text or stretch-text (text that can be collapsed into a short heading or expanded when selected). User navigation can also be adapted using techniques such as direct guidance and adaptive ordering and hiding of links. Adaptive techniques have been applied in many domains. Marucci and Paternò [2002] describe a Web system supporting an adaptive museum guide that provides virtual visitors with different types of information related to the domain objects presented (introduction, summary, comparison, difference, curiosity) according to their profile, knowledge level, preferences and history of interactions.

To date, only a few publications have considered user modelling to support the design of multi-platform applications, and there are still many issues that need to be solved in this context. For example, Hippie [Oppermann and Specht 2000] describes a prototype that applies user modelling techniques to aid users in accessing museum information through either a web site or a PDA while they are in the museum. In our approach we also consider the use of mobile outdoor technologies and provide user models integrated with task models developed in the design phase.

In this chapter we first introduce a detailed scenario to illustrate the type of support that we have designed. Then we describe our method, how the user model is structured and how such information is used to obtain an adaptive user interface. We also discuss the types of rules for using information in the user model to drive adaptive behaviour. Finally, we provide some concluding remarks.

11.2. AN ILLUSTRATIVE SCENARIO

In this section we provide the reader with a concrete example of the type of support that our approach provides. The scenario describes an application that provides an interactive guide to a town.

From a desktop computer at the hotel, John visits the Carrara web site. He finds it interesting. In particular, he is interested in marble sculptures located close to Piazza Garibaldi. He spends most of his time during the virtual visit (Figure 11.2a) accessing the related pages and asking for all the available details on such artworks.

The next day, John leaves the hotel and goes to visit the historic town center. When he arrives he accesses the town's web site through his phone.

The system inherits his preferences and levels of knowledge from the virtual visits performed in the hotel. Thus, it allows him to access information on the part of the town

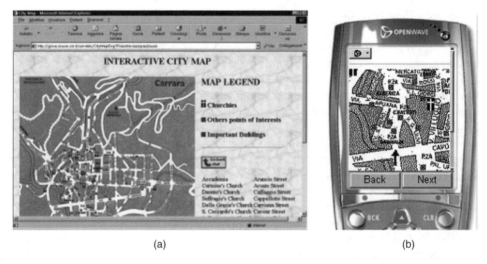

(a)   (b)

Figure 11.2. Spatial information provided through (a) the desktop and (b) the cell phone.

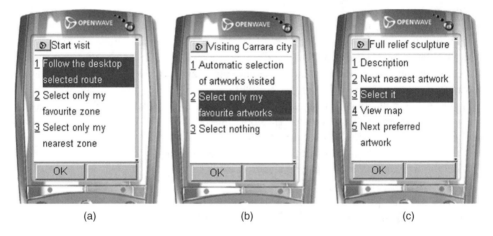

(a)   (b)   (c)

Figure 11.3. Cell phone support during the visit to the historic town.

that interests him most (Figure 11.3a) and navigation is supported through adaptive lists. These lists are based on a ranking determined by the interests expressed in the previous visit using the desktop system. During the physical visit he sees many works of art that impress him, but there is no information available nearby, so he annotates them through the phone interface (Figures 11.3b and 11.3c).

In the evening, when he is back in the hotel, John accesses the town web site again through his login. The application allows him to access an automatically generated guided tour of the town (Figure 11.4) with an itinerary based on the locations of the works of art

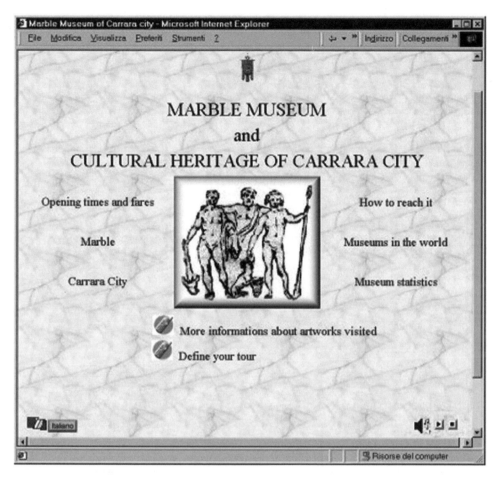

Figure 11.4. User interface to the desktop system after access through phone.

that impressed him. He can modify the tour if he no longer finds some of the proposed works of art interesting. In this way, he can perform a new visit of the most interesting works of art, receiving detailed information about them.

11.3. GENERAL DESCRIPTION OF THE APPROACH

In order to support the development of systems that adapt to the current user, device and context, we use the models shown in Figure 11.5. In this diagram we consider both the static development of interactive systems for a set of platforms and the dynamic adaptation of interactive systems to changes in context in real time.

- In the static case, a system specification in terms of the supported tasks is used to create an abstract version of a user interface (including both abstract presentation and dialogue

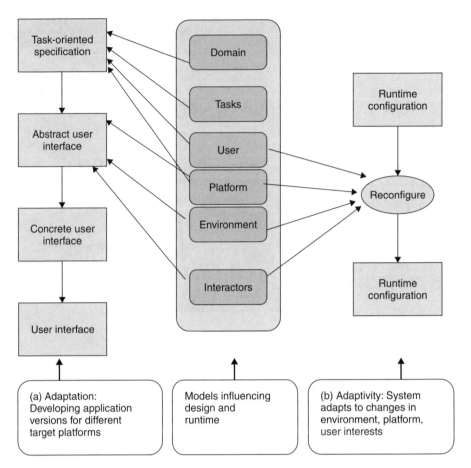

Figure 11.5. Models considered at design time and run-time.

design); from the abstract user interface, a concrete user interface is derived, in which the abstract interaction mechanisms are bound to platform-specific mechanisms.

- In the dynamic case, external triggers lead to the real-time reconfiguration of the interactive system during use. These triggers can be user actions (e.g., connecting a PDA to a mobile phone to provide a network connection) or events in the environment (e.g., changing noise and light level as a train enters a tunnel, or network failure).

In order to better describe the static development and the dynamic reconfiguration of systems, we refer to a number of models:

- The *Task Model* describes a set of activities that users intend to perform while interacting with the system. We can distinguish two types of task models: the system task model, which is how the designed system requires tasks to be performed, and the user task model, which is how users expect to perform their activities. A mismatch between these two models can generate usability problems.

- The *Domain Model* defines the objects that a user can access and manipulate in the user interface. This model also represents the object attributes and relationships according to semantically rich expressions.
- *Interactors* [Paternò and Leonardi 1994] describe the different interaction mechanisms independent of platform (e.g., the basic task an interactor is able to support). The interactor model operates primarily at the level of the abstract description of the user interface.
- The Platform Model describes the physical characteristics of the target platforms, for example characteristics of the available interactive devices such as pen, screen, voice input and video cameras.
- The *Environment Model* specifies the user's physical environment.
- The *User Model* describes information such the user's knowledge, interests, movements and personal preferences.

We have identified a set of design criteria for using logical information in the models to generate multimedia user interfaces adapted to a specific user, platform and context of use. For a given task and device, these design criteria indicate, for example, which interaction and presentation techniques are the most effective in a specific configuration setting, and how the user interface should adapt to a change of device or environmental conditions.

In the following sections we will show how these models take part in the overall design process. The discussion will focus on the task model and user model, describing the role they play in the generation of user interfaces that adapt to changes in context. Afterward, we will describe their relationships in more detail, and we will present an example that helps to explain the approach.

11.4. ROLE OF THE TASK MODEL IN DESIGN

The design of multi-platform applications can employ different approaches. It is possible to support the same type of tasks with different devices. In this case, what has to be changed is the set of interaction and presentation techniques to support information access while taking into account the resources available in the device considered. However, in some cases designers should consider different devices also with regard to the choice of tasks to support. For example, phones are more likely to be used for quick access to limited information, whereas desktop systems better support browsing through large amounts of information. Since the different devices can be divided in clusters sharing a number of properties, the vast majority of approaches considers classes of devices rather than single devices. On the one hand, this approach tends to limit the effort of considering all the different devices. On the other hand, different device types might be needed because of the heterogeneity of devices belonging to the same platform.

The fact that devices and tasks are so closely interwoven in the design of multiplatform interactive applications is a central concern running through our method, which is composed of a number of steps allowing designers to start with an overall envisioned

Figure 11.6. Deriving multiple user interfaces from a single task model.

task model of a nomadic application and then derive effective user interfaces for multiple devices (see Figure 11.6). The approach involves four main steps:

1. High-level task modelling of a multi-context application: In this phase, designers define the logical activities to be supported and the relationships among them. They develop a single model that addresses the various contexts of use and roles; they also develop a domain model to identify the objects manipulated in tasks and the relationships among such objects. Such models are specified using the ConcurTaskTrees (CTT) notation. The CTT Environment tool [Mori *et al.* 2002] publicly available at http://giove.cnuce.cnr.it/ctte.html supports editing and analysis of task models using this notation. The tool allows designers to explicitly indicate the platforms suitable to support each task.
2. Developing the system task model for the different platforms: Here designers filter the task model according to the target platform and, if necessary, further refine the task model for specific devices. In this filter-and-refine process, tasks that cannot be supported on a given platform are removed and the navigational tasks necessary to interact with the platform are added. In other cases it is necessary to add supplementary details on how a task is decomposed for a specific platform.
3. From system task model to abstract user interface: Here the goal is to obtain an abstract description of the user interface. This description is composed of a set of abstract presentations that are identified through an analysis of the task relationships. These abstract presentations are then structured by means of interactors (see Section 3 for definition). Then we identify the possible transitions among the user interface

presentations as a function of the temporal relationships in the task model. Analysing task relationships can be useful for structuring the presentation. For example, the hierarchical structure of the task model helps to identify interaction techniques and objects to be grouped together, as techniques and objects that have the same parent task are logically more related to each other. Likewise, concurrent tasks that exchange information can be better supported by highly integrated interaction techniques.

4. User interface generation: This phase is platform-dependent and device-dependent. For example, if the platform is a cellular phone, we also need to know the type of micro-browser supported and the number and types of soft-keys available in the specific device considered.

In the following sections we discuss these steps in detail. We have defined XML versions of the language for task modelling (ConcurTaskTrees) and the language for modelling abstract interfaces; we have also developed automatic transformations among these representations.

11.4.1. FROM THE TASK MODEL TO THE ABSTRACT USER INTERFACE

The task model is the starting point for defining an abstract description of the user interface. This abstract description has two components: a presentation component (the static structure of the user interface) and a dialogue component (the dynamic behaviour).

The shift from task to abstract interaction objects is performed through three steps:

1. Calculation of Enabled Task Sets (ETS): the ETSs are sets of tasks enabled over the same period of time according to the constraints indicated in the task model. They are automatically calculated through an algorithm that takes as input (i) the formal semantics of the temporal operators of the CTT notation and (ii) a task model. For example, if two tasks t1 and t2 are supposed to be concurrently performed, then they belong to the same ETS; they can be performed in any order so their execution will be enabled over the same period of time. If they are supposed to be carried out following a sequential order (first t1 then t2), they cannot belong to the same ETS since the performance of t2 will be enabled only after the execution of t1; thus they will never be enabled during the same interval of time. The need for calculating ETSs is justified by the fact that the interaction techniques supporting the tasks belonging to the same enabled task set are logically candidates to be part of the same presentation. In this sense, the ETS calculation provides a first set of potential presentations. Furthermore, the calculation of the ETS implies the calculation of the conditions that allow passing from ETS to ETS—we called them 'transitions'.

2. Heuristics for optimizing presentation sets and transitions: these heuristics help designers reduce the number of presentations considered in the final user interface. This is accomplished by grouping together tasks belonging to different ETSs. In fact, depending on the task model the number of ETSs can be rather high. As a rule of thumb, the number of ETSs is of the same order as the number of enabling operators in the task model. So, in this phase we specify rules (heuristics) to reduce their number by merging two or more ETSs into new sets, called Presentation Task Sets (PTS).

3. Mapping presentation task sets and their transitions onto sets of abstract interaction objects and dialogue: a number of rules have been identified in order to perform the mapping between a task and a suitable abstract user interface object. These rules are based on the analysis of the multi-dimensional information associated with tasks – for example, the goal, the objects manipulated and the frequency of the task. Each dimension functions as a sort of condition during the visit of the tree-like structure of the language describing interactors (see Figure 11.7), in order to select the most suitable one. The transitions between different presentation sets are directly mapped into connections linking the different presentations of the abstract user interface.

All of these transformations are supported by our TERESA tool [Mori *et al.* 2003] publicly available at http://giove.cnuce.cnr.it/teresa.html.

11.4.2. THE LANGUAGE FOR ABSTRACT USER INTERFACES

The set of presentation sets obtained in the previous step is the initial input for building the abstract user interface specification. This specification is composed of interactors or Abstract Interaction Objects (AIOs) associated with the basic tasks. Such interactors are high-level interaction objects that are classified first by type of basic task supported, then by type and cardinality of the associated objects and lastly by presentation aspect.

Figure 11.7 shows that an interface is composed of one or more presentations and each presentation is characterised by an aio or an aio_composition and 0 or more connections. There are two main types of objects in the abstract user interface: elementary abstract interaction objects (aio) and complex expressions (aio_composition) derived from applying the operators to these interaction objects. While the operators describe the static organisation of the user interface (in the next section we provide more detail on them),

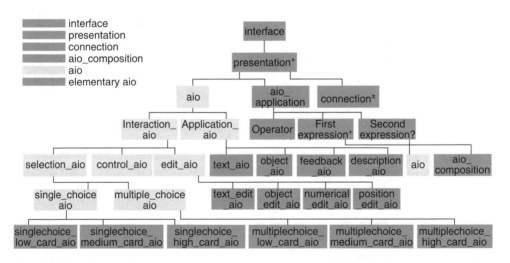

Figure 11.7. The tree-like representation of the language for specifying the abstract user interface.

the set of connections describes how the user interface evolves over time, namely its dynamic behaviour.

11.4.2.1. From Presentation Task Sets to Abstract User Interface Presentations

The abstract user interface is mainly defined by a set of interactors and the associated composition operators. The type of task supported, the type of objects manipulated and their cardinality are useful elements for identifying the interactors. In order to compose such interactors we have identified a number of composition operators for designing usable interfaces. These composition operators are associated with communication goals that designers aim to achieve [Mullet and Sano 1995]:

- *Grouping (G)*: The objective is to group together two or more elements, so this operator should be applied when the involved tasks share some characteristics. A typical situation is when the tasks have the same parent task. This is the only operator for which the position of the different operands is irrelevant.
- *Ordering (O)*: This operator is applied when some kind of sequential order exists among elements. The most typical sequential order is the temporal order. The order in which the different elements appear within this operator reflects the order within the group.
- *Relation (R)*: This operator is applied when a relation exists between n elements y_i, $i = 1, \ldots, n$ and one element x. In the task model, a typical situation is when a leaf task t is at the right-hand side of a disabling operator. In this case all the tasks that could be disabled by t (at whatever task tree level) are in relation to t. This operator is not commutative.
- *Hierarchy (H)*: This operator means that a hierarchy exists among the involved interactors. The importance level associated with the operands identifies the degree of visual prominence that the associated interaction objects should have in the user interface. The degree of importance can be derived from the frequency of access or from details of the application domain. Various techniques can be used to convey importance. In graphical user interfaces, one method is allotting more screen space to objects that are hierarchically more important.

These operators are applied to tasks belonging to the same PTS, depending on the temporal relationships among those tasks. The temporal relationships are derived from the task model in the following manner: if the two concerned tasks are siblings, the temporal relationship is represented by the CTT operator existing between them; if this is not the case (e.g. the two tasks have different parent tasks) the temporal relationship is easily derived because temporal relationships between tasks are inherited by their subtasks.

11.4.2.2. The Dialogue Component

In specifying the dynamic behaviour of the abstract user interface, an important role is played by abstract interaction objects associated with the transitions. For each presentation task set P, *transition(P)* specifies the conditions allowing for the transition of the abstract user interface from the current presentation task set P into another presentation task set P'. The transitions can directly correspond to tasks, or, alternatively, can be expressed by means of a Boolean expression. For example, when we want to express that more than one task has to be executed in order to trigger the activation of a different presentation, an AND operator combines the tasks.

11.4.3. FROM THE ABSTRACT USER INTERFACE TO ITS IMPLEMENTATION

Once the elements of the abstract user interface have been identified, each interactor is mapped onto interaction techniques supported by the specific device configuration (operating system, toolkit, etc.). For example if the object of the abstract user interface allows for a single selection from a set of objects, various implementations are available to the designer depending on the capabilities of the platform or device in question; these can include radio button menus, pull-down menus, list menus, etc.

In addition, since relationships between interactors are expressed with composition operators, they have to be appropriately implemented in order to convey their logical meaning in the final user interface. Several techniques are available for this purpose. For instance, in graphical user interfaces, a typical example is the set of techniques for conveying groupings by using classical presentation patterns such as proximity, similarity and continuity. If a different modality is used, the meaning of the same operators should be conveyed through different mechanisms. For example, in audio user interfaces, we would convey groupings with aural attributes such as pitch and volume.

As another example, a hierarchy operator for textual objects in a graphical user interface could represent important objects with larger fonts, whereas in an audio-based user interface, the hierarchy operator could represent important verbal information with a higher volume.

11.5. RELATIONS BETWEEN TASK MODEL AND USER MODEL

In our approach we assume that a model-based method has been followed in the design of the multi-platform application. As noted earlier in this chapter, the ConcurTaskTrees notation [Paternò 1999] allows designers to develop task models of nomadic applications. This means that in the same model, designers can describe tasks to be performed on different platforms and their interrelationships [Paternò and Santoro 2002]. From this high level description it is possible to obtain first the system task model associated with each platform and then the corresponding device-level user interface. The task model can also be expressed in XML format.

In our case we use the XML specification as input for the creation of the user model that will be used for adaptivity at run-time. The two models share some information, but also contain different elements. This means that some elements of the task model are removed and others added in order to make the two models compatible. In addition, the user model is mainly characterised by values that are updated dynamically based on users' interactions with the interface. For each user, the user model is updated when the user interacts with any of the available platforms. A run-time support algorithm uses the

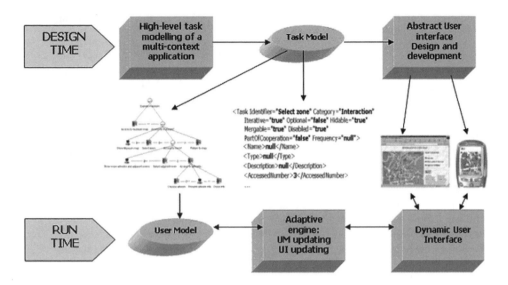

Figure 11.8. Relationships between task model and user model.

user model to modify the user interface presentation, navigation and content by applying previously defined adaptivity rules.

One advantage of this approach is that the task model developed at design time already provides useful information for run-time adaptive support (Figure 11.8). This information from the task model at design time includes:

- The temporal dependencies among tasks performed on different platforms;
- The tasks that can be performed through multiple platforms;
- The association of tasks with domain objects and related attributes;
- The definition of objects and attributes accessible through a given platform.

The performance of some tasks (from either phone or desktop) can change the level of interest associated with some domain objects (for example the preferred city zone), and this information can also be used to adapt the presentation support for a platform different from that currently in use (for example, the order of the links in a list).

11.6. THE USER MODEL

In our approach, the user model is structured in such a way as to indicate user preferences and acquired knowledge depending on the user's access to the application. Referring to the scenario of use of the Carrara Web site in section 2:

- User preferences can include, for example, the preferred city zone, navigation style, theme or features of an artwork.

- Acquired knowledge can include, for example, the level of knowledge about an author, a historical period or a material.
- The general format of the user model (in XML file format) includes:

As we can see in Figure 11.9, the user model is tightly related to the task model. It contains information that is dynamically updated such as the number of times that a task has been performed or that an object has been accessed. It also contains fields that allow dynamic modification of the availability of performing a specific task: *Mergeable* indicates whether to merge the execution of a task with a different task, *Hideable* indicates whether to hide its performance in another, more general task, and *Disabled* indicates whether to completely disable it for the current user.

For each task, all the attributes listed above can be defined (through a dedicated tool), including the properties related to adaptive support (Figure 11.10). After this step, the tool generates the XML file in the following general form (Figure 11.11):

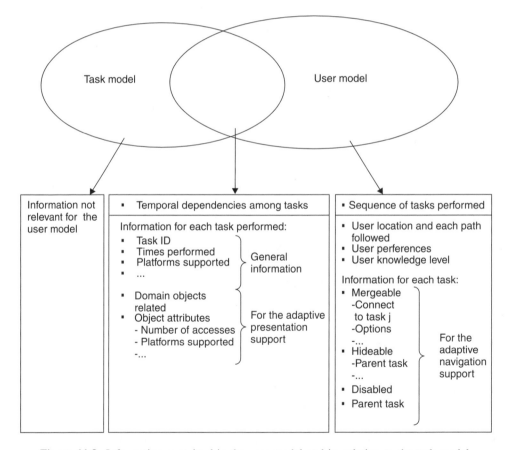

Figure 11.9. Information contained in the user model and its relation to the task model.

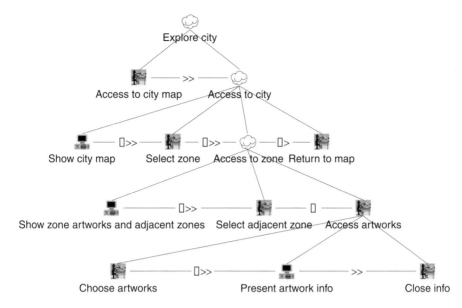

Figure 11.10. Example of a task model.

```
<Task Identifier="Select zone" Category="Interaction"
     Iterative="true" Optional="false" Hidable"="true"
     Mergable="true" Disabled="true"
     PartOfCooperation="false" Frequency="null">
<Name>null</Name>
<Type>null</Type>
<Description>null</Description>
<AccessedNumber>3</AccessedNumber>
...
```

Figure 11.11. An excerpt of the XML file containing information about task models.

This file is updated during the user session. From analysis of the file, the system is able to determine the tasks performed by the user and their sequence, as well as the object classes and related subclasses. From this user input, the system computes the navigation preferences by analysing information such as the sequence of tasks, the tasks never performed, and the tasks most frequently performed. The system also evaluates the presentation preferences by analysing the objects' classes and subclasses.

The location is an attribute related only to mobile interactive platforms. For example if the user has accessed the system by mobile phone, then after the user selects an item from the *Materials* list, the system offers the option of displaying only the artworks made of local materials.

The domain model is structured in terms of object classes and related subclasses that are manipulated during task performance. The relationships between tasks and domain objects are represented in the user model. The association between tasks and object instances can be either static or dynamic. For example, in the task of selecting an element from a list of predefined values, the association is static, whereas in the task of displaying information on a work of art whose name is provided by the user, the association is dynamic.

The domain objects that can be accessed and manipulated vary by device. In general, domain objects that can be manipulated by phone are more limited than those accessible via desktop computers and have different spatial attributes related to the user position, such as *closeness*.

Likewise, the supported tasks depend on the interaction platform. For example, some tasks are associated with a virtual visit on a desktop computer and others are associated with access by mobile phone. In addition, performance of certain tasks on one platform may depend on the accomplishment of other tasks through other devices (for example the desktop task of reviewing an itinerary previously annotated with a phone device).

In response to the user's behaviour in real time, the user model dynamically updates user knowledge and preferences. This has the effect of updating objects, attributes and task performance frequencies. The application can dynamically change the supported navigation according to the frequency of performance of certain tasks and the frequency of use of certain objects.

11.7. ADAPTIVE RULES

This section describes the rules that are used to drive the adaptivity of the user interface. In the next subsection we will explain how these rules are handled, and how they result in adaptive navigation and presentation as a function of the users' interactions with the system on different platforms. In particular we will examine examples of the adaptation of navigation, presentation and content of the user interface. The following tables (Tables 11.1, 11.2 and 11.3) show when a rule comes into force and the effect on interactive system behaviour. It is possible to relate such rules to the identification of interaction patterns directly from the end-user experience [Seffah and Javahery 2003].

11.7.1. NAVIGATION AS A FUNCTION OF TASK FREQUENCY

Here we discuss how the system handles the situations where the user always repeats the same sequence of tasks. For example, we can consider when the user selects a set of domain objects associated with a general topic and then a more refined subset iteratively (see Figure 11.12).

The recurrent selection of a specific type of artwork (e.g. made of bronze, defined as full relief sculpture, etc.), followed by a more specific selection (e.g. bronze artworks from the 20th century, full relief sculpture by the artist Vatteroni, etc.) causes the appearance in the interface of a new link for direct access to the subclass: 'Bronze artworks in XX Century' or 'Vatteroni's full relief sculpture'. This link will appear until the user has visited all the artworks belonging to that subset or until the system detects different preferences.

We can follow the corresponding changes in the user model: for each task there is an attribute that represents the possibility of that task's being *merged*, an indication of the

Table 11.1. Rules for adaptive navigation.

If...	Then...
The user always performs the same sequence of tasks in order to reach a goal	Change the navigation support so as to reduce the time required to achieve the goal
The user performs a task on one platform and then accesses the application through another platform	Change the user model state to enable or disable certain tasks
The user never selects a task (for example, a link selection) during one or more sessions on any platform	Hide the task support from all platforms (for example, remove link)
Mobile context: The user is near an object of interest in the physical world	Advise users through their mobile device
Mobile context: The user is following a physical path in the environment	Determine the next object of interest for users based on their preference and location
The user often selects a domain object set that satisfies a given rule (for example belonging to a city zone, with the same characteristics, etc.)	Change navigation modality so as to enable the related tasks.
The user never selects a domain object set that satisfies a given rule (for example belonging to a city zone, with the same characteristics, etc.)	Change navigation modality so as to disable the related tasks.

Table 11.2. Rules for adaptive presentation.

If...	Then...
The user often selects a domain object (independently of the task order and platform)	Provide access to this object or attribute in a high priority position.
The user never selects a domain object (independently of the task order and platform)	Provide access to this object or attribute in a low priority position.
The user often performs the same type of tasks	Change the presentation according to the most frequently used task types

task to which it can be connected and the new name to be given to this unified task as well as the number of accesses, object instance and object subsets selected.

In the previous example (the recurrent selection of bronze artworks and then bronze artworks from the XX Century), this will generate a link *Bronze artworks in XX Cen.*, in both the desktop and phone interfaces in which the user can select the material. During dynamic generation of the user interface, the system first analyses the XML file content and then generates the links.

Table 11.3. Rules for adaptive content.

If...	Then...
The user has already seen a specific domain object and then accesses a similar object	Change the content so as to explain the difference or similarity as compared to the previously seen object
The user shows advanced knowledge of a certain topic	Increase the detail in the description of the elements of interest, within the constraints of the current device

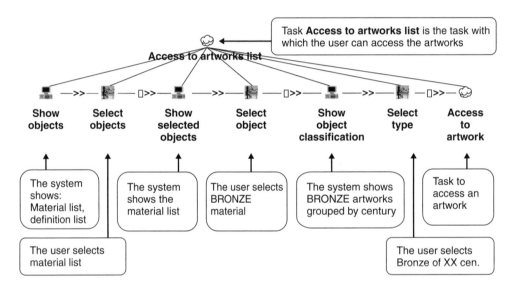

Figure 11.12. A task model for two-stage selection of objects.

Another example is when the user never performs certain tasks during a session or during different sessions. In this case the system will remove the tasks in question. Thus, if the task is never performed over one or more sessions in any platform (assuming that it is defined for multiple platforms), it can be disabled by setting the corresponding attribute.

11.7.2. NAVIGATION AS A FUNCTION OF TASK PERFORMANCE

Tasks performed in a specific platform can generate a change in the task model for another platform. For example, let us consider a scenario where the user previously selects a tour on a desktop computer, indicates preferences for a city zone and then accesses the application through a cell phone.

When the user selects a tour on the desktop computer, via either a map or the predefined link, the task *Follow the desktop selected route* in the user model will be modified (see Figure 11.13). The corresponding *Disabled* attribute, previously set to true, will be set to false, and the corresponding object instance will be the tour chosen by the user.

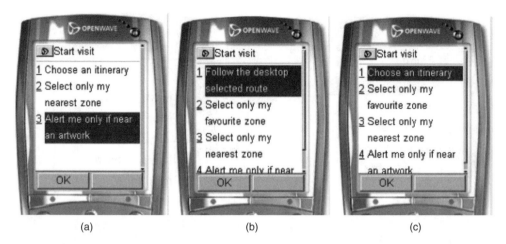

(a)   (b)   (c)

Figure 11.13. Access to the application (a) for the first time, (b) after a desktop visit that selected a tour and (c) after a desktop visit that did not select a tour.

For the reverse case, from the mobile platform the user chooses the option of selecting the artworks seen during the visit, so as to view descriptions and details later on at home using the desktop computer. This will enable the task *More Information about artworks visited* in the desktop platform, and each of the artworks selected will be added as the objects corresponding to that task.

11.7.3. MODIFICATION OF PRESENTATION

The following example demonstrates a change in presentation for a task whose objects are the artworks located in the historic city center. The user can access these artworks by choosing one of the following alternatives: Streets, Buildings, Churches, or Squares. Suppose that the user often chooses 'Streets'. The user model contains the choice task whose objects correspond to the artworks of the city, along with the specific platforms from which each object can be accessed.

More generally, the user model also contains the objects manipulated by each task as well as the platforms supporting each object. For each user choice, the system stores the objects selected in the user model. In the example mentioned above, the user first selects 'artworks in Carrara city' and then the object 'Streets'. The recurrent choice of this attribute will cause a change in the order of items in the corresponding list (see Figure 11.14).

In summary, if the user selects an object on one platform, this will cause a change in the sequence of all lists containing that object, across all platforms.

11.7.4. MODIFICATION OF CONTENT PRESENTATION

In one of the rules in Table 11.1, if the user frequently accesses a domain object, this causes a modification of the content presentation and an updating of the user's knowledge

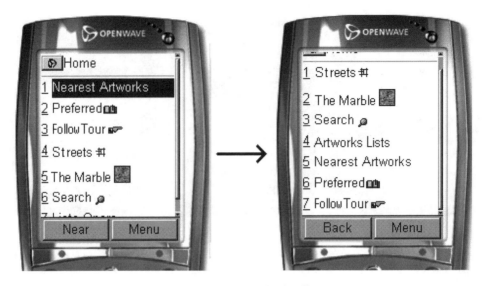

Figure 11.14. An adaptive list.

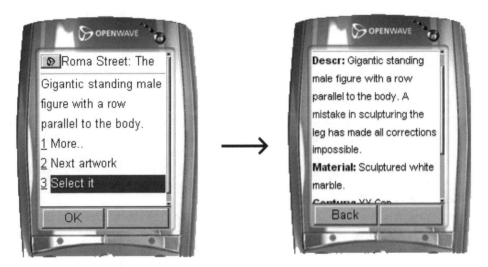

Figure 11.15. The user frequently asks for more information; the system generates it automatically.

level (while maintaining the same navigation path). For example, we can consider a scenario where the user accesses the description of an artwork and frequently asks for more information about that work.

The solution consists of introducing the ability to perform the same low-level task in the task hierarchy without performing the intermediate tasks. In the example in Figure 11.15, this means that the user can directly access detailed information about a work of art.

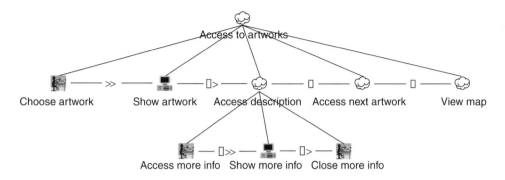

Figure 11.16. The task model for adaptive content.

Figure 11.16 shows the resulting task model. For each task that can be *Hideable* and is performed multiple times, we can conceal that task within the *Show Artwork* task. In the example, this means that when the system shows information on the artwork, it already includes detailed information. At the same time, the knowledge level of the user is updated (the user always accesses more information). When the user accesses the system from any platform, the knowledge level will be inherited.

In order to avoid user misunderstandings or confusion because of the adaptive support, it is possible to clearly indicate what part of the user interface is adaptive. For example, in a cell phone, a soft key can highlight the individual adaptive links or call up an adaptive list of frequently accessed links.

11.8. CONCLUSIONS

This chapter discussed how to provide adaptive support for multiple platforms based on task and user modelling techniques. The method was illustrated through a case study in the museum application domain.

In particular, this chapter addressed the use of task models at design time and their relationships with user models. A set of rules was introduced, based on the user model, for modifying presentation and dialogue as a function of users' interactions on different platforms. These rules allow applications to better support users' goals.

Future work will be dedicated to analysing in more detail whether adaptivity, especially in mobile phones, can sometimes disorient users. We will perform studies to determine how to introduce adaptivity in a way that avoids disorientation.

ACKNOWLEDGEMENTS

This work has been partially supported by the CAMELEON project (http://giove.cnuce. cnr.it/cameleon.html).

REFERENCES

Booch, G., Rumbaugh, J., and Jacobson, I. (1999) *Unified Modeling Language Reference Manual*. Addison Wesley.

Brusilovsky, P. (1996) Methods and techniques of adaptive hypermedia. *User Modelling and User Adapted Interaction*, 6(2–3), 87–129. URL: http://www.cntrib.andrew.cmu.edu/plb/UMUAI.ps.

Mori, G., Paternò, F., and Santoro, C. (2002) CTTE: Support for Developing and Analysing Task Models for Interactive System Design. *IEEE Transactions on Software Engineering*, 28 (8), 797–813.

Mori, G., Paternò, F., and Santoro, C. (2003) Tool Support for Designing Nomadic Applications, *Proceedings of ACM IUI 2003 International Conference on Intelligent User Interfaces*, January 12–15, 2003, Miami, FL, USA, 141–8. ACM Press.

Mullet, K., and Sano, D. (1995) *Designing Visual Interfaces*. Prentice Hall.

Oppermann, R., and Specht, M. (2000) A Context-Sensitive Nomadic Information System as an Exhibition Guide. *Proceedings of the Handheld and Ubiquitous Computing Second International Symposium*, Bristol, UK, September 25–27, 2000, 127–42.

Paternò, F. (1999) *Model-based Design and Evaluation of Interactive Applications*. Springer Verlag.

Paternò, F., and Leonardi, A. (1994) A Semantics-based Approach to the Design and Implementation of Interaction Objects. *Computer Graphics Forum*, 13(3), 195–204.

Paternò, F., and Santoro, C. (2002) One Model, Many Interfaces. *Proceedings of the 4th International Conference on Computer-Aided Design of User Interfaces CADUI 2002*, May 15–17, 2002, Valenciennes, Belgium, 143–54. Kluwer Academics, Dordrecht.

Seffah, A., and Javahery, H. (2003) Multiple User Interfaces: Definitions, Challenges and Research Opportunities, Chapter 2 in this book.

Part V

Architectures, Patterns, and Development Toolkits

Migrating User Interfaces Across Platforms Using HCI Patterns

Homa Javahery, Ahmed Seffah, Daniel Engelberg, and Daniel Sinnig

Human-Centered Software Engineering Group Department of Computer Science,
Concordia University, Canada

12.1. INTRODUCTION

User interfaces are an important part of any interactive software system. In fact, the user interface (UI) occupies a large share of the total size of a typical system [Myers 1993]. A major milestone in interactive system evolution was the shift from text-based interfaces to more complex graphical user interfaces. The web, as a vital medium for information transfer, has also had a major impact on UI design. It emphasized the need for more usable interfaces that are accessible by a wider range of people. Recently, the introduction of new platforms and devices, in particular mobile phones and Personal Digital Assistants (PDAs), has added an extra layer of complexity to UI system changes. In the migration of interactive systems to these new platforms and architectures, modifications have to be made to the UI while ensuring the application of best design practices. We will demonstrate how this can be achieved through the use of Human-Computer Interaction (HCI) Patterns.

Multiple User Interfaces. Edited by A. Seffah and H. Javahery
© 2004 John Wiley & Sons, Ltd ISBN: 0-470-85444-8

As a starting point, let us take the example of web applications, which are usually designed for a standard desktop computer with a web browser. With the rapid shift toward wireless computing, these web applications need to be customized and migrated to different devices with different capabilities. We need to rethink the strategies for displaying information in the context of devices with smaller and lower-resolution screens. As an illustration, an airline reservation system might separate the tasks of choosing a flight and buying the ticket into two separate screens for a small PDA. However, this separation is not required for a large screen. Furthermore, the PDA interface might eliminate images or it might show them in black and white. Similarly, text might be abbreviated on a small display, although it should be possible to retrieve the full text through a standardized command. For all these situations, HCI patterns facilitate the transition to different devices while ensuring that constraints are taken into account, and that usability is not compromised.

Figure 12.1 illustrates how HCI patterns can be applied to display the CNN site (www.cnn.com) on different devices. Although the basic functionality and information content are the same for all three platforms (desktop, PDA and mobile phone), they have been adapted according to the context of use and the limitations of each platform. Depending on the device and the constraints imposed, the presentation of the site will be different. HCI patterns help designers choose appropriate presentations for the design of each interface. In Figure 12.1, different pattern implementations are used to address the same navigation problem, which is how to assist the user in reaching specific and frequently-visited pages.

Figure 12.1. HCI patterns in a MUI framework.

In this chapter, we will address existing UIs that require migration to different platforms. Two approaches can be used when migrating UIs to other platforms: reengineering and redesign. The emphasis of this chapter is not on the differences between the methods, but rather on how both methods can benefit from the use of HCI patterns:

- Reengineering is a technique that reuses the original system with the goal of maintaining it and adapting it to required changes. It has a fundamental goal of preserving the knowledge contents of the original system through the process of evolving it to its new state. In the process of concretely applying the reengineering, HCI patterns can be used to abstract and redeploy the UI onto different platforms. Reengineering in itself is a complex undertaking, and therefore tools are needed to support the transition so as to limit time and costs.
- Redesign is a simplified version of reengineering, and can be more practical than reengineering in certain contexts. Redesign using patterns involves a direct transformation of patterns, as an example, from a desktop-based set of HCI patterns to a PDA-based set of patterns. In contrast to reengineering, there is no intermediate step of creating a platform-independent UI model. The consequences of this simplification are described later in this chapter.

The following section gives a brief overview of HCI patterns. In Section 12.3, we summarize the steps involved in redesigning UIs with pattern mapping and present a case study. In Section 12.4 we discuss research directions for the use of HCI patterns in reengineering, the differences between reengineering and redesign, and for which context of development each approach is best suited. Finally, we discuss future investigations and conclude our work in Section 12.5.

12.2. A BRIEF OVERVIEW OF HCI PATTERNS

The architect Christopher Alexander introduced the idea of patterns in the early 1970s [Alexander 1979]. His idea stemmed from the premise that there was something fundamentally incorrect with 20th century architectural design methods and practices. He introduced patterns as a three-part rule to help architects and engineers with the design of buildings, towns, and other urban structures. His definition of a pattern was as follows: 'Each pattern is a three-part rule, which expresses a relation between a certain context, a problem, and a solution' [Alexander 1979]. The underlying objective of Alexander's patterns was to tackle architectural problems that occurred over and over again in a particular environment, by providing commonly accepted solutions. The concept of patterns became very popular in the software engineering community with the wide acceptance of the Gang of Four's book '*Design Patterns: Elements of Re-usable Object-Oriented Software*' [Gamma *et al.* 1995]. During the last three years, the HCI community has been a forum for discussion on patterns for user interface design. An HCI pattern is an effective way to transmit experience about recurrent problems in the HCI domain, which includes UI design issues. A pattern is a named, reusable solution to a recurrent problem in a particular context of use.

The following are the motivations for using patterns as a **tool** for redesigning or reengineering an existing user interface.

First, there exist a number of HCI pattern catalogues that carry a significant amount of reusable design knowledge. Many groups and individuals have devoted themselves to the development of these catalogues, and examples include *Common Ground, Experiences*, and *Interaction Design Patterns* [Tidwell 1999; Coram and Lee 1998; Welie 2002]. Some suggest a classification of their pattern catalogues according to the type of application [Tidwell 1999], while others tailor their catalogue of patterns for a specific platform [Welie 2002]. In addition, Mahemoff and Johnston [1999] propose the *Planet Pattern Language* for internationalizing interactive systems. This language addresses the high-level issues that developers encounter when specifying requirements for international software. It helps developers document and access information about target cultures and shows them how these resources can help them to customize functionality and user interface design.

Secondly, HCI patterns have the potential to drive the entire UI design process [Borchers 2000; Lafrenière and Granlund 1999; Javahery and Seffah 2002]. HCI patterns deal with all types of issues relating to the interaction between humans and computers, and apply to different levels of abstraction. Depending on the type of application, they can be categorized according to different UI facets, such as Navigation, Information/Content, and Interaction (which includes forms and other input components) for web applications. For software developers unfamiliar with newly emerging platforms, patterns provide a thorough understanding of context of use and examples that show how the pattern applies to different types of applications and devices. Some researchers have also suggested adding implementation strategies and information on how a pattern works, why it works (rationale), and how it should be coded [Javahery and Seffah 2002; Welie *et al.* 2000].

Table 12.1. A description of the Quick Access pattern.

Pattern Name	Quick Access
Type	Navigation support in small and medium web sites
Context of use	• Useful for the novice and expert user
	• Assists the user to reach specific pages from any page and at any time
	• Menu to reflect important web site content
Consequences	• Decreases memory (cognitive) load
	• Increases web page accessibility
	• Increases subjective user satisfaction and trust
Solution	• Groups most convenient action links such as *Top Stories, News, Sports*, etc. for a News site
	• Uses meaningful metaphors and accurate phrases as labels
	• Places it consistently throughout the whole web site
Implementation strategy	• Implemented as a GUI toolbar for traditional desktop applications
	• Implemented as a combo-box or a pop-up menu for small size screens such as PDA and some mobile phones (depends on screen size and device capabilities)
	• Implemented as a selection for mobile phones

Thirdly, HCI patterns are an interesting reengineering tool because the same pattern can be implemented differently on various platforms. For example, the Quick Access pattern in Table 12.1 helps the user reach specific pages, which reflect important web site content, from any location on the site. For our news example (Figure 12.1), it can provide direct and quick access to central pages such as *Top Stories*, *News*, *Sports*, and *Business*. For a web browser on a desktop, it is implemented as an *index browsing toolbar* using embedded scripts or a Java applet in HTML. For a PDA, the Quick Access pattern can be implemented as a *combo box* using the Wireless Markup Language (WML). For a mobile phone, the Quick Access pattern is implemented as a *selection* [Welie 2002] using WML. Pattern descriptions should provide advice to pattern users for selecting the most suitable implementation for a given context.

For the sake of simplicity, we will refer to HCI patterns simply as 'patterns' for the duration of this chapter.

12.3. REDESIGNING USER INTERFACES WITH PATTERN MAPPING

As illustrated in Figure 12.2, using the traditional GUI as a starting point, it is possible to redesign the UI for platform migration, by using *pattern mapping*. The patterns of the existing GUI are transformed or replaced in order to redesign and re-implement the user interface. Since patterns hold information about design solutions and context of use, platform capabilities and constraints are implicitly addressed in the transformed patterns.

To illustrate the use of patterns in the redesign process, in what follows, we describe the fundamentals of the pattern-based redesign process as illustrated in Figure 12.2.

12.3.1. THE EFFECT OF SCREEN SIZE ON REDESIGN
Different platforms use different screen sizes, and these different screen sizes afford different types and variants of patterns. In this section we will address how the change

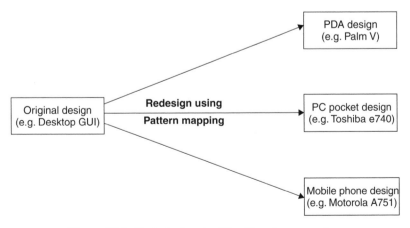

Figure 12.2. Redesigning the UI with pattern mapping.

Table 12.2. Screen size of PDAs and desktops.

Device	Screen Area (cm^2)	Pixels
Standard PDA	35–45	25,000–100,000
Standard desktop computer	600–900	480,000–786,000

in screen size between two platforms affects redesign at the pattern level. We will focus on the redesign of desktop architectures to PDA architectures, as a function of their difference in screen size. This section provides a framework for the redesign of navigation architectures at the presentation layer of design. The amount of information that can be displayed on a given platform screen is determined by a combination of area and number of pixels, as illustrated in Table 12.2.

Comparing area, a standard desktop monitor offers approximately 20 times the area of a typical PDA. If we compare pixels, the same desktop monitor has approximately 10 times the pixels of the PDA.

The total difference in information capacity between platforms will be somewhere between these two measures of 20 times the area and 10 times the pixels. We can conclude that to transform a desktop display architecture to a PDA display architecture, the options are as follows:

1. To reduce architecture size, it is necessary to significantly reduce both the number of pages and the quantity of information per page.
2. To hold constant the architecture size (i.e. topics or pages), it is necessary to significantly reduce the quantity of information per page (by a factor of about 10 to 20).
3. To retain the full amount of information in the desktop architecture, it is necessary to significantly increase the size of the architecture, since the PDA can hold less information per page.

The choice of transformation strategy will depend on the size of the larger architecture and the value of the information:

- For small desktop architectures, the design strategy can be weighted either toward reducing information if the information is not important, or toward increasing the number of pages if the information is important.
- For medium or large desktop architectures, it is necessary to weight the design strategy heavily toward reducing the quantity of information, since otherwise the architecture size and number of levels would rapidly explode out of control.

Finally, we can consider transformation of patterns and graphical objects in the context of the amount of change that must be applied to the desktop design or architecture to fit it into a PDA format. The list is ordered from the most direct to the least direct transformation:

1. *Identical.* For example, drop-down menus can usually be copied without transformation from a desktop to a PDA.

2. *Scalable* changes to the size of the original design or to the number of items in the original design. For example, a long horizontal menu can be adapted to a PDA by reducing the number of menu elements.
3. *Multiple* of the original design, either simultaneously or sequentially in time. For example, a single long menu can be transformed into a series of shorter menus.
4. *Fundamental* change to the nature of the original design. For example, permanent left-hand vertical menus are useful on desktop displays but are not practical on most PDAs. In transformation to a PDA, left-hand menus normally need to be replaced with an alternative such as a drop-down menu.

This taxonomy of transformation types is especially relevant to the automation of cross-platform design transformation since the designs that are easiest to transform are those that require the least transformation. The taxonomy therefore identifies where human intervention will be needed for design decisions in the transformation process. In addition, when building a desktop design for which a PDA version is also planned, the taxonomy indicates which patterns to use in the desktop design to allow easy transformation to the PDA design.

12.3.2. PATTERN-BASED REDESIGN: A CASE STUDY WITH NAVIGATION PATTERNS

In this section, we discuss the use of patterns in design transformations from desktop to PDA platforms. The method transforms a core set of patterns based on various factors, including screen size and architecture size.

For our case study, we will consider transformations for the patterns for navigation outlined in Table 12.3. This list is far from exhaustive, but helps to communicate the flavour and abstraction level of patterns for navigation that we are targeting. Due to space limitations, we can only provide the title and a brief description, rather than the full description format as described in [Borchers 2002].

Figure 12.3 illustrates some of the navigation patterns from Table 12.3 as extracted from the existing home page of a desktop-based Web portal (www.cbc.ca). Once these patterns are extracted from the desktop-based architecture, they can be transformed and re-applied in a PDA architecture.

Table 12.4 describes the types of cross-platform transformations that are recommended for the HCI patterns in Table 12.3, and which can be used to redesign the CBC News site. These transformations offer the closest and simplest equivalent in the corresponding platform. In the third column, the suffix 's' after a pattern indicates 'scaled (down)', and the suffix 'm' indicates 'multiple (sequence)'.

Figure 12.4 demonstrates the redesigned interface of the CBC site for migrating to a PDA platform. The permanent horizontal menus at the top (P5) in the original desktop UI were redesigned to a shorter horizontal menu (P5s). In order to accommodate this change on the small PDA screen, the three different horizontal menus had to be shortened, and only important navigation items were used. The keyword search pattern (P13) remains as a keyword search. The permanent vertical menu at the left (P6) is redesigned to a drop-down menu (P15). The drop-down menu in the PDA design also includes the menu headings 'What's on today?' and 'Online features' from the temporary vertical menu (P3)

Table 12.3. Examples of HCI patterns.

	HCI Pattern	Definition or comments
P.1	Bread crumbs	Navigation trail from home page down to current page; see [Welie 2002].
P.2	Temporary horizontal menu bar at top	Displayed in a specific context (not permanent). Typically called up by an item in a left-hand vertical menu.
P.3	Temporary (contextual) vertical menu at right in content zone	Called up by a higher-level menu or a link. Might be permanent on a single page, but not repeated across the site.
P.4	Information portal	Broad first and second level on home page. Same principle as the "Directory" pattern [Welie 2002].
P.5	Permanent horizontal menu bar at top	Standard, single-row menu bar
P.6	Permanent vertical menu at left	Vertical menu repeated across all pages of a site. Can have one or multiple levels of embedding.
P.7	Progressive filtering	Allows user to reach target by applying sequential filters [Welie 2002].
P.8	Shallow embedded vertical menu	A single-level menu or a 2-level embedded menu
P.9	Sub-site	Shallow main menu or broad portal leading to smaller sub-sites with simple navigation architectures
P.10	Container navigation	Different levels of menu displayed simultaneously in separate zones (e.g. Outlook Express or Netscape Mail)
P.11	Deeply embedded vertical menu	e.g. File manager menu
P.12	Alphabetical index	Index contains hyperlinks to pages containing or describing the indexed terms
P.13	Key-word search	Search engine
P.14	Intelligent agent	Human-machine interfaces that aim to improve the efficiency, effectiveness and naturalness of human-machine interaction by representing, reasoning and acting on models of the user, domain, task, discourse and media
P.15	Drop-down menu	A menu of commands or options that appears when the user selects an item with a mouse
P.16	Hybrid navigation	Start with key-word search, then present details of search target in menu format

in the original desktop design. Finally, the information portal (P4), which is the first thing that captures the user's attention, is redesigned to a smaller information portal (P4s).

12.3.3. ARCHITECTURE SIZE AS AN ADDED VARIABLE IN REDESIGN

Up to this point the role of only one factor affecting MUI redesign has been analysed, namely screen size. Another relevant and analogous factor is architecture size. As discussed in [Engelberg and Seffah 2002], different sizes of architecture require different HCI patterns for navigation. The factor of architecture size can be added to (or rather

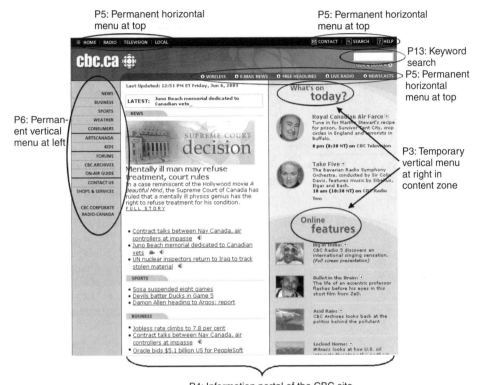

Figure 12.3. Patterns extracted from the CBC News site.

multiplied by) the factor of screen size to obtain a more general framework for MUI redesign at the level of patterns.

Table 12.5 outlines the redesign approach from an existing large display to a PDA platform, taking into account both the size of architecture and the size of the display. This table is based partly on an analysis performed in [Engelberg and Seffah 2002]. In the first column, the architecture size of the desktop UI is approximated by the number of hierarchical levels in the information architecture and assumes the typical case of a tree-shaped architecture. The patterns in the third column of Table 12.5 refer to the first column of Table 12.4. The solutions in the last column of Table 12.5 refer to the transformations in the last column of Table 12.4.

Usability compromises in migration: The last column of Table 12.5 displays the results of the three possible size-based transformations to an architecture referred to earlier in this chapter. In the migration from a desktop architecture to a PDA architecture, there is less room to display information, and therefore a usability compromise must be made:

1. To reduce architecture size, it is necessary to significantly reduce both the number of pages, hence the number of topics, and the quantity of information per page. This is

Table 12.4. Examples of HCI pattern transformations for different screen sizes.

HCI pattern in desktop display	Type of transformation	Replacement pattern in small PDA display
P.1 Bread crumbs	Scalable or fundamental	P.1s – Shorter bread crumb trail; P.15 – Drop-down 'History' menu.
P.2 Temporary horizontal menu	Scalable or fundamental	P.2s – Shorter menu; P.3 – Link to full-page display of menu options ordered vertically
P.3 Temporary vertical menu in content zone	Identical, scalable or fundamental	P.3 – Temporary vertical menu in content zone; P.3s – Shorter temporary vertical menu; or P.15 – Drop-down menu
P.4 Information portal	Scalable	P.4s – Smaller information portal
P.5 Permanent horizontal menu at top	Scalable or fundamental	P.3 – Link to full-page display of menu options ordered vertically P.5s – Shorter horizontal menu at top;
P.6 Permanent vertical menu at left	Fundamental	P.5s – Shorter horizontal menu at top P.15 – Drop-down menu
P.7 Progressive filtering	Identical	P.7 – Progressive filtering
P.8 Shallow embedded vertical menus	Identical or fundamental	P.3m – Sequence of temporary vertical menus in content zone; P.8 – Shallow embedded vertical menus
P.9 Sub-site	Scalable or fundamental	P.3 – Temporary vertical menu in content zone; P.9s – Smaller sub-site;
P.10 Container navigation (3 containers)	Scalable or fundamental	P.10s – Container navigation (2 containers); P.7 – Progressive filtering
P.11 Deeply embedded vertical menus	Multiple or fundamental	P.3m – Sequence of single-level menus; P.8m – Sequence of shallow embedded menus
P.12 Alphabetical index	Scalable	P.12s – Alphabetical index (less items per page or smaller index)
P.13 Key-word search	Identical	P.13 – Key-word search
P.14 Intelligent agents	Identical	P.14 – Intelligent agents
P.15 Drop-down menu	Identical, scalable or fundamental	P.15 – Drop-down menu; P.15s – Shorter drop-down menu; Hyperlink to P.3 – Temporary vertical menu in content zone
P.16 Hybrid navigation: Key-word search	Identical or scalable	P.16s – Hybrid approach with smaller or fewer deeply embedded menus

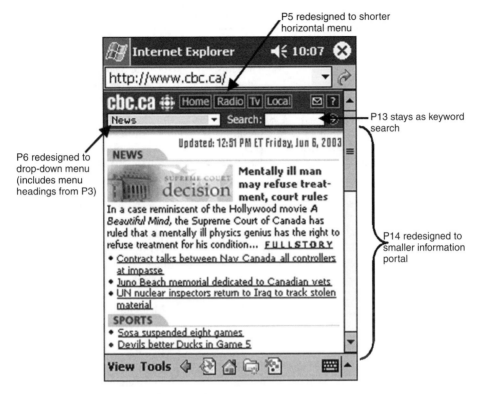

Figure 12.4. Migration of the CBC site to a PDA platform using pattern mapping.

often the preferred choice for migration to a PDA platform and reflects a high priority placed on ease of navigation.

2. To maintain a constant architecture size (i.e. pages or number of topics), it is necessary to significantly reduce the quantity of information per page; however the number of pages (topics) can be kept constant. (It would not normally be sensible to employ the other option, i.e. reduce the number of pages and keep quantity of information per page constant). This choice reflects a balance of priorities between contents and ease of navigation.

3. To retain the full amount of information in the desktop architecture, it is necessary to significantly increase the size of the architecture (number of pages), since the PDA can hold less information per page. This option is rarely desirable, but could be necessary in certain contexts.

Thus the last column of Table 12.5 presents the set of HCI design patterns to be used for navigation in a PDA architecture, as a combined function of:

1. Screen size: How the HCI design patterns used for navigation in desktop systems are transformed into PDA-compatible patterns (see Table 12.4),

Table 12.5. Redesign as a function of architecture size.

Architecture size of desktop UI	Best approach for navigation design	HCI navigation patterns (desktop display)	HCI navigation patterns for use in a PDA display (*Note that Small, Medium and Large refer to size of the PDA architecture, not the size of the PDA display*)
Small (2 levels)	Model (menu) based approach: Simple 1- and 2-level main menus	**Small (desktop)** = {P.3, P.5, P.6, P.8}	To reduce architecture size: Reduce both info per page and number of pages, and use the **Small (PDA)** pattern set. To hold architecture size constant: Reduce info per page and transform the Small (desktop) set, resulting in **Small (PDA)** = {P.3, P.3m, P.3s, P.5s, P.8, P.15}. To hold information constant: Architecture size increases, so transform the next larger (i.e. Medium) desktop set, resulting in **Medium (PDA)** = {P.1s, P.2s, P.3, P.3m, P.3s, P.4s, P.5s, P.7, P.8, P.10s, P.12s, P.15, P.15s}.
Medium (4 levels)	Model (menu) based approach: Simple 1- and 2-level menus calling up contextual (temporary) menus for the deeper levels; Alphabetical index	**Medium (desktop)** = {P.1, P.2, P.3, P.4, P.5, P.6, P.7, P.8, P.10, P.12, P.15}	To reduce architecture size: Reduce both info per page and number of pages, and transform the Small desktop set, resulting in **Small (PDA)** = {P.3, P.3m, P.3s, P.5s, P.8, P.15}. To hold architecture size constant: Reduce info per page and then transform the Medium desktop set, resulting in **Medium (PDA)** = {P.1s, P.2s, P.3, P.3m, P.3s, P.4s, P.5s, P.7, P.8, P.10s, P.12s, P.15, P.15s}. To hold information constant: Architecture size increases, so transform the next larger (i.e. Large) desktop set, resulting in **Large (PDA)** = {P.1s, P.2s, P.3, P.3m, P.3s, P.4s, P.5s, P.7, P.8, P.8m, P.9s, P.10s, P.12s, P.13, P.14, P.15, P.15s, P.16s}.

Table 12.5 (*continued*)

Archi-tecture size of desktop UI	Best approach for navigation design	HCI navigation patterns (desktop display)	HCI navigation patterns for use in a PDA display (*Note that Small, Medium and Large refer to size of the PDA architecture, not the size of the PDA display*)
Large (6+ levels)	Menu-based approaches adapted to deep menus;	**Large (desktop) =** {P.1, P.2, P.3, P.4, P.5, P.6, P.7, P.8, P.9, P.10, P.11, P.12, P.13, P.14, P.15, P.16}	To reduce architecture size: Reduce both info per page and number of pages, and then transform the Medium desktop set, resulting in **Medium (PDA)** = {P.1s, P.2s, P.3, P.3m, P.3s, P.4s, P.5s, P.7, P.8, P.10s, P.12s, P.15, P.15s}.
	Data-based (bottom-up) approaches (depending on efficiency and context of use), e.g. key-word search and alphabetical index;		To hold architecture size constant: Reduce info per page and then transform the Large desktop set, resulting in **Large (PDA)** = {P.1s, P.2s, P.3, P.3m, P.3s, P.4s, P.5s, P.7, P.8, P.8m, P.9s, P.10s, P.12s, P.13, P.14, P.15, P.15s, P.16s}.
	Hybrid approach: (see pattern P.16 in Table 12.3)		To hold information constant: Architecture size will need to increase, however essentially the same patterns are used as for **Large (PDA).**

2. The architecture size of the desktop (source) architecture, and
3. The three possible usability compromises in migration listed in the previous paragraph.

Up to this point, we have demonstrated how patterns can be used as a redesign tool for migrating a web portal from a desktop platform to a PDA. The problem with redesign, however, is that the same exercise has to be repeated for each platform. If we were to have a generic UI model of the application or web site, migrating to different platforms would be facilitated. In such a case, design strategies and content-related information would be separate from presentation issues. The next section introduces our ideas for using patterns in UI reengineering to try to come up with a UI model, which can then be instantiated to different platforms. What we propose in the following section is a future perspective on how patterns can be applied to reengineering user interfaces.

12.4. RESEARCH DIRECTIONS FOR THE USE OF PATTERNS IN REENGINEERING USER INTERFACES

Reengineering consists of a reverse engineering phase, a transformation phase, and a forward engineering phase [Moore 1996]. Figure 12.5 illustrates how reengineering can be performed with HCI patterns. The process begins with a user interface (e.g. desktop in Figure 12.5) to be reengineered, or migrated to different platforms. The reengineering steps are as follows:

1. *Reverse engineering*: Consists of a pattern extraction and abstraction phase, resulting in the creation of a platform-independent UI model.
2. *Transformation*: Patterns and design strategies in the platform-independent UI model are analysed, and transformed if appropriate.
3. *Forward engineering*: The platform-independent UI model is first instantiated to different platforms based on constraints and capabilities; the patterns are then implemented on different platforms.

In what follows, we will further clarify the above points and detail the process of our proposed pattern-assisted reengineering method.

Figure 12.5. Pattern-assisted UI reengineering.

12.4.1. PATTERN-ASSISTED REENGINEERING

In pattern-assistedReengineering reengineering, we create a platform-independent UI model that can be instantiated to different platforms. The main benefit of such a model is that it captures content-related information, design strategies, and context of use attributes, independent of the device. If we want to add features or somehow change the design, this will be reflected on all devices. Applying certain presentation rules and methods which are platform-dependent onto this model results in components suited for defined platforms. Once the platform components are defined, the layout can be defined according to screen size, resolution, and other device-specific constraints. To facilitate the reengineering process, patterns can be used as a tool since they encapsulate design knowledge with different platform-specific implementation schemes. Applying appropriate HCI patterns in reengineering can make the process of design easier, and will result in fewer usability errors. In addition, since patterns are context-oriented, their use will ensure that the best solution has been applied.

The first step of reverse engineering consists of extracting patterns from the original UI. In the process of pattern extraction, we match this knowledge against known patterns and identify which patterns were used in the original interface. If we take the example of a web portal or any web application, the extracted patterns can be logically grouped into the following descriptive UI facets: (1) Navigation (2) Information or Content (3) Interaction (such as forms and other input components). This step is identical to the pattern extraction step in redesign (Figure 12.3).

During the abstraction step of reverse engineering, extracted patterns are abstracted into higher levels of design concepts and goals, generally referred to as design strategies. The aim is to create a platform-independent UI model that can then be instantiated to different platforms. Other techniques and artefacts such as domain analysis, personae, and use cases can be applied to create a more complete UI model. The platform-independent UI model has the following characteristics:

- Includes design strategies. An example of a design strategy is query-based navigation versus conceptual model-based navigation.
- Includes patterns that are abstracted sufficiently to become platform-independent. An example is the Quick Access pattern (Table 12.1).
- This step clearly separates content and design from presentation issues.

Moore [1996] defines the UI transformation phase as consisting of transforming an 'abstract model' into a 'restructured abstract model', with human analyst input. In our approach, during the transformation step, patterns and design strategies in the model are analysed to determine suitability to any new design requirements. Inappropriate patterns and design strategies are replaced by appropriate patterns, or removed. In addition, new patterns can be added to the model based on user requirements and task-based changes. It is important to differentiate between transformation at a higher level of design, which is platform and implementation-independent, and transformation at a lower level of design, which deals with presentation issues. Since our objective is to create a generic UI model, presentation issues and implementation strategies are not taken into account during the transformation phase, but rather, during forward engineering. If we consider the Quick

Access pattern, a design decision could include the addition of this pattern during the transformation phase. However, presentation and implementation details of this pattern are abstracted away until platform constraints and capabilities are taken into account.

In the first step of forward engineering, the platform-independent UI model is instantiated to target devices based on their constraints and capabilities. Dialogue style, look-and-feel, and presentation issues are considered at this level. Depending on the device, different presentation components (or implementation strategies) may apply for each pattern and design strategy in the UI model. For example, if we go back to the Quick Access pattern and want to apply it to a PDA, the combo box will be used. However, for the PC, the toolbar is the appropriate implementation strategy.

Another example is the progress pattern, which can be applied if 'the user wants to know whether or not the operation is still being performed as well as how much longer the user will need to wait' [Welie 2002]. The instantiation of this pattern for a desktop application can include parameters such as the estimated time left, the transfer rate, and the progress bar. A mobile phone may use the same process, but some parameters have to be left out due to platform constraints. In such a case, only the percentage of time left to complete the task may be displayed.

A final example is the search pattern. On the PC, this pattern can be instantiated using a complex dialogue box including different text fields, checkboxes and radio buttons, whereas on the PDA, a simple search input box is appropriate. Simple search can be integrated in most devices since it does not need much space. Having rules that highlight such details can make the process of MUI migration easier.

Another relevant consideration for MUI migration is the ability to preview images. For instance, when running an application on a mobile device, such as a mobile telephone, we may not have the ability to preview images due to the platform constraints of smaller screen size and lower resolution. The preview pattern is appropriate for the desktop, but not for the mobile phone, and should therefore not be implemented during forward engineering.

The second step of forward engineering is pattern implementation. In this step, instantiated patterns for the target device are applied, combined and coded, resulting in the new UI. The new UI will be better suited to new requirements, since patterns are context-oriented. Pattern-assisted reengineering simplifies the process of reengineering the UI for a new context of use. However, this method does not cover all aspects and is not complete. Currently, we are far from the point where defined patterns cover all possible situations and where a model can be created purely by combining existing patterns.

12.4.2. COMPARING REENGINEERING TO REDESIGN

In contrast to redesign, which was introduced in Section 12.3, reengineering is useful for the following reasons:

1. By creating a platform-independent UI model, reengineering facilitates forward engineering to a broad family of different platforms. The platform-independent UI model encourages design reuse, reduces the total project workload and ensures coherence between the different members of the application family. In comparison, direct redesign

without such a model could be less efficient in the context of a large family of platforms, and does not include safeguards for maintaining coherence between applications.

2. Software systems inevitably change over time. Original requirements have to be modified to account for changing user needs, changes in system environment, as well as advances in technology. To manage the process of system change, we need to maintain the system [Sommerville 2000]. Reengineering is a technique that explores the idea of reusing the original system with the goal of maintaining it and adapting it to required changes. The creation of a platform-independent UI model can facilitate future maintenance, since elements of the model can be changed, rather than the design of each specific platform.

3. In the reverse engineering phase, the process of reengineering allows the designer to re-evaluate the user's task-goals in a broader context of use. This re-evaluation can improve the usability and utility of the system. In contrast, direct redesign does not question the task goals, and therefore can result in repeating old mistakes or missing opportunities for optimization.

Although reengineering is advantageous as outlined in the above points, it has a number of disadvantages:

1. It is time-consuming and complex in today's context of rapid software development.
2. The lack of a clearly-defined pattern taxonomy can be problematic during the abstraction stages and description of the UI model.
3. The forward engineering phase requires a set of defined rules with regard to implementation strategies of each pattern, depending on the device constraints and capabilities.

12.5. CONCLUSION AND FUTURE INVESTIGATIONS

In the current technological context, required changes to already-existing UIs are inevitable, such as migrating applications to multiple platforms. Writing code from scratch is no longer a viable solution when applying changes since it requires a large amount of resources, or is simply too risky to perform as the original knowledge may get lost. An interactive system that is up and running is often a fundamental asset to the company using it, as it carries with it a certain amount of domain knowledge and experience. To manage the process of system change, we need to maintain the system [Sommerville 2000].

To address these issues of reuse and maintenance, this chapter discussed the idea of using HCI patterns as an approach for UI redesign and migration to different platforms. The reengineering ideas presented in this chapter are a starting point and continued research will be needed to validate them. Some work in the area of reengineering patterns has begun [Ismail and Keller 1999; Ducasse et al. 2000; Beedle 1997], however there is still much that needs to be explored, especially in UI reengineering.

Using patterns can effectively fill the gap in existing methods for migrating UIs since they capture best design practices and can play a role throughout the complete redesign process. The application of patterns has a number of advantages: first, they can reduce the time required for redesign since for the most common usability and UI design problems, a pattern solution already exists; secondly, usability errors are reduced since most of the patterns have already been tested on other systems; finally, patterns help in the comprehension of the system for future maintenance.

In this chapter we have introduced redesign through pattern mapping as a simplified version of reengineering. This approach is a significant improvement over non-structured migration methods currently in use, for the following reasons:

- The method provides a standardized table of pattern transformations, thereby reducing the redesign effort and ensuring consistency in redesign.
- The standardized transformations formalize best practices in design, thereby ensuring optimal quality of the migrated user interface.
- The method helps designers in design choices associated with (1) the size of the source architecture and target architecture and (2) the amount of information to maintain in migrating from the source platform to the target platform.
- The method is simple enough to be used easily by novice designers, as compared to reengineering which currently requires a considerable degree of expertise and abstract reasoning ability.

As a prospective outlook, we have introduced pattern-assisted UI reengineering as an advancement of UI redesign. Similarly to UI redesign, HCI patterns are extracted from the user interface. However, these patterns, which are associated with various UI facets, are then abstracted into a platform-independent UI model. In order to adapt the UI to new platform-specific requirements, the platform-independent model is instantiated for various platforms, by using patterns.

For the purpose of MUI migration, UI reengineering offers the following advantages over UI redesign:

- by creating a platform-independent UI model, reengineering facilitates forward engineering to a broad family of different platforms;
- reengineering reuses the original system with the goal of maintaining it and adapting it to required changes, which facilitates future maintenance;
- through reverse engineering the designer can re-evaluate the user's task-goals in a broader context of use. This re-evaluation can improve the usability and utility of the system and avoids repeating old mistakes.

Pattern-assisted reengineering offers the very useful ability of easily extracting multiple platform-specific designs from a single generic (platform-independent) UI model. However, the current state of the art in HCI patterns and MUI research is not yet mature enough to handle all the requirements of pattern-assisted reengineering. Before generic UI pattern-based models can be defined, more research must be addressed to define the multiple levels of abstraction of patterns and to create a clear, well-structured taxonomy of HCI patterns. Thus, within a pattern-based framework, the simplified 'redesign' method proposed here is currently the most practical approach for migration of UIs between platforms.

Whether they are used in redesign or in reengineering, HCI patterns facilitate the UI migration process by encapsulating high-level design choices and rationales and allowing the designer to operate at a higher level of abstraction.

ACKNOWLEDGEMENTS

We are grateful to Dr. Peter Forbrig, Ashraf Gaffar, and Jovan Strika for their contribution to the MUI reengineering effort.

REFERENCES

Alexander, C. (1979) *The Timeless Way of Building*. Oxford University Press.

Beedle, M. (1997) Pattern based reengineering. *Object Magazine*, January, 1997.

Borchers, J.O. (2000) *A Pattern Approach to Interaction Design*, Proceedings of the DIS 2000 *International Conference on Designing Interactive Systems*, August 16–19, 2000, New York, 369–78. ACM Press.

Coram, T., and Lee, J. (1998) *Experiences: A Pattern Language for User Interface Design*, http://www.maplefish.com/todd/papers/experiences.

Ducasse, S., Debbe, R., and Richner, T. (eds) (2000) *Type-Check Elimination: Two Reengineering Patterns*. *Software Composition Group*, Bern University, http://www.iamunibe.ch/~scg.

Engelberg, D., and Seffah, A. (eds) (2002) *Design Patterns for the Navigation of Large Information Architectures*. *Proceedings of 11th Annual Usability Professional Association Conference*, July 8–12, 2002, Orlando, Florida.

Gamma, E., Helm, R., Johnson, R. and Vlissides, J. (eds) (1995) *Design Patterns: Elements of Reusable Object Oriented Software*. Addison Wesley.

Ismail K., and Keller, R. (eds) (1999) *Transformations for Pattern-based Forward Engineering*. Universite de Montreal, http://www.iro.umontreal.ca/~labgelo/Publications/Papers/sts99.pdf.

Javahery, H., and Seffah, A. (eds) (2002) *A Model for Usability Pattern-Oriented Design*. *Proceedings of TAMODIA* 2002, 18–19 July 2002, Bucharest, Romania, 104–110.

Lafrenière, D., and Granlund, A. (eds) (1999) *A Pattern-Supported Approach to the User Interface Design*. UPA Workshop Report, http://www.gespro.com/lafrenid/Workshop_Report.pdf.

Mahemoff, M., and Johnston, L. (1999) *The Planet Pattern Language for Software Internationalization*, Proceedings of Pattern Languages of Program Design (PLOP), September 15–18, 1999, Monticello, IL.

Moore, M. (1996) *Representation Issues for Reengineering Interactive Systems*. *ACM Computing Surveys*, 28(4), December 1996.

Myers, B. (1993) *Why are Human-Computer Interfaces Difficult to Design and Implement?* Carnegie Mellon University School of Computer Science Technical Report, no. CMU-CS-93-183 July 1993.

Sommerville, I. (2000) *Software Engineering*, 6th Edition. Addison-Wesley, Boston, MA.

Tidwell, J. (1999) *Common Ground: A Pattern Language for Human-computer Interface Design*. http://www.mit.edu/~jtidwell/commond_ground.html.

Welie, M.V. (2002) *Interaction Design Patterns*. http://www.welie.com/patterns.

Welie, M.V., Van der Veer, G.C., and Eliëns, A. (eds) (2000) *Patterns as Tools for User Interface Design*. *International Workshop on Tools for Working with Guidelines*, October 7–8, 2000, Biarritz, France.

Support for the Adapting Applications and Interfaces to Context

Anind K. Dey[1] and Gregory D. Abowd[2]

[1] *Intel Research, Berkeley USA*
[2] *College of Computing Georgia Institute of Technology USA*

13.1. INTRODUCTION

The typical user is no longer facing a desktop machine in the relatively predictable office environment. Rather, users have to deal with diverse devices, mobile or fixed, sporting diverse interfaces and used in diverse environments. In appearance, this phenomenon is a step towards the realization of Mark Weiser's ubiquitous computing paradigm, or 'third wave of computing', where specialized devices outnumber users [Weiser 1991]. However, many important pieces necessary to achieve the ubiquitous computing vision are not yet in place. Most notably, interaction paradigms with today's devices fail to account for a major difference with the static desktop interaction model. Devices are now often used in changing environments, yet their interfaces and services do not adapt to those changes very well. Although moving away from the desktop brings up a new variety of situations in which an application may be used, computing devices are left unaware of

Multiple User Interfaces. Edited by A. Seffah and H. Javahery
© 2004 John Wiley & Sons, Ltd ISBN: 0-470-85444-8

their surrounding environment. One hypothesis that a number of ubiquitous computing researchers share is that enabling devices and applications and their interfaces to automatically adapt to changes in their surrounding physical and electronic environments, will lead to an enhancement of the user experience.

The information in the physical and electronic environments creates a *context* for the interaction between humans and computational services. Context is any information that characterizes a situation related to the interaction between users, applications, and the surrounding environment. A growing research activity within ubiquitous computing deals with the challenges of *context-aware computing* [Moran and Dourish 2001], understanding and handling context that can be sensed automatically in a physical environment and treated as implicit input to positively affect the behavior of an application.

Apart from the demonstration of a variety of prototype context-aware applications, the majority of which are location-based services, there has been relatively little advancement in context-aware computing over the past five years. There are technology- and human-centered challenges that stem from a poor understanding of what constitutes context and how it should be represented. We lack conceptual models and tools to support the rapid development of rich context-aware applications that might better inform the empirical investigation of interaction design and the social implications of context-aware computing. The work presented here attempts to enable a new phase of context-aware application development. We want to help application developers understand what context is, what it can be used for, and to provide concepts and practical support for the software design and construction of context-aware applications.

Difficulties arise in the design, development and evolution of context-aware applications. Designers lack conceptual tools and methods to account for context-awareness. As a result, the choice of context information used in applications is very often driven by the context acquisition mechanisms available – hardware and software sensors. This approach entails a number of risks. First, the initial choice of sensors may not be the most appropriate. The details and shortcomings of the sensors may be carried up to the application level and hinder the flexibility of the interaction and further evolution of the application. More importantly, a sensor-driven approach constrains the possibilities of designers by limiting the kind of applications they are able to design and the context of uses that they can imagine.

Developers face another set of problems related to distribution, modifiability and reusability. Context-aware applications are often distributed because they acquire or provide context information in a number of different places. For example, in the Conference Assistant scenario, each room and venue of the conference must be able to sense the presence and identity of users. Although mechanisms for distribution are now mainstream on desktops and servers, they are not always appropriate for distributed networks of sensors. Indeed, context-awareness is most relevant when the environment is highly dynamic, such as when the user is mobile. Thus, context-aware applications may be implemented on very diverse kinds of computing platforms, ranging from handheld devices to wearable computers to custom-built embedded systems. As a result, context-aware applications require lightweight, portable and interoperable mechanisms across a wide range of platforms.

As with graphical user interfaces, and more crucially so given the overall lack of experience with context-aware applications, iterative development is key to creating usable context-aware applications. Thus, applications must be implemented in a way that makes it easy to modify context-related functions. There are currently few guidelines, models or tools that support this requirement. Finally, application developers should be able to reuse satisfactory context-aware solutions. However, there are no methods or tools available to make this task any easier.

Before going into the details of infrastructure requirements for handling context, we will present a more formal definition of context and describe different ways of using context to enhance or augment multiple user interfaces. Taken from our survey on context-aware computing [Dey and Abowd 2000], we define *context* as:

> *Any information that can be used to characterize the situation of entities (i.e. whether a person, place or object) that are considered relevant to the interaction between a user and an application, including the user and the application themselves. Context is typically the location, identity and state of people, groups and computational and physical objects.*

A goal of context acquisition is to determine what a user is trying to accomplish. Because the user's objective is difficult to determine directly, context cues can be used to help infer this information and to inform an application on how to best support the user. The definition we have provided is quite general. This is because context-awareness represents a generalized model of input, including both implicit and explicit input, allowing almost *any* application to be considered context-aware as long as it reacts to input.

Having presented a definition of context, we must now examine how applications can effectively use context information. In our survey on context-aware computing, we proposed a classification of context-aware functions that a context-aware application may implement. All of these functions are directly applicable to multiple user interfaces. This classification introduces three categories of functions, related to the presentation of information, the execution of services, and the storage of context information attached to other captured information for later retrieval.

The first category, *presenting information and services*, refers to applications that either present context information to the user or use context to propose appropriate selections of actions to the user. There are several examples of this class of functions in the literature and in commercially available systems: showing the user's or his/her vehicle's location on a map and possibly indicating nearby sites of interest [Abowd *et al.* 1997; Bederson 1995; Davies *et al.* 1998; Feiner *et al.* 1997; Fels *et al.* 1998; Lamming and Flynn 1994; McCarthy and Anagost 2000]; presenting a choice of printers close to the user [Schilit *et al.* 1994]; sensing and presenting in/out information for a group of users [Salber 1999]; ambient information displays [Heiner *et al.* 1999; Ishii and Ullmer 1997; Mynatt *et al.* 1998; Weiser and Brown 1997]; and providing remote awareness of others [Schmidt *et al.* 1999].

The second category, *automatically executing a service*, describes applications that trigger a command or reconfigure the system on behalf of the user according to context changes. Examples include: the Teleport system in which a user's desktop environment follows the user as he/she moves from workstation to workstation [Bennett *et al.* 1994]; car navigation systems that recompute driving directions when the user misses a turn [Hertz

1999]; a recording whiteboard that senses when an informal and unscheduled encounter of individuals occurs and automatically starts recording the ensuing meeting [Brotherton *et al.* 1999]; mobile devices enhanced with sensors that determine their context of use to change their settings and actions [Harrison *et al.* 1998; Hinckley *et al.* 2000; Schmidt *et al.* 1999]; a camera that captures an image when the user is startled as sensed by biometric sensors [Healey and Picard 1998]; and devices that deliver reminders when users are at a specified location [Beigl 2000; Lamming and Flynn 1994].

In the third category, *attaching context information for later retrieval*, applications tag captured data with relevant context information. For example, a zoology application tags notes taken by the user with the location and the time of the observation [Pascoe *et al.* 1998]. The informal meeting capture system mentioned above provides an interface to access informal meeting notes based on who was there, when the meeting occurred and where the meeting was located. In a similar vein are Time-Machine Computing [Rekimoto 1999] and Placeless Documents [Dourish *et al.* 2000], two systems that attach context to desktop or networked files to enable easier retrieval. Some of the more complex examples in this category are memory augmentation applications such as Forget-Me-Not [Lamming and Flynn 1994] and the Remembrance Agent [Rhodes 1997].

In this section, we have established a definition of context, and we have further elicited the dimensions of context information, as well as its possible uses in context-aware applications. In the remainder of this chapter, we will discuss the difficulties in using context and present three different forms of architectural support for building context-aware applications. This includes using a component-based approach called the Context Toolkit, an extension to the Context Toolkit that uses a more holistic situation-based approach, and a more specialized approach for dealing with individual types of context.

13.2. WHY CONTEXT IS DIFFICULT TO USE AND WHY SUPPORT IS NEEDED FOR IT

With the perspective of an application designer in mind, we will describe an infrastructure that automatically supports all the tasks that are common across applications, requiring the designer to only provide support for the application-specific tasks. To this end, we have identified a number of requirements that the framework must fulfill to enable designers to more easily deal with context. These requirements are: separation of concerns; context interpretation; transparent, distributed communications; constant availability of context acquisition; context storage; and resource discovery. We will now discuss each of these requirements.

13.2.1. SEPARATION OF CONCERNS

One of the main reasons why context is not used more often in applications is that there is no common way to acquire and handle context. In general, context is handled in an improvised fashion. Application developers choose whichever technique is easiest to implement, at the expense of generality and reuse. Two common ways in which context has been handled include connecting sensor drivers directly into applications and using servers to hide sensor details. With some applications [Harrison *et al.* 1998; Orr and Abowd 2000], the drivers for sensors used to detect context are directly hardwired into the applications themselves. In this situation, application designers are forced to write code that deals with the sensor details, using whatever protocol the sensors dictate. As well, there is no separation between application semantics and the sensor details, leading to a loss of generality and making it difficult to reuse sensors.

Ideally, we would like to handle context in the same manner as we handle user input. User interface toolkits support application designers in handling input. They provide the valuable widget abstraction to enable designers to use input without worrying about how the input was collected. This abstraction provides many benefits and has been used not only in standard keyboard and mouse interaction, but also with pen and speech input [Arons 1991] and with the unconventional input devices used in virtual reality [MacIntyre and Feiner 1996]. It facilitates the separation of application semantics from low-level input handling details. It supports reuse by allowing multiple applications to create their own instances of a widget. It contains not only a querying mechanism but also possesses a notification, or *callback*, mechanism to allow applications to obtain input information as it occurs. Finally, in a given toolkit, all the widgets have a common external interface, allowing applications to treat all widgets in a similar fashion, not having to deal with differences between individual widgets.

There are systems that support event management [Bauer *et al.* 1998; Schilit 1995], through the use of querying mechanisms, notification mechanisms, or both, to acquire context from sensors. However, this previous work has suffered from the design of specialized servers, which do not share a common interface, forcing an application to deal with each server in a distinct manner, resulting in a minimal range of server types being used. By separating how context is acquired from how it is used, applications can now use contextual information without worrying about the details of a sensor and how to acquire context from it. These details are not completely hidden and can be obtained if needed.

13.2.2. CONTEXT INTERPRETATION

There may be multiple layers that context data go through before reaching an application, due to the need for additional abstraction. This can be as simple as needing to abstract a smart card id into its owner's name, but can be much more complex in taking all of the context available in a room to determine that a meeting is occurring. From an application designer's perspective, the use of these multiple layers should be transparent. In order to support this transparency, context must often be interpreted before it can be used by an application. To be easily reusable by multiple applications, interpretation needs to be provided by the architecture. There are a number of systems that provide mechanisms to perform transparent recursive interpretation [Dey *et al.* 1998; Kiciman and Fox 2000].

13.2.3. TRANSPARENT, DISTRIBUTED COMMUNICATIONS

Traditional user input comes from the keyboard and mouse, connected directly to the computer with which they are being used. However, devices used to sense context are commonly physically distributed and not connected to the same computer running an application that will react to that context. The fact that communication is distributed should be transparent to both sensors and applications, to simplify the design and building of both sensors and applications, relieving the designer of having to build a communications framework.

13.2.4. CONSTANT AVAILABILITY OF CONTEXT ACQUISITION

With GUI applications, user interface components such as buttons and menus are instantiated, controlled and used by only a single application (with the exception of some groupware applications). In contrast, context-aware applications should not instantiate individual components that provide sensor data, but must be able to access existing ones, when they require it. This eases the programming burden on the application designer by not requiring him/her to instantiate, maintain or keep track of components that acquire context, while allowing him/her to easily communicate with them. Because these components run independently of applications, there is a need for them to be persistent, available all the time.

13.2.5. CONTEXT STORAGE AND HISTORY

A requirement linked to the need for constant availability is the desire to maintain historical information. User input widgets maintain little, if any, historical information. For example, a file selection dialog box keeps track of only the most recent files that have been selected and allows a user to select those easily. In general though, if a more complete history is required, it is left up to the application to maintain it. In contrast, a component that acquires context information, should maintain a history of all the context it obtains. Context history can be used to establish trends and predict future context values.

13.2.6. RESOURCE DISCOVERY

In order for an application to communicate with a sensor (or rather its software interface), it must know what kind of information the sensor can provide, where it is located and how to communicate with it (protocol, language and mechanisms to use). For distributed sensors, this means knowing, at a minimum, both the hostname and port of the computer the sensor is running on. To be able to effectively hide these details from the application, the architecture needs to support a form of resource discovery [Schwartz et al. 1992]. With a resource discovery mechanism, when an application is started, it could specify the type of context information required. The mechanism would be responsible for finding any applicable components and for providing the application with ways to access them.

In this section, we have illustrated why an infrastructure is needed to support the building of context-aware applications and presented a number of infrastructure requirements. In the following three sections, we will present three different forms of infrastructure support: a generic component-based system modelled after GUI systems (Section 13.3); a generic situation-based system modelled after blackboards (Section 13.4); and a specialized system for dealing with specific types of context (Section 13.5).

13.3. BASIC COMPONENT-BASED ARCHITECTURE AND THE CONFERENCE ASSISTANT APPLICATION

Our initial object-oriented infrastructure, *the Context Toolkit*, was designed to address the requirements from the previous section. The infrastructure consists of five main types of objects:

- Widget, implements the widget abstraction;
- Server, responsible for aggregation of context;
- Interpreter, responsible for interpretation of context;
- Service, responsible for performing actions;
- Discoverer, responsible for supporting resource discovery.

Figure 13.1 shows the relationship between the objects in the Context Toolkit and an application. Each of these objects is autonomous in execution. They are instantiated independently of each other and execute in their own threads, supporting our requirement for independence. These objects can be instantiated on a single or on multiple computing devices. Although our base implementation is written in Java, the mechanisms used for communication (HTTP over XML) are programming language independent, allowing implementations in other languages.

A context widget is a software component that provides applications with access to context information from their operating environment. In the same way that GUI widgets mediate between the application and the user, context widgets mediate between the application and its operating environment. As a result, just as GUI widgets insulate applications from some presentation concerns, context widgets insulate applications from context acquisition concerns. To address context-specific operations, we introduce four additional categories of components in our infrastructure – interpreters, aggregators, services and discoverers.

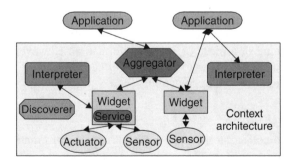

Figure 13.1. Data flow between applications and the Context Toolkit.

13.3.1. CONTEXT WIDGETS

GUI widgets hide the specifics of the input devices being used from the application programmer (allowing changes with minimal impact on applications), manage interaction to provide applications with relevant results of user actions, and provide reusable building blocks. Similarly, context widgets provide the following benefits:

- They provide a separation of concerns by hiding the complexity of the actual sensors used from the application. Whether the location of a user is sensed using Active Badges, floor sensors, an RF (radio frequency) based indoor positioning system or a combination of these, they should not impact the application.
- They abstract context information to suit the expected needs of applications. A widget that tracks the location of a user within a building or a city notifies the application only when the user moves from one room to another, or from one street corner to another, and doesn't report less significant moves to the application. Widgets provide abstracted information that we expect applications to need the most frequently.
- They provide reusable and customizable building blocks of context sensing. A widget that tracks the location of a user can be used by a variety of applications, from tour guides to car navigation to office awareness systems. Furthermore, context widgets can be tailored and combined in ways similar to GUI widgets. For example, a meeting sensing widget can be build on top of a presence sensing widget.

From the application's perspective, context widgets encapsulate context information and provide methods to access it in a way very similar to a GUI widget. Context widgets provide callbacks to *notify* applications of significant context changes and attributes that can be *queried* or *polled* by applications. As mentioned earlier, context widgets differ from GUI widgets in that they live much longer, they execute independently from individual applications, they can be used by multiple applications simultaneously, and they are responsible for maintaining a complete history of the context they acquire. Example context widgets include presence widgets that determine who is present in a particular location, temperature widgets that determine the temperature for a location, sound level widgets that determine the sound level in a location, and activity widgets that determine what activity an individual is engaged in.

From a designer's perspective, context widgets provide abstractions that encapsulate acquisition and handling of a piece of context information. However, additional abstractions are necessary to handle context information effectively. These abstractions embody two notions – interpretation and aggregation.

13.3.2. CONTEXT AGGREGATORS

Aggregation refers to collecting multiple pieces of context information that are logically related into a common repository. The need for aggregation comes in part from the distributed nature of context information. Context must often be retrieved from distributed sensors, via widgets. Rather than have an application query each distributed widget in turn (introducing complexity and making the application more difficult to maintain), *aggregators* gather logically related information relevant for applications and make it available within a single software component. Our definition of context given earlier describes the need to collect related context information about the relevant entities (people, places, and objects) in the environment. Aggregators aid the architecture in supporting the delivery of specified context to an application, by collecting related context about an entity in which the application is interested.

An aggregator has similar capabilities to a widget. Applications can be notified of changes in the aggregator's context, can query/poll for updates, and access stored context about the entity the aggregator represents. Aggregators provide an additional separation of concerns between how context is acquired and how it is used.

13.3.3. CONTEXT INTERPRETERS

Context interpreters are responsible for implementing the interpretation abstraction discussed in the requirements section. *Interpretation* refers to the process of raising the level of abstraction of a piece of context. For example, location may be expressed at a low level of abstraction such as geographical coordinates or at higher levels such as street names. Simple inference or derivation transforms geographical coordinates into street names using, for example, a geographic information database. Complex inference using multiple pieces of context also provides higher-level information. As an illustration, if a room contains several occupants and the sound level in the room is high, one can guess that a meeting is going on by combining these two pieces of context. Most often, context-aware applications require a higher level of abstraction than what sensors provide. *Interpreters* transform context information by raising its level of abstraction. An interpreter typically takes information from one or more context sources and produces a new piece of context information.

Interpretation of context has usually been performed by applications. By separating the interpretation out from applications, reuse of interpreters by multiple applications and widgets is supported. All interpreters have a common interface so other components can easily determine what interpretation capabilities an interpreter provides and will know how to communicate with any interpreter. This allows any application, widget or aggregator to send context to an interpreter to be interpreted.

13.3.4. SERVICES

The three components we have discussed so far, widgets, interpreters and aggregators, are responsible for acquiring context and delivering it to interested applications. If we examine the basic idea behind context-aware applications, that of acquiring context from the environment and then performing some action, we see that the step of taking an action is not yet represented in this architecture. *Services* are components that execute actions on behalf of applications.

From our review of context-aware applications, we have identified three categories of context-aware behaviors or services. The actual services within these categories are quite diverse and are often application-specific. However, for common context-aware services that multiple applications could make use of (e.g. turning on a light, delivering or displaying a message), support for that service within the architecture would remove the need for each application to implement the service. This calls for a service building block from which developers can design and implement services that can be made available to multiple applications.

A context service is an analog to the context widget. Whereas the context widget is responsible for retrieving state information about the environment from a sensor (i.e. input), the context service is responsible for controlling or changing state information in the environment using an actuator (i.e. output). As with widgets, applications do not need to understand the details of how the service is being performed in order to use them.

13.3.5. DISCOVERERS

Discoverers are the final component in the Context Toolkit. They are responsible for maintaining a registry of the capabilities that exist in the framework. This includes knowing what widgets, interpreters, aggregators and services are currently available for use by applications. When any of these components are started, it notifies a discoverer of its presence and capabilities, and how to contact that component (e.g. language, protocol, machine hostname). Widgets indicate what kind(s) of context they can provide. Interpreters indicate what interpretations they can perform. Aggregators indicate what entity they represent and the type(s) of context they can provide about that entity. Services indicate what context-aware service they can provide and the type(s) of context and information required to execute that service. When any of these components fail, it is a discoverer's responsibility to determine that the component is no longer available for use.

Applications can use discoverers to find a particular component with a specific name or identity (i.e. white pages lookup) or to find a class of components that match a specific set of attributes and/or services (i.e. yellow pages lookup). For example, an application may want to access the aggregators for all the people that can be sensed in the local environment. Discoverers allow applications to not have to know *a priori* where components are located (in the network sense). They also allow applications to more easily adapt to changes in the context-sensing infrastructure, as new components appear and old components disappear.

13.3.6. CONFERENCE ASSISTANT APPLICATION

We will now present the Conference Assistant, the most complex application that we have built with the Context Toolkit. It uses a large variety of context including user location, user interests and colleagues, the notes that users take, interest level of users in their activity, time, and activity in the space around the user. A separate sensor senses each type of context, thus the application uses a large variety of sensors as well. This application spans the entire range of context types and context-aware features we identified earlier.

13.3.6.1. Application Description

We identified a number of common activities that conference attendees perform during a conference, including identifying presentations of interest to them, keeping track of colleagues, taking and retrieving notes, and meeting people that share their interests. The Conference Assistant application currently supports all but the last conference activity and was fully implemented and tested in a scaled-down simulation of a conference. The following scenario describes how the application supports these activities.

A user is attending a conference. When she arrives at the conference, she registers, providing her contact information (mailing address, phone number, and email address), a list of research interests, and a list of colleagues who are also attending the conference. In return, she receives a copy of the conference proceedings and a Personal Digital Assistant (PDA). The application running on the PDA, the Conference Assistant, automatically displays a copy of the conference schedule, showing the multiple tracks of the conference, including both paper tracks and demonstration tracks. On the schedule (Figure 13.2a), certain papers and demonstrations are highlighted (light gray) to indicate that they may be of particular interest to the user.

The user takes the advice of the application and walks towards the room of a suggested paper presentation. When she enters the room, the Conference Assistant automatically displays the name of the presenter and the title of the presentation. It also indicates whether audio and/or video of the presentation are being recorded. This impacts the user's behavior, taking fewer or greater notes depending on the extent of the recording available. The presenter is using a combination of PowerPoint and Web pages for his presentation. A thumbnail of the current slide or Web page is displayed on the PDA. The Conference Assistant allows the user to create notes of her own to 'attach' to the current slide or Web page (Figure 13.3). As the presentation proceeds, the application displays updated information for the user. The user takes notes on the presented slides and Web

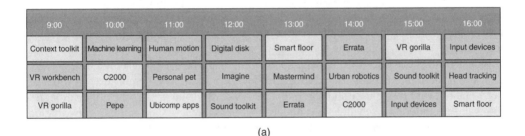

(a)

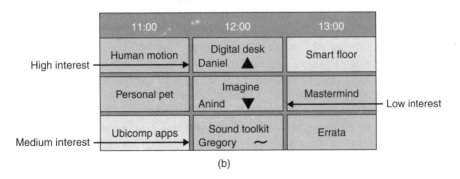

(b)

Figure 13.2. (a) Schedule with suggested papers and demos highlighted (light-colored boxes) in the three (horizontal) tracks; (b) Schedule augmented with users' location and interests in the presentations being viewed.

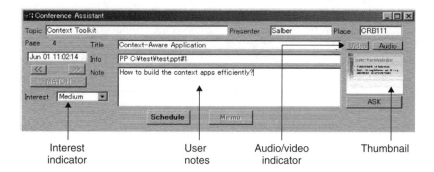

Figure 13.3. Screenshot of the Conference Assistant note-taking interface.

pages using the Conference Assistant. The presentation ends and the presenter opens the floor for questions. The user has a question about the presenter's tenth slide. She uses the application to control the presenter's display, bringing up the tenth slide, allowing everyone in the room to view the slide in question. She uses the displayed slide as a reference and asks her question. She adds her notes on the answer to her previous notes on this slide.

After the presentation, the user looks back at the conference schedule display and notices that the Conference Assistant has suggested a demonstration to see based on her interests. She walks to the room where the demonstrations are being held. As she walks past demonstrations in search of the one she is interested in, the application displays the name of each demonstrator and the corresponding demonstration. She arrives at the demonstration she is interested in. The application displays any PowerPoint slides or Web pages that the demonstrator uses during the demonstration. The demonstration turns out not to be relevant to the user and she indicates her level of interest to the application. She looks at the conference schedule and notices that her colleagues are in other presentations (Figure 13.2b). A colleague has indicated a high level of interest in a particular presentation, so she decides to leave the current demonstration and to attend that presentation. The user continues to use the Conference Assistant throughout the conference for taking notes on both demonstrations and paper presentations.

She returns home after the conference and wants to retrieve some information about a particular presentation. The user executes a retrieval application provided by the conference. The application shows her a timeline of the conference schedule with the presentation and demonstration tracks (Figure 13.4a). It provides a query interface that allows the user to populate the timeline with various events: her arrival and departure from different rooms, when she asked a question, when other people asked questions or were present, when a presentation used a particular keyword, or when audio or video were recorded. By selecting an event on the timeline (Figure 13.4a), the user can view (Figure 13.4b) the slide or Web page presented at the time of the event, audio and/or video recorded during the presentation of the slide, and any personal notes she may have taken on the presented information. She can then continue to view the current presentation, moving back and forth between the presented slides and Web pages.

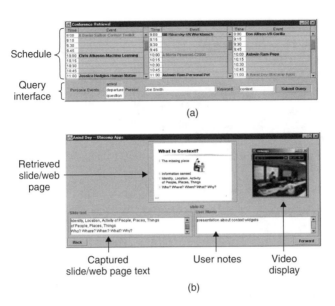

Schedule

Query
interface

Retrieved
slide/web
page

(a)

Captured
slide/web page text

User notes

Video
display

(b)

Figure 13.4. Screenshots of the retrieval application: (a) query interface and timeline annotated with events and (b) captured slideshow and recorded audio/video.

Similarly, a presenter can use a third application with the same interface to retrieve information about his/her presentation. The application displays a presentation timeline, populated with events about when different slides were presented, when audience members arrived and left the presentation, the identities of questioners and the slides relevant to the questions. The presenter can 'relive' the presentation, by playing back the audio and/or video, and moving between presentation slides and Web pages.

The Conference Assistant is the most complex context-aware application we have built. It uses a wide variety of sensors and a wide variety of context, including real-time and historical context. This application supports all three types of context-aware features: presenting context information, automatically executing a service, and tagging of context to information for later retrieval.

13.3.6.2. Application Design

The application features presented in the above scenario have all been implemented. The Conference Assistant makes use of a wide range of context. In this section, we discuss the application architecture and the types of context used, both in real time during a conference and after the conference, as well as how they were used to provide benefits to the user.

During registration, a User Aggregator is created for the user, shown in the architecture diagram of Figure 13.5. It is responsible for aggregating all the context information about the user and acts as the application's interface to the user's personal context information. It subscribes to information about the user from the public registration widget, the user's

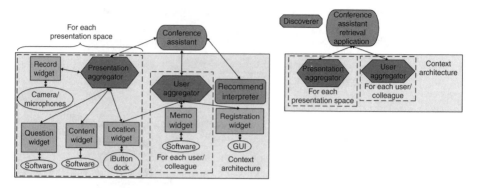

Figure 13.5. Context architecture for the Conference Assistant application during and after the conference.

memo widget and the location widget in each presentation space. When the user is attending the conference, the application first uses information about what is being presented at the conference and her personal interests (registration widget) to determine what presentations might be of particular interest to her (the recommend interpreter). The application uses her location (location widget), the activity (presentation of a Web page or slide) in that location (content and question widgets) and the presentation details (presenter, presentation title, whether audio/video is being recorded) to determine what information to present to her. The text from the slides is being saved for the user, allowing her to concentrate on what is being said rather than spending time copying down the slides. The memo widget captures the user's notes and any relevant context to aid later retrieval. The context of the presentation (presentation activity has concluded, and the number and title of the slide in question) facilitates the user's asking of a question. The context is used to control the presenter's display, changing to a particular slide for which the user had a question.

There is a Presentation Aggregator for each physical location where presentations/demos are occurring, responsible for aggregating all the context information about the local presentation and acting as the application's interface to the public presentation information. It subscribes to the widgets in the local environment, including the content widget, location widget and question widget. The content widget uses a software sensor that captures what is displayed in a PowerPoint presentation and in an Internet Explorer Web browser. The question widget is also a software widget that captures what slide (if applicable) a user's question is about, from their Conference Assistant application. The location widget used here is based on Java iButton technology.

The list of colleagues provided during registration allows the application to present other relevant information to the user. This includes both the locations of colleagues and their interest levels in the presentations they are currently viewing. This information is used for two purposes during a conference. First, knowing where other colleagues are helps an attendee decide which presentations to see herself. For example, if there are two interesting presentations occurring simultaneously, knowing that a colleague is attending one of the presentations and can provide information about it later, a user can choose to attend the other presentation. Secondly, as described in the user scenario, when a user is attending a presentation that is not relevant or interesting to her, she can use the context of her colleagues to decide which presentation to move to. This is a form of social or collaborative information filtering [Shardanand and Maes 1995].

After the conference, the retrieval application uses the conference context to retrieve information about the conference. The context includes public context such as the time when presentations started and stopped, whether audio/video was captured at each presentation, the names of the presenters, the rooms in which the presentations occurred, and any keywords the presentations mentioned. It also includes the user's personal context such as the times at which she entered and exited a room, the rooms themselves, when she asked a question, and what presentation and slide or Web page the question was about. The application also uses the context of other people, including their presence at particular presentations and questions they asked, if any. The user can use any of this context information to retrieve the appropriate slide or Web page and any recorded audio/video associated with the context.

The Conference Assistant does not communicate with any widget directly, but instead communicates only with the user's user aggregator, the user aggregators belonging to each colleague and the local presentation aggregator. It subscribes to the user's user aggregator for changes in location and interests. It subscribes to the colleagues' user aggregators for changes in location and interest level. It also subscribes to the local presentation aggregator for changes in a presentation slide or Web page when the user enters a presentation space and unsubscribes when the user leaves. It also sends its user's interests to the recommend interpreter to convert them to a list of presentations in which the user may be interested. The interpreter uses text matching of the interests against the title and abstract of each presentation to perform the interpretation.

Only the memo widget runs on the user's handheld device. The registration widget and associated interpreter run on the same machine. The user aggregators are all executing on the same machine for convenience, but can run anywhere, including on the user's device. The presentation aggregator and its associated widgets run on any number of machines in each presentation space. The content widget needs to be run on only the particular computer being used for the presentation.

In the conference attendee's retrieval application, all the necessary information has been stored in the user's user aggregator and the public presentation aggregators. The architecture for this application (Figure 13.5) is much simpler, with the retrieval application only communicating with the user's user aggregator and each presentation aggregator. As shown in Figure 13.4, the application allows the user to retrieve slides (and the entire presentation including any audio/video) using context via a query interface. If personal context is used as the index into the conference information, the application polls the user aggregator for the times and location at which a particular event occurred (user entered or left a location, or asked a question). This information can then be used to poll the correct presentation aggregator for the related presentation information. If public context is used as the index, the application polls all the presentation aggregators for the times at which a particular event occurred (use of a keyword, presence or question by a certain person). As in the previous case, this information is then used to poll the relevant presentation aggregators for the related presentation information.

13.3.7. SUMMARY

The Conference Assistant, as mentioned earlier, is our most complex context-aware application. It supports interaction between a single user and the environment, and between multiple users. Looking at the variety of context it uses (location, time, identity, activity) and the variety of context-aware services it provides (presentation of context information, automatic execution of services, and tagging of context to information for later retrieval), we see that it completely spans our categorization of both context and context-aware services. This application would have been extremely difficult to build if we did not have the underlying support of the Context Toolkit. We have yet to find another application that spans this feature space.

Figure 13.5 demonstrates quite well the advantage of using aggregators. Each presentation aggregator collects context from four widgets. Each user aggregator collects context from the memo and registration widgets plus a location widget for each presentation space. Assuming 10 presentation spaces (three presentation rooms and seven demonstration spaces), each user aggregator is responsible for 12 widgets. Without the aggregators, the application would need to communicate with 42 widgets, obviously increasing the complexity. With the aggregators and assuming three colleagues, the application just needs to communicate with 14 aggregators (10 presentation and four user), although it would only be communicating with one of the presentation aggregators at any one time.

Our component-based architecture greatly eases the building of both simple and complex context-aware applications. It supports each of the requirements from the previous section: separation of concerns between acquiring and using context, context interpretation, transparent and distributed communications, constant availability of the infrastructure, context storage and history and resource discovery. Despite this, some limitations remain:

- Transparent acquisition of context from distributed components is still difficult.
- The infrastructure does not deal with the dynamic component failures or additions that would be typical in environments with many heterogeneous sensors.
- When dealing with multiple sensors that deliver the same form of information, it is desirable to fuse information. This sensor fusion should be done without further complicating application development.

In the following sections we will discuss additional programming support for context that addresses these issues.

13.4. SITUATION SUPPORT AND THE CYBREMINDER APPLICATION

In the previous section, we described the Context Toolkit and how it helps application designers to build context-aware applications. We described the context component abstraction that used widgets, interpreters and aggregators, and showed how it simplified thinking about and designing applications. However, this context component abstraction has some flaws that make it harder to design applications than it needs to be. The extra steps are:

- locating the desired set of interpreters, widgets and aggregators;
- deciding what combination of queries and subscriptions are necessary to acquire the context the application needs;
- collecting all the acquired context information together and analyzing it to determine when a situation interesting to the application has occurred.

A new abstraction called the *situation abstraction*, similar to the concept of a blackboard, makes these steps unnecessary. Instead of dealing with components in the infrastructure individually, the situation abstraction allows designers to deal with the infrastructure as a single entity, representing all that is or can be sensed. Similar to the context component abstraction, designers need to specify what context their applications are interested in. However, rather than specifying this on a component-by-component basis and leaving it up to them to determine when the context requirements have been satisfied, the situation abstraction allows them to specify their requirements at one time to the infrastructure and leaves it up to the infrastructure to notify them when the request has been satisfied, removing the unnecessary steps listed above and simplifying the design of context-aware applications.

In the context component abstraction, application programmers have to determine what toolkit components can provide the needed context using the discoverer and what combination of queries and subscriptions to use on those components. They subscribe to these components directly and when notified about updates from each component, combine them with the results from other components to determine whether or not to take some action. In contrast, the situation abstraction allows programmers to specify what information they are interested in, whether that be about a single component or multiple components. The Context Toolkit infrastructure determines how to map the specification onto the available components and combine the results. It only notifies the application when the application needs to take some action. In addition, the Context Toolkit deals automatically and dynamically with components being added and removed from the infrastructure. On the whole, using the situation abstraction is much simpler for programmers when creating new applications and evolving existing applications.

13.4.1. IMPLEMENTATION OF THE SITUATION ABSTRACTION

The main difference between using the context component abstraction and the situation abstraction is that in the former case, applications are forced to deal with each relevant component individually, whereas in the latter case, while applications can deal with individual components, they are also allowed to treat the context-sensing infrastructure as a single entity.

Figure 13.6 shows how an application can use the situation abstraction. It looks quite similar in spirit to Figure 13.1. Rather than the application designer having to determine what set of subscriptions and interpretations must occur for the desired context to be acquired, it hands this job off to a connector class (shown in Figure 13.6, sitting between

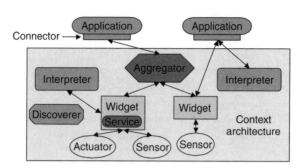

Figure 13.6. Typical interaction between applications and the Context Toolkit using the situation abstraction.

the application and the context architecture). This connector class determines what subscriptions and interpretations are required (with the help of a Discoverer) and interacts with the infrastructure to make it happen. More details on the algorithm behind this determination can be found in [Dey 2000].

13.4.2. CYBREMINDER: A COMPLEX EXAMPLE THAT USES THE SITUATION ABSTRACTION

We will now describe the CybreMinder application, a context-aware reminder system, to illustrate how the situation abstraction is used in practice. The CybreMinder application is a prototype application that was built to help users create and manage their reminders more effectively [Dey and Abowd 2000]. Current reminding techniques such as post-it notes and electronic schedulers are limited to using only location or time as the trigger for delivering a reminder. In addition, these techniques are limited in their mechanisms for delivering reminders to users. CybreMinder allows users to specify more complex and appropriate situations or triggers and associate them with reminders. When these situations are realized, the associated reminder will be delivered to the specified recipients. The recipient's context is used to choose the appropriate mechanism for delivering the reminder.

13.4.2.1. Creating the Reminder and Situation

When users launch CybreMinder, they are presented with an interface that looks quite similar to an e-mail creation tool. As shown in Figure 13.7, users can enter the names of the recipients for the reminder. The recipients could be themselves, indicating a personal reminder, or a list of other people, indicating that a third party reminder is being created. The reminder has a subject, a priority level (ranging from lowest to highest), a body in which the reminder description is placed, and an expiration date. The expiration date indicates the date and time at which the reminder should expire and be delivered, if it has not already been delivered.

In addition to this traditional messaging interface, users can select the context tab and be presented with the situation editor (Figure 13.8a). This interface allows dynamic

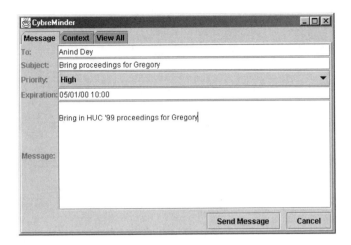

Figure 13.7. CybreMinder reminder creation tool.

construction of an arbitrarily rich situation, or context, that is associated with the reminder being created. The interface consists of two main pieces for creating and viewing the situation. Creation is assisted by a dynamically generated list of valid sub-situations that are currently supported by the CybreMinder infrastructure (as assisted by the Context Toolkit described later). When the user selects a sub-situation, they can edit it to fit their particular situation. Each sub-situation consists of a number of context types and values. For example, in Figure 13.8a, the user has just selected the sub-situation that a particular user is present in the CRB building at a particular time. The context types are the user's name, the location (set to CRB) and a timestamp.

In Figure 13.8b, the user is editing those context types, requiring the user name to be 'Anind Dey' and not using time. This sub-situation will be satisfied the next time that Anind Dey is in the location 'CRB'. The user indicates which context types are important by selecting the checkbox next to those attributes. For the types that they have selected, users may enter a relation other than '='. For example, the user can set the timestamp after 9 p.m. by using the '>' relation. Other supported relations are '>=', '<', and '<='. For the value of the context, users can either choose from a list of pre-generated values, or enter their own.

At the bottom of the interfaces in Figure 13.8, the currently specified situation is visible. The overall situation being defined is the conjunction of the sub-situations listed. Once a reminder and an associated situation have been created, the user can send the reminder. If there is no situation attached, the reminder is delivered immediately after the user sends the reminder. However, unlike e-mail messages, sending a reminder does not necessarily imply immediate delivery. If a situation is attached, the reminder is delivered to recipients at a future time when all the sub-situations can be simultaneously satisfied. If the situation cannot be satisfied before the reminder expires, the reminder is delivered both to the sender and recipients with a note indicating that the reminder has expired.

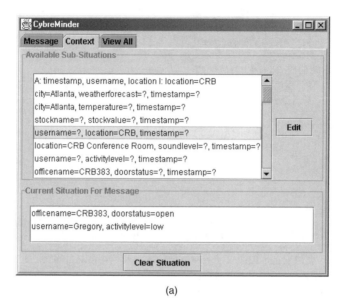

(a)

(b)

Figure 13.8. CybreMinder (a) situation editor and (b) sub-situation editor.

13.4.2.2. Delivering the Reminder

Thus far, we have concentrated on the process of creating context-aware reminders. We will now describe the delivery process. When a reminder is delivered, either because its associated situation was satisfied or because it has expired, CybreMinder determines what is the most appropriate delivery mechanism for each reminder recipient. The default signal is to show the reminder on the closest available display, augmented with an audio cue. However, if a recipient wishes, they can specify a configuration file that will override this default.

A user's configuration file contains information about all of the available methods for contacting the user, as well as rules defined by the user on which method to use in which situation. If the recipient's current context and reminder information (sender identity and/or priority) matches any of the situations defined in his/her configuration file, the specified delivery mechanism is used. Currently, we support the delivery of reminders via SMS on a mobile phone, e-mail, displaying on a nearby networked display (wearable, handheld, or static CRT) and printing to a local printer (to emulate paper to-do lists).

For the latter three mechanisms, both the reminder and associated situation are delivered to the user. Delivery of the situation provides additional useful information to users, helping them understand why the reminder is being sent at this particular time. Along with the reminder and situation, users are given the ability to change the status of the reminder (Figure 13.9a left). A status of 'completed' indicates that the reminder has been addressed and can be dismissed. The 'delivered' status means the reminder has been delivered but still needs to be addressed. A 'pending' status means that the reminder should be delivered again when the associated situation is next satisfied. Users can explicitly set the status through a hyperlink in an e-mail reminder or through the interface shown in Figure 13.9b.

The CybreMinder application is the first application we built that used the situation abstraction. It supports users in creating reminders that use simple situations based on time or location, or more complex situations that use additional forms of context. The situations that can be used are only limited by the context that can be sensed. Table 13.1 shows natural language and CybreMinder descriptions for some example situations.

13.4.2.3. Building the Application

The Context Toolkit-based architecture used to build CybreMinder is shown in Figure 13.10. For this application, the architecture contains a user aggregator for each user of CybreMinder and any available widgets, aggregators and interpreters. When CybreMinder launches, it makes use of the discovery protocol in the Context Toolkit to query for the context components currently available to it. It analyzes this information and determines what sub-situations are available for a user to work with. The sub-situations are simply the collection of subscription callbacks that all the context widgets and context aggregators provide. For example, a presence context widget contains information about the presence of individuals in a particular location (specified at instantiation time). The callback it provides contains three attributes: a user name, a location, and a timestamp. The location is a constant, set to 'home', for example. The constants in each callback are used to populate the menus from which users can select values for attributes.

When the user creates a reminder with an associated situation, the reminder is sent to the aggregator responsible for maintaining context about the recipient – the user aggregator. CybreMinder can be shut down any time after the reminder has been sent to the recipient's aggregator. The recipient's aggregator is the logical place to store all reminder information intended for the recipient because it knows more about the recipient than any other

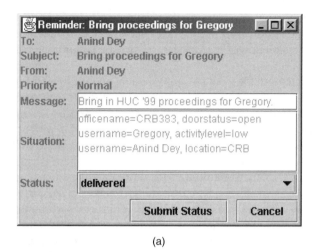

(a)

(b)

Figure 13.9. CybreMinder display of (a) a triggered reminder and (b) all reminders.

component and is always available. This aggregator analyzes the given situation and creates subscriptions to the necessary aggregators and widgets (using the extended Context Toolkit object) so that it can determine when the situation has occurred. In addition, it creates a timer thread that awakens when the reminder is set to expire. Whenever the aggregator receives a subscription callback, it updates the status of the situation in question. When all the sub-situations are satisfied, the entire situation is satisfied and the reminder can be delivered.

The recipient's aggregator contains the most up-to-date information about the recipient. It tries to match this context information along with the reminder sender and priority level with the rules defined in the recipient's configuration file. The recipient's context and the

Table 13.1. Natural language and CybreMinder descriptions of example situations.

Situation	Natural Language Description	CybreMinder Description
Time	9:45 am	Expiration field: 9:45 am
Location	Forecast is for rain and Bob is leaving his apartment	City = Atlanta, WeatherForecast = rain Username = Bob, Location = Bob's front door
Co-Location	Sally and colleague are co-located	Username = Sally, Location = *1 Username = Bob, Location = *1
Complex #1	Stock price of X is over $50, Bob is alone and has free time	StockName = X, StockPrice > 50 Username = Bob, Location = *1 Location = *1, OccupantSize = 1 Username = Bob, FreeTime > 30
Complex #2	Sally is in her office and has some free time, and her friend is not busy	Username = Sally, Location = Sally's office Username = Sally, FreeTime = 60 Username = Tom, ActivityLevel = low

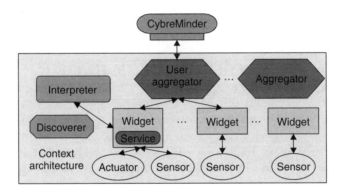

Figure 13.10. Architecture diagram for the CybreMinder application using the situation abstraction.

rules consist of collections of simple attribute name – value pairs, making them easy to compare. When a delivery mechanism has been chosen, the aggregator calls a widget service that can deliver the reminder appropriately.

13.4.3. SUMMARY

Use of the situation abstraction allows end-users to attach reminders to arbitrarily complex situations that they are interested in, which the application then translates into a system specification of the situations. Users are not required to use templates or hardcoded situations, but can use any context that can be sensed and is available from their environment. This application could have been written to use widgets, aggregators and interpreters directly, but instead of leveraging off the Context Toolkit's ability to map between user-specified situations and these components, the application programmer would have to provide this ability making the application much more difficult to build.

The situation abstraction allows application designers to program at a higher level and alleviates the designer from having to know about specific context components. It allows designers to treat the infrastructure as a single component and not have to deal with the details of individual components. In particular, this supports the ability to specify context requirements that bridge multiple components. This includes requirements for unrelated context that is acquired by multiple widgets and aggregators. It also includes requirements for interpreted context that is acquired by automatically connecting an interpreter to a widget or aggregator. Simply put, the situation abstraction allows application designers to simply describe the context they want and the situation they want it in, and to have the context infrastructure provide it. This power comes at the expense of additional abstraction. When designers do not want to know the details of context sensing, the situation abstraction is ideal. However, if the designer wants greater control over how the application acquires context from the infrastructure or wants to know more about the components in the infrastructure, the context component abstraction may be more appropriate. Note that the situation abstraction could not be supported without context components. The widgets, interpreters and aggregators with their uniform interfaces and ability to describe themselves to other components makes the situation abstraction possible.

13.5. FUSION SUPPORT AND THE IN/OUT BOARD APPLICATION

While the Context Toolkit does provide much general support for building arbitrarily complex context-aware applications, sometimes its generality is a burden. The general abstractions in the Context Toolkit are not necessarily appropriate for novice context-aware programmers to build simple applications. In particular, support for fusing multiple sources of context is difficult to support in a general fashion and can be much more appropriately handled by focusing on specific pieces of context. Location is far and away the most common form of context used for ubiquitous computing applications. In this section, we explore a modified programming infrastructure, motivated by the Context Toolkit but consciously limited to the specific problems of location-aware programming. This Location Service is further motivated by the literature on location-aware computing, where we see three major emphases:

- deployment of specific location sensing technologies (see [Hightower and Borriello 2001] for a review);
- demonstration of compelling location-aware applications; and
- development of software frameworks to ease application construction using location [Moran and Dourish 2001]

In this section, we present a specialized construction framework, the location service, for handling location information about tracked entities. Our goal in creating the location service is to provide a uniform, geometric-based way to handle a wide variety of location technologies for tracking interesting entities while simultaneously providing a simple and extensible technique for application developers to access location information in a form most suitable for their needs. The framework we present divides the problem into three specific activities:

- acquisition of location data from any of a number of positioning technologies;
- collection of location data by named entities; and
- monitoring of location data through a straightforward and extensible query and translation mechanism.

We are obviously deeply influenced by our earlier work on the Context Toolkit. After a few years of experience using the Context Toolkit, we still contend that the basic separation of concerns and programming abstractions that it espouses are appropriate for many situations of context-aware programming, and this is evidenced by a number of internal and external applications developed using it. However, in practice, we did not see the implementation of the Context Toolkit encouraging programmers to design context-aware applications that respected the abstractions and separation of concerns. Our attempt at defining the location service is not meant to dismiss the Context Toolkit but to move toward an implementation of its ideas that goes further toward directing good application programming practices.

This work is an explicit demonstration of the integration of multiple different location sensing technologies into a framework that minimizes an application developer's requirement to know about the sensing technology. We also provide a framework in which more complicated fusion algorithms, such as probabilistic networks [Castro *et al.* 2001], can be used. Finally, we provide an extensible technique for interpreting and filtering location information to meet application-specific needs.

We provide an overview of the software framework that separates the activities of acquisition, collection and application-specific monitoring. Each of these activities is then described in detail, emphasizing the specific use of location within the Aware Home Research Initiative at Georgia Tech [Aware Home 2003]. We conclude with a description of some applications developed with the aid of the location service.

13.5.1. THE ARCHITECTURE OF THE LOCATION SERVICE

Figure 13.11 shows a high-level view of the architecture of the location service. Any number of location technologies acquire location information. These technologies are augmented with a software wrapper to communicate a geometry-based (i.e., three-dimensional coordinates in some defined space) XML location message, similar in spirit to the widget abstraction of the Context Toolkit. The location messages are transformed into Java objects and held in a time-ordered queue. From there, a collation algorithm attempts to use separate location objects that refer to the same tracked entity. When a location object relates to a known (i.e., named) entity, then it is stored as the current location for that entity. A query subsystem provides a simple interface for applications to obtain location information for both identified and unidentified entities. Since location information is stored as domain-specific geometric representations, it is necessary to transform location to a form desirable for any given application. This interpretation is done by means

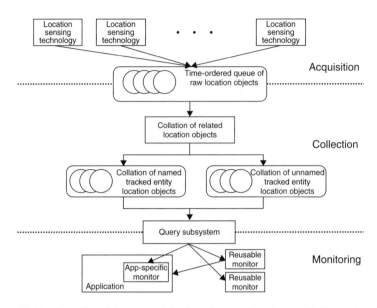

Figure 13.11. Overall architecture of the location service. Arrows indicate data flow.

of monitor classes, reusable definitions of spatially significant regions (e.g., rooms in a house) that act as filters to signal important location information for any given application.

There are several important properties of this service. First, the establishment of a well-defined location message insulates the rest of the location service from the details of the specific location sensing technologies. The overall service will work whether or not the constituent sensing technologies are delivering location objects. Second, the collation, or fusion, algorithm within the collection layer can be changed without impacting the sensing technologies or the location-aware applications. Third, the monitor classes are reusable and extensible, meaning simple location requirements don't have to be recreated by the application developer each time and complex location needs can be built up from simpler building blocks.

13.5.2. REPRESENTING LOCATION

The location service assumes that raw positioning data is delivered as geometric data within one of a set of known reference frames. The raw positioning data object consists of:

- a four-tuple, (x,y,z,d), consisting of a three-dimensional positional coordinate and a reference frame identifier, d, which is used to interpret the positional coordinate;
- an orientation value as a three-tuple, if known;
- the identity of the entity at that location, if known;
- a timestamp for when the positioning data was acquired; and
- an indication of the location sensing technology that was the source of the data.

Not every sensing technology can provide all of this information. The Collector attempts to merge multiple raw location objects in order to associate a location value with a collection of named and unnamed entities. This results in a new location object for tracked entities that is stored within the collector and made available via the query subsystem for applications to access and interpret.

13.5.3. DETAILS ON POSITIONING SYSTEMS

To exercise the framework of the location service, we have instantiated it with a variety of location sensing technologies. We describe them briefly here. The first two location sensing technologies existed prior to the development of the location service, and the latter two were developed afterwards to validate the utility of the framework.

13.5.3.1. The RFID Floor Mat System

For a while, we have been interested in creating natural ways to track people indoors. While badge technologies have been very popular in location-aware research, they are unappealing in a home environment. Several researchers have suggested the instrumentation of a floor for tracking purposes [Addlesee *et al*. 1997; Orr and Abowd 2000]. These are very appealing approaches, but require somewhat abnormal instrumentation of the floor and are computationally heavyweight. Prior to this work on the location service, we were very much driven by the desire to have a single location sensing technology that would deliver room-level positioning throughout the house. As a compromise between the prior instrumented floors work and badging approaches, we arrived at a solution of floor mats that act as a network of RFID antennae (see Figure 13.12). A person wears a passive RFID tag below the knee (usually attached to a shoe or ankle) and the floor mat antenna can then read the unique ID as the person walks over the mat. Strategic placement of the floor mats within the Aware Home provided us with a way to detect position and identity as individuals walked throughout the house.

13.5.3.2. Overhead Visual Tracking

Although room level location information is useful in many applications, it remains very limiting. More interesting applications can be built if better location information can be provided. Computer vision can be used to infer automatically the whereabouts and activities of individuals within the home. The Aware Home has been instrumented with cameras in the ceiling, providing an overhead view of the home. The visual tracking system, in the kitchen, attempts to coordinate the overlapping views of the multiple cameras in a given space (see Figure 13.13). It does not try to identify moving objects, but keeps track of the location and orientation of a variety of independent moving 'blobs' over time and across multiple cameras.

13.5.3.3. Fingerprint Detection

Commercial optical fingerprint detection technology is now currently available and affordable. Over the span of one week, two undergraduates working in our lab created a fingerprint detection system that, when placed at doorways in the Aware Home, can

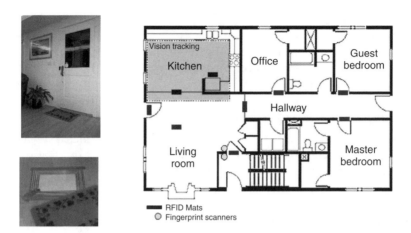

Figure 13.12. The RFID floor mat positioning system. On the left are a floor mat placed near an entrance to the Aware Home and an RFID antenna under the mat. Strategic placement of mats around the floor plan of the house provides an effective room-level positioning system. Also shown are the locations of the vision and fingerprint systems.

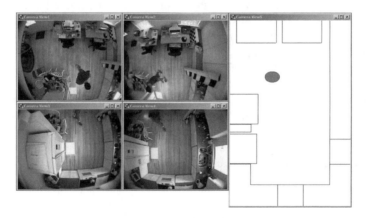

Figure 13.13. The visual tracking system. Four overhead cameras track moving 'blobs' in the kitchen of the Aware Home.

be another source of location information. The fingerprint detection system reports the identity of the person whose finger was scanned along with the spatial coordinates for the door and the orientation of the user.

13.5.3.4. Open-Air Speaker ID

Speaker ID technology developed by digital signal processing experts at Georgia Tech can be used as another source of location information. An 'always on' microphone records five-second samples that are compared against the known population of the house. If there is a close enough match, the identity of the user along with the location of the microphone is provided, functioning similarly to the RFID floor mat system.

13.5.4. FUSION AND AGGREGATION OF LOCATION

Location data from the independent location sensing technologies are sent as messages. As long as the individual sensing technology allows for a socket communication, it is straightforward to extract these messages and incorporate them into the location service. In the collector subsystem, these messages are stored in a time-sequenced queue. Within the collector, the main task is to consume the raw location objects and update two collections of tracked objects, one for identified or named entities and one for currently unidentified or unnamed entities.

The collector has access to the agreed namespace for entities to be tracked. If a label for a location object does not match a name in this namespace, the Collector attempts to translate the label to one of these names using sensor-specific translation tables. So, for example, an RFID location event will contain a label that is a unique integer that is then mapped to the name of an individual who is the declared owner of that tag. Recall that some of the location sensing technologies, such as the visual tracker, produce anonymous location objects that are not assumed to be known entities.

When the Collector reads a location object that does not contain a known identity, it searches the queue of raw location objects to see if there is one that can be mapped to a known entity and that corresponds to the anonymous one. If so, then the two are merged and given the name of the known entity. Each named tracked entity has a special storage area to contain its current location data. Another special storage area contains location data for currently unidentified tracked objects. Figure 13.14 shows a listing of the contents of the two collections and a simple graphical depiction of location that is derived from the collector subsystem.

Currently, the algorithm for merging location data uses a fairly straightforward temporal and spatial heuristic. For example, when a location object with no identity is consumed, the merge algorithm tries to find a location object with an identity around the same time and place. While this relatively naïve algorithm has its own problems, it is incorporated into the collector system in such a way that it can be replaced relatively easily with a more sophisticated routine. For example, with an increased number of location-sensing technologies feeding the queue of raw location objects, scalability of the fusion algorithm may be a concern. Currently, the visual tracking system creates the highest bandwidth of location objects and we are able to handle 15 location object updates per second. The most important feature of the collector is that it consumes the raw location data objects and produces what is effectively a database of named and unnamed location objects. Application programmers need only consider this repository of current locations for tracked entities when constructing their applications.

13.5.5. ACCESSING, INTERPRETING AND HANDLING LOCATION DATA
WITHIN AN APPLICATION

Applications need to access location data for relevant entities. Since the collector provides a model of location data as a repository indexed by identities of known and unknown entities, we need to provide ways for applications to query this repository and trigger events

(a)

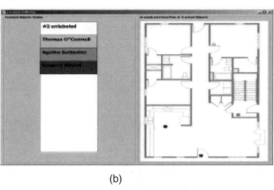

(b)

Figure 13.14. (a) A listing of the contents of the two collections of tracked entity location objects (named and unnamed); (b) a graphical depiction of the same information, showing location relative to the floor plan of the Aware Home.

based on application-relevant interpretations of that location information. The location service provides a Monitor class in Java that application developers can reuse and extend to perform all three tasks of querying, interpreting and providing application-specific location event triggers. The kinds of application requests we want to make easy to program are of the following types:

- Where is a particular individual? The answer should be in a form that is relevant to that application (for example, room names for a home).
- Who is in a particular location? Again, the location is application-specific.
- Alerting when a person moves into or out of a particular location.
- Alerting when individuals are collocated.

We will show how these kinds of application questions are supported in our discussion of monitors.

13.5.5.1. Querying for Current Location Information

The query subsystem presents a straightforward API to request information about named and unnamed tracked entities. Query requests represent Boolean searches on all of the attributes for the tracked location objects. A Java RMI server receives query requests and returns matched records from named/unnamed repositories. This query system makes it very easy to request information about any number of entities that are being tracked.

13.5.5.2. Interpreting Location for Application-Specific Needs

Up to this point, all of the location information in the location service is geometric and relative to a set of pre-defined domains. Application designers want location information in many different forms and it is the role of the monitoring layer to provide mechanisms for interpreting geometric location data into a variety of alternate geometric and symbolic forms. However, to make this translation possible, knowledge of the physical space needs to be encoded, and this knowledge may be very application-specific. For example, one application may want to know on what floor within a home various occupants are located, whereas another application may want to know the rooms, and another is interested in knowing whether someone is facing the television. Each of these spatial interpretations is encoded as a translation table from geometric information to the appropriate symbolic domain. Furthermore, these translation tables are dependent on the coordinate reference frame used for a particular location object, that is, the translation would be different for a home versus an office building.

To facilitate a variety of spatial interpretations, the Monitor class is constructed with a specific instance of a spatial translation table. These translation tables can be reused across different instances of the Monitor class and any of its subclass extensions. Any geometric position data returned from a query is automatically translated with respect to the spatial interpretation and can be delivered to an application.

13.5.5.3. Filtering and Delivering Information to Applications

Even when it has been translated into a meaningful representation, not all location data is relevant to an application. The other objective of a monitor is to provide ways to filter the location events that can trigger application-specific behavior. By setting up an application as a listener for a given instance of a monitor, the monitor can then control which location events are delivered to the application. We have created several examples of monitors that provide the capabilities suggested earlier. All monitors extend the base class Monitor that provides the ability to create, send and receive queries and process the results through a selected spatial interpretation look-up table and send selected events to applications that subscribe as listeners to the monitor.

13.5.6. SAMPLE APPLICATION DEVELOPMENT

A canonical indoor location application is the In/Out board, which indicates the location of a set of normal occupants of a building. Figure 13.15 shows a screenshot of a simple In/Out board. This application was originally built within the Context Toolkit to react to location changes from the whole-house RFID. It was rewritten to take advantage of the multiple location-sensing technologies.

Another motivating application for us is to use context to facilitate human–human communication [Nagel *et al.* 2001]. Within an environment like a home or office, we would like to automatically route text, audio and video communications to the most

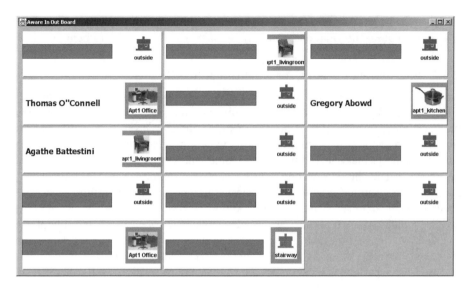

Figure 13.15. An In/Out Board application developed using the Location Service. Names of people other than the authors have been suppressed for publication.

appropriate place based on knowledge of the recipient's context. The location service makes it fairly straightforward to determine in which room the recipient is located, and in some spaces even the orientation. We have implemented an instant messaging client that will send a text-based message to a display that is nearest to the recipient. A more interesting variation of this messaging can occur if we monitor not only for the location of the recipient but also determine if that person is alone. While location is not always sufficient context to infer human activity, it is often necessary. The location service is designed to make it easier for inferences on location information to be shared across multiple applications.

13.5.7. SUMMARY

There is no universal location-sensing technology to support location anywhere–anytime. As a result, for the foreseeable future, operational location systems will have to combine separate location-sensing technologies in order to increase the scale over which location can be delivered. The contribution of the location service is as a framework to facilitate the construction of location-aware applications that insulates the application programmer from the details of multiple location-sensing technologies. Motivated by the initial separation of concerns espoused by the Context Toolkit, the location service separates problems of sensor acquisition, collection/aggregation, and monitoring of location information. Unlike the Context Toolkit, the implementation of the location service presents a cleaner programming interface for the development of location-aware applications; the programmer need only reuse or extend existing monitors to access, interpret and filter location data for entities of interest.

An important general topic for future research is the extent to which the layered solution presented for the location service will actually apply to other context types, such as activity. Information about tracked entities is contained within the collector layer of the location service. While this solution is reasonable when we consider only location context, we still must decide whether this approach can be used to address other forms of context.

13.6. CONCLUSIONS

An important component of multiple user interfaces is their ability to adapt to changing context, whether that be the actual users, their capabilities, location, or any other environmental cues. In this chapter, we have illustrated the value of using context and detailed the difficulties in using it. To address these difficulties, we presented the Context Toolkit, a component-based infrastructure that greatly eases the development of context-aware applications and provided the example of the Conference Assistant application, a compelling and complex context-aware application. The initial Context Toolkit had short-comings, forcing application developers to deal with new components being added and components failing along with the more time-consuming issue of dealing with contextual information that crossed multiple components. To address these shortcomings, we introduced the extended Context Toolkit which supports the situation abstraction, allowing application developers to specify their context interests at a high level and to have this information collected and filtered by the infrastructure. We presented the situation-aware reminder system, CybreMinder, to illustrate how the situation abstraction can be used. Finally, we presented the location service, a specialized system that deals strictly with providing applications with access to location context. It uses the same general notions as the Context Toolkit in terms of context acquisition, collection and monitoring (or subscription), but better supports application programmers in separating the application from the context infrastructure. An In/Out Board was used to illustrate the use of the location service. Note that while the example applications discussed in this chapter used simple forms of context such as location, time, identity and activity, the frameworks discussed are just as suited to more complex forms of context such as user goals and plans. Unfortunately, we do not have sensors or interpreters that can reliably produce these forms of context.

There is still much research to be performed in the area of context-aware computing. An interesting area to explore is how to make context even easier to use. One aspect of this is investigating how to support complex features such as fusion of homogeneous and heterogeneous context, as described in the previous section. Another aspect is to expand on existing efforts like Phidgets [Greenberg and Fitchett 2001], which make it very easy to build prototype applications that use simple sensors and actuators via a USB connection. A third aspect of making context easier to use is to provide techniques for dealing with sensors or interpretations of context information that is not completely accurate. It is unreasonable to assume that sensors or inferencing techniques will ever be 100% reliable, so our approach to this problem has been to bring users into the loop, allowing them to manually make corrections [Dey *et al.* 2002].

Another related area of interest is in how to provide additional control to users. Most context-aware applications are built assuming that the developer builds the application with no ability for the end-user to modify how the application operates. One aspect of this is how to support end-user privacy in a sensor-rich environment. This includes providing feedback to users about what context is being sensed and how that context is being used. It also includes supporting users in controlling how that sensed context is used in a built application and the further goal of allowing end-users to create their own context-aware applications from scratch.

Finally, as with the development of any software toolkit or architecture that supports the building of user interfaces, there is tension between evaluating the infrastructure independently of the interfaces and applications it supports, and evaluating the interfaces and applications themselves [Edwards *et al.* 2003]. When designing and evaluating infrastructure, we need to look critically at whether it supports core application features, whether it supports the development of applications as well as novel extensions to the architecture itself, and how it naturally and beneficially constrains application developers in leading them down an appropriate application design path.

ACKNOWLEDGEMENTS

We would like to thank Daniel Salber for his contributions in formulating the requirements and implementing the Context Toolkit. We would also like to thank Agathe Battestini and Thomas O'Connell for their contributions in building the location service, the discoverer service and other extensions to the toolkit. Finally, we thank all the members of the Future Computing Environments group that have used the toolkit and provided valuable feedback on it.

REFERENCES

Abowd, G.D., Atkeson, C.G., Hong, J., Long, S., Kooper, R., and Pinkerton, M. (1997) Cyberguide: A mobile context-aware tour guide. *ACM Wireless Networks*, 5(3), 421–433.

Addlesee, M., Jones, A., Livesey, F., and Samaria, F. (1997) The ORL Active Floor. *IEEE Personal Communications*, 4 (5), 35–41.

Arons, B. (1991) The Design of Audio Servers and Toolkits for Supporting Speech in the User Interface. *Journal of the American Voice I/O Society*, 27–41.

Aware Home Research Initiative (2003). http://www.awarehome.gatech.edu.

Bauer, M., Heiber, T., Kortuem, G., and Segall, Z. (1998) A Collaborative Wearable System with Remote Sensing. *Proceedings of the 2nd International Symposium on Wearable Computers (ISWC'98)*, 10–17. IEEE: Los Alamitos, CA.

Bederson, B.B. (1995) Audio Augmented Reality: A prototype automated tour guide. *Proceedings of CHI '95*, 210–211. ACM Press: New York, NY.

Beigl, M. (2000) MemoClip: A location based remembrance appliance. *Personal Technologies*, 4(4), 230–34.

Bennett, F., Richardson, T., and Harter, A. (1994) Teleporting: Making applications mobile. *Proceedings of the 1994 Workshop on Mobile Computing Systems and Applications*.

Brotherton, J., Abowd, G.D., and Truong, K. (1999) Supporting Capture and Access Interfaces for Informal and Opportunistic Meetings. *GIT-GVU-99-06 Tech. Rep.* Atlanta, GA: Georgia Institute of Technology, GVU Center.

Castro, P., Chiu, P., Kremeneck, T., and Muntz, R. (2001) A Probabilistic Room Location Service for Wireless Networked Environments. *Proceedings of Ubicomp 2001*, 18–34.

Davies, N., Mitchell, K., Cheverst, K., and Blair, G. (1998) Developing a Context-Sensitive Tour Guide. *Proceedings of the 1st Workshop on Human Computer Interaction for Mobile Devices*.

Dey, A.K. (2000). *Providing Architectural Support for Building Context-Aware Applications*. Georgia Institute of Technology dissertation.

Dey, A.K., and Abowd, G.D. (2000) CybreMinder: A context-aware system for supporting reminders. *Proceedings of the 2nd International Symposium on Handheld and Ubiquitous Computing (HUC2K)*, 172–186.

Dey, A.K., Abowd, G.D., and Wood, A. (1998) CyberDesk: A framework for providing self-integrating context-aware services. *Knowledge Based Systems*, 11(1), 3–13.

Dey, A.K., Mankoff, J., Abowd, G.D., and Carter, S. (2002) Distributed Mediation of Ambiguous Context in Aware Environments. *Proceedings of UIST 2002*, 121–30.

Dourish P., Edwards, W.K., LaMarca, A., Lamping, J., Petersen, K., Salisbury, M., Thornton, J.. and Terry., D.B. (2000) Extending Document Management Systems with Active Properties. *ACM Transactions on Information Systems* 18(2), 140–170.

Edwards, W.K., Bellotti, V., Dey, A.K., and Newman, M. (2003). Stuck in the Middle: The challenges of user-centered design and evaluation for infrastructure. To appear in the *Proceedings of CHI 2003*.

Feiner, S., MacIntyre, B., Hollerer, T., and Webster, T. (1997) A Touring Machine: Prototyping 3D mobile augmented reality systems for exploring the urban environment. *Personal Technologies*, 1(4), 208–217.

Fels, S., Sumi, Y., Etani, T., Simonet, N., Kobayshi, K. and Mase, K. (1998) Progress of C-MAP: A context-aware mobile assistant. *Proceedings of the AAAI 1998 Spring Symposium on Intelligent Environments*, 60–67.

Greenberg, S., and Fitchett, C. (2001) Phidgets: easy development of physical interfaces through physical widgets. In *Proceedings of UIST 2001*, 209–218.

Harrison, B.L., Fishkin, K.P., Gujar, A., Mochon, C., and Want, R. (1998) Squeeze Me, Hold Me, Tilt Me! An exploration of manipulative user interfaces. *Proceedings of the CHI'98*, 17–24.

Healey, J., and Picard, R.W. (1998) StartleCam: A cybernetic wearable camera. *Proceedings of the 2nd International Symposium on Wearable Computers (ISWC'98)*, 42–49.

Heiner, J.M., Hudson, S.E., and Tanaka, K. (1999) The Information Percolator: Ambient information display in a decorative object. *Proceedings of UIST'99*, 141–8.

Hertz (1999) NeverLost. Available at: http://www.hertz.com/serv/us/prod_lost.html.

Hightower, J., and G. Borriello. (2001) Location systems for ubiquitous computing. *IEEE Computer* 34(8), 57–66.

Hinckley K., Pierce, J., Sinclair, M., and Horvitz, E. (2000) Sensing techniques for mobile interaction. *Proceedings of UIST 2000*, 91–100.

Ishii, H. and Ullmer, B. (1997) Tangible Bits: Towards seamless interfaces between people, bits and atoms. *Proceedings of CHI '97*, 234–41.

Kiciman, E., and Fox, A. (2000) Using dynamic mediation to integrate COTS entities in a ubiquitous computing environment. *Proceedings of the 2nd International Symposium on Handheld and Ubiquitous Computing (HUC2K)*, 211–26.

Lamming, M., and Flynn, M. (1994) Forget-me-not: Intimate computing in support of human memory. *Proceedings of the FRIEND 21: International Symposium on Next Generation Human Interfaces*, 125–8.

McCarthy, J.F., and Anagost, T.D. (2000) EventManager: Support for the peripheral awareness of events. *Proceedings of the 2nd International Symposium on Handheld and Ubiquitous Computing (HUC2K)*, 227–35.

MacIntyre, B., and Feiner, S. (1996) Language-level support for exploratory programming of distributed virtual environments. *Proceedings of UIST'96*, 83–94.

Moran, T., and Dourish, P. (eds) (2001) *Context-aware computing*. Special triple issue of the *Human-Computer Interaction (HCI) Journal*, 16 (2–4–). Lawrence Erlbaum, Mahwah, NJ.

Mynatt, E.D., Back, M., Want, R., Baer, M., and Ellis, J.B. (1998) Designing Audio Aura. *Proceedings of CHI '98*, 566–73.

Nagel, K., Kidd, C., O'Connell,, T. Dey, A.K., and Abowd, G.D. (2001) The Family Intercom: Developing a context-aware audio communication system. In *Proceedings of Ubicomp 2001*, Technical Note, 176–83.

Orr, R.J., and Abowd, G.D. (2000) The Smart Floor: A mechanism for natural user identification and tracking. Short paper in *Proceedings of CHI 2000*, 275–60.

Pascoe, J., Ryan, N.S., and Morse, D.R. (1998) Human-Computer-Giraffe Interaction: HCI in the field. *Proceedings of the Workshop on Human Computer Interaction with Mobile Devices*.

Rekimoto, J. (1999) Time-Machine Computing: A time-centric approach for the information environment. *Proceedings of UIST'99*, 45–54.

Rhodes, B.J. (1997) The Wearable Remembrance Agent. *Proceedings of the 1st International Symposium on Wearable Computers (ISWC'97)*, 123–8.

Salber D., Dey, A.K., and Abowd, G.D. (1999) The Context Toolkit: Aiding the development of context-enabled applications. *Proceedings of CHI '99*, 434–41.

Schilit B., Adams, N., and Want, R. (1994) Context-aware computing applications. *Proceedings of the 1st International Workshop on Mobile Computing Systems and Applications*, 85–90.

Schilit, B. (1995) *System architecture for context-aware mobile computing*. Columbia University Dissertation.

Schmidt, A., Aidoo, K.A., Takaluoma, A., Tuomela, U., Laerhoven, K.V., and Velde, W.V.D. (1999) Advanced interaction in context. *Proceedings of the 1st International Symposium on Handheld and Ubiquitous Computing (HUC '99)*, 89–101.

Schwartz, M.F., Emtage, A., Kahle, B., and Neuman, B.C. (1992) A Comparison of Internet Resource Discovery Approaches. *Computer Systems*, 5(4), 461–93.

Shardanand, U., and Maes, P. (1995) Social information filtering: Algorithms for automating "word of mouth". In the *Proceedings of CHI '95*, 210–17.

Weiser, M. (1991) The Computer for the 21st Century. *Scientific American*, 265(3), 66–75.

Weiser, M., and Brown, J.S. (1997) The Coming Age of Calm Technology, in *Beyond Calculation: The Next Fifty Years of Computing*, 1st edn (eds P.J. Denning and R.M. Metcalfe) Springer Verlag, New York, 75–86.

A Run-time Infrastructure to Support the Construction of Distributed, Multi-User, Multi-Device Interactive Applications

Simon Lock[1] and Harry Brignull[2]

[1] *Computing Department, Lancaster University, UK*
[2] *School of Cognitive and Computing Sciences University of Sussex, UK*

14.1. INTRODUCTION

Many predict that we will one day be immersed in a world surrounded by devices, both visible and invisible, which we will be able to use collectively to help in our day-to-day activities [Cooperstock *et al.* 1997; Streitz *et al.* 1999]. For example, if we want an application to make use of the devices we have to hand (such as a PDA, a GPS device, or a hands-free cell-phone), we should be able to do so fluidly and easily. We have already witnessed an explosion of the number of devices in our environment and on our persons.

Multiple User Interfaces. Edited by A. Seffah and H. Javahery
© 2004 John Wiley & Sons, Ltd ISBN: 0-470-85444-8

Past models of user interaction have concentrated upon single PCs with interaction taking place through the traditional combination of keyboard, screen and mouse. However in the future, this model may no longer be appropriate for representing multiple interface systems. Such environments can be characterised as consisting of numerous devices used concurrently in highly dynamic arrangements, with devices routinely entering and leaving various configurations. In addition to this, user interaction is often 'spread' across several different devices, with both input and output being achieved through any number of these devices.

Currently the user faces massive problems if they want such devices to work together. Installation and configuration is a time-consuming task, which can in some circumstances require low-level development knowledge. This is a long way from the vision of being able to dynamically choose the devices you want to work together on the fly, in a fluid and easy interaction. The Dynamo infrastructure aims to provide the capability to support such activities, helping make this future vision of multi-device interaction a reality.

Imagine a room or building filled with devices: desktop PCs, interactive whiteboards, handheld devices, mobile phones, speakers, webcams, pointing devices, keyboards and so on. Add to this a constant ebb and flow of devices as they are switched on and off, brought in and out of range, break down or are repaired, and we have a particularly interesting and challenging domain. Now imagine people wandering in and out of this area, some with personal devices, some without, all wanting to interact with a range of applications through the available devices. Some people want to work together, some want to check their e-mail, some want to enter into a discussion, some want to find other people, others want to find their lunch. There is a vast array of interactive, collaborative and multimedia applications which they may wish to use. How can we support a developer in building useful applications to fulfil the needs of all of these people?

In this chapter, we first introduce a simple possible interaction scenario which involves a multi-user interface configuration. Although the scenario is a relatively simple one, it provides a basis for identifying some of the key MUI issues as we see them. Using this scenario as a target, we go on to describe an infrastructure which would allow the scenario to be realised. To aid this, we will derive a set of requirements for an infrastructure to support the realisation of the scenario. These form the requirements for the proposed Dynamo support infrastructure which we then go on to describe in some detail. This description begins with an initial outline of the infrastructure's design, a more in-depth consideration of its implementation and then an illustration of some of the features of its operation. Finally, we provide an illustration of the infrastructure in operation for the original MUI interaction scenario.

14.2. MUI INTERACTION SCENARIO

The scenario chosen to help derive requirements for a support infrastructure is a public/private electronic message board which contains a set of messages for particular users. These messages may either be presented with the contents visible (i.e. an open message) or the contents hidden (i.e. a folded message) A user may select a message from the message board in order to attempt to view its contents.

For example, let us consider the case of Vic and Bob who both work in the same research group at a particular university. Their research department has a large-scale, public, electronic message-board in the reception of an office building. Bob leaves home one morning with his cell-phone in his pocket. Sitting on the train on the way to the office, he decides to send a message to his colleague Vic. He switches on his phone and sends an SMS to the message-board's call-service. When Vic arrives in the office reception, she notices that a message is waiting for her on the message-board. Using her PDA, she downloads and opens up the message to privately view the contents. Vic decides to reply to the message, but rather than using the PDA's small keyboard which can be slow and difficult to use accurately, she makes use of a full-size wireless QWERTY keyboard which is available in reception. Using this, she creates a reply message on her PDA and posts it on the message board for Bob to see. Bob, who is still on the train, wonders if Vic has seen the message, so he dials up the message-board's call-centre and listens to a synthesised version of the reply Vic has posted.

Although this example is relatively simplistic, it is possible to extract and extrapolate a number of requirements for a MUI support infrastructure from its features and implications. These requirements are discussed in the following section.

14.3. REQUIREMENTS FOR INFRASTRUCTURE

The previous example indicates a number of key considerations which must be taken into account when constructing MUI applications. These are often departures from the acknowledged wisdom regarding single user, single interface systems. We must take all of these factors into account when constructing a support infrastructure of MUI applications. We can summarise the individual requirements for the proposed MUI support infrastructure as follows:

- *Spreading of interaction across devices*: A key requirement of any MUI support infrastructure is that instead of being tethered to a single device, a user interface must be 'spreadable' across a number of different devices. This capability offers numerous benefits to the user, including the ability to use multiple devices together in an application, which means that they can exploit the unique characteristics of each device. For example, a PDA is suited to the display of private information and for small, detailed and accurate types of interaction for a single user, whereas a large-screen 'whiteboard'-type device is suited to displaying public information and for broad-brush, large-scale interaction and for co-operative, multi-user interaction. Similarly, a QWERTY keyboard device is suited to prolonged text input whereas a stylus and touch-screen is more suited to sketching and annotating. For the mobile user, such an infrastructure allows them to take all the devices they have available on their person (and in the immediate environment) and create a 'bricolage', i.e. an assemblage improvised from the devices ready to hand, combining them to produce a far greater functionality than would be available from the devices as disparate entities. Benefits are particularly obvious for groupware users who wish to bring an eclectic mix of devices to a meeting and have them work together easily and efficiently.

- *Heterogeneity of devices*: Any potential developed infrastructure should support a wide range of heterogeneous devices with a variety of physical properties. This versatility will allow the widest possible range of, and maximise the possible combinations of, interactive devices. The devices from the described interaction scenario which require support from the MUI infrastructure include SMS (text message) gateways, large scale displays, PDAs, speech recognition and speech synthesis devices. The infrastructure should be flexible enough to be able to deal with a vast range of sensory and sub-conscious interaction devices, including displays, audio input and output, bio-sensing, mechanical input (e.g. mouse, stylus, keyboard) and so on.
- *Dynamic device integration*: In order to support the fluctuating nature of modern computing environments, support is provided for the dynamic addition, removal and update of maintained services. This is to allow applications to make use of all of the available devices at any one time, thus permitting the richest possible interaction to take place. New devices which become available should be able to be integrated into an already executing application in real-time. Outdated or unavailable devices should be updated or removed from those available to an application. An application should be able to use any available device with the minimum of overhead and without need for the developer to know of its existence or exact nature at construction time.
- *Device suitability determination*: So as to make best use of the interaction devices which are currently available, a successful support infrastructure should provide some mechanism for an application to determine the most suitable of the currently available devices. In order to do this, the application must provide some high level description of the features of a required device and the infrastructure should take on the responsibility of identifying the most appropriate device that is available. The device description mechanism must be such that the developer requires no knowledge of the devices currently available.
- *Adaptation of interaction*: Due to the different properties of the devices available for interaction in a particular situation, a support infrastructure should provide the ability for elements of the user interface to be suitably adaptive. This is to allow the user interface of an application to be appropriately displayed on whichever devices are being used to present it. We can split these concerns into the adaptation of both view and control elements of interactive interfaces. Both of these aspects must be adaptable for an infrastructure to be successful.
- *Accounting for privacy of information*: When working with shared applications, interfaces and devices, the privacy of data manipulated and displayed is of particular importance. It would be most undesirable for private information to be presented on a public device (e.g. a large display or via a public speaker system). For these reasons, it is desirable for any development infrastructure to include facilities to control and manage the presentation of private data and the utilisation of devices.
- *Data synchronisation across devices*: Due to the potentially large number of devices being used to present a user interface within a MUI configuration, some mechanism for managing the synchronisation of the data between the devices is required. Such a mechanism would be responsible for ensuring that the data being presented or manipulated on one device was consistent with other instances of that data at other locations

within the configuration. For these reasons, a scalable data synchronisation mechanism is essential to ensure the success of any support infrastructure.

14.4. EXISTING APPROACHES

The Pebbles project [Myers *et al.* 1998] has undertaken some interesting work in the use of handheld devices to extend the interaction capabilities of desktop PCs and large scale projected displays. Pebbles consists of a number of static, fixed applications with purpose built 'remote control' software components installed on handhelds, with augmented or specially constructed applications on desktops or large scale displays. Configurations which have been built include the SlideshowCommander PowerPoint controller, the PebblesDraw multi-user whiteboard application, the ShortCutter desktop remote control, the RemoteCommander keyboard and mouse event simulator, and so on.

A basic infrastructure which uses multiple interaction devices to support the construction of applications has been developed by Rekimoto [Rekimoto 1998]. This system uses fixed configurations of handheld devices to extend the capabilities of an interactive whiteboard. The handheld devices are used to present a fixed set of interactive components such as tool palettes, text entry boxes and web or filestore conduits.

The iRoom infrastructure [Fox *et al.* 2000] connects interactive whiteboards, projection tables, wireless mice and keyboards, and handheld devices to support distributed user interaction. The focus in this system is on the augmentation of existing windows applications, such as PowerPoint, so that they may utilise the additional interaction properties of handheld devices. The infrastructure is based around a centralised 'event heap' which handles the reception and transfer of interaction events between devices. Data synchronisation in this infrastructure is achieved using the third party 'Coast' mechanism developed by the German National Research Centre for Information Technology [Schuckmann *et al.* 1996]. This infrastructure relies on pre-defined, loose confederations of devices based on 'soft' linkages and asynchronous messaging.

Another infrastructure which has been developed to provide support for distributed user interaction is the BEACH system [Tandler 2000]. BEACH uses the concepts of segments (display areas) and overlays (windows) to abstract over the resources of interaction devices. Each device is represented as a BEACH client and communicates with other devices and BEACH components using a multi-user event handling mechanism. As with the iRoom above, BEACH is dependent upon the Coast system for data synchronisation. Within this infrastructure, display segments may be integrated together to form larger display areas. This approach also introduces scaling and rotation mechanisms to allow graphical user interfaces to be transformed to suit a particular display device.

A situated computing framework which has been developed by Pham *et al.* [2000] uses combinations of hardware devices from a user's environment for the presentation of multimedia content such as still images, audio and video. Possible presentation devices include handhelds, PCs, workstations, TVs, telephones, and so on. A novel feature of this particular infrastructure is the support it provides for the dynamic ebb and flow of interaction devices within the user's vicinity. A key feature of the infrastructure is the automated selection of the most appropriate available output device to present a particular type of media. If no suitable devices are available then adaptations to the media may be applied, such as splitting, conversion and filtering, to try to match the media to devices.

The SharedNotes system [Greenberg *et al.* 1999] is a single application, rather than a support infrastructure, but one which demonstrates a number of interesting features. This application utilises handheld devices, desktop PCs and a public electronic whiteboard to provide interaction with a group note-taking and discussion system. SharedNotes supports the creation of private notes on handheld devices, the selective publicisation of these notes on the whiteboard, the collaborative editing of notes and the taking away of a static record of notes at the end of a discussion session. The application also supports the automatic synchronisation of all instances of a note on the various devices upon which it may be present.

The Manifold architecture [Marsic 2001] (based on the DISCIPLE framework [Wang *et al.* 1999]) was created for developing groupware applications. Manifold performs the automatic run-time mapping of application data content to suit available presentation devices. XML documents are used to describe application data, with style sheets being used to perform the dynamic mapping. Manifold, however, only employs hardwired device-side application elements with no run-time selection of MVC control elements. The described infrastructure appears relatively heavyweight and encounters various difficulties when utilised with less powerful devices such as handhelds. The main focus of this work is on providing different views of the same synchronised data on alternative devices (e.g. 3D workstation versus handheld). No spreading of user interface is performed and no run-time discovery of devices is employed, with all devices having to be explicitly added to a DISCIPLE session.

WebSplitter [Han *et al.* 2000] is a mechanism which can be used to spread the content of a web page across a number of different devices. In this way, it is possible to utilise nearby multimedia devices if the current browsing device is not able to handle all of the content (e.g. if it were a PDA). So, for example, a nearby speaker could be used to present audio content or a large display could be used to present pages to a number of people at once. XML tags within the web pages are used to selectively filter and partition the page between the different devices. WebSplitter also allows devices such as PDAs to be used as remote controls to facilitate page navigation. WebSplitter uses a service discovery component to facilitate devices and applications in discovering each other. This is also used to ensure the most suitable device is used to present a particular element of a web page. A policy file is used to define the mapping rules which govern which XML tags should be distributed to which types of device. This is done using the concept of 'capabilities' which are special tags that describe properties of devices. All of these facilities permit the ad-hoc integration of devices and the subsequent run-time distribution of web page content. Finally, WebSplitter provides a mechanism to allow a user to manually alter the distribution of components among the available devices.

The system developed by Petrovski [Petrovski and Grundy 2001] provides a distributed, multi-user health information system. This system uses a variety of devices including PDAs, WAP phones, desktop machines and generic web browsers in a relatively static configuration. Jini, surrogate architecture, Java server pages and CORBA middleware components are used to provided connectivity between devices, application logic and the commercial back-end database. No support for discovery is provided and devices must be integrated manually into the system configuration. The system does provide multiple representations of the same centralised data, but these are either through pre-defined user interfaces or take the form of inflexibly generated web pages. The whole system is based around a client server model where all data is passed to and from a centralised machine.

Table 14.1 shows to what extent the existing approaches meet the previously identified requirements from our interaction scenario.

It is clear from the above table that no one approach fulfils all of the requirements derived from our interaction scenario. The following sections describe the design, implementation and operation of an infrastructure which does in fact meet all of these requirements.

14.5. DESIGN OF INFRASTRUCTURE AND DEVELOPMENT FRAMEWORK

In this section we introduce the design for an application infrastructure and development framework which will fulfil all of the specified requirements for the interaction scenario.

In Figure 14.1, the various entities are as follows:

- User – A person who interacts with the system
- Device – A physical device which permits user–system interaction
- Service – A component which provides access to a particular aspect of a device
- Consumer – A component which provides a programmatic interface to a remote service
- Application – A purpose-built application which utilises the infrastructure and device services

In the infrastructure, a device is represented by a service or set of services that encapsulate the total functionality of the particular device. For example, we can decompose a standard PC device into offering a display service, a text input service, a pointing service, and perhaps audio in and out services. The physical devices are controlled by these services, which are accessed by device consumer classes across some form of interconnect (e.g. a TCP/IP network). The infrastructure provides numerous device services and device consumers for commonly used interaction devices, including those shown in Table 14.2.

The Dynamo infrastructure utilises a centralised registry of currently available device services to support discovery by applications. This registry supports the real-time registration, reconfiguration and removal of device services. When an application requires interaction with a user, a device service which can provide suitable interaction is first identified from the registry. To aid in the identification and selection of device services, each one is associated with a number of description parameters which describe their specific abilities and constraints. Identified devices can then be dynamically linked into and utilised by the application at run-time. Once a device service has been selected, a device consumer is then created and linked to it in order to allow an application access to the features of the service. Once a device has been linked to an application, it is then possible

Table 14.1. How existing approaches satisfy the requirements.

Criteria / Approach	Smearing input	Smearing output	Full Heterogeneity	Dynamic integration	Suitability determination	Adaptation of view	Adaptation of control	Accounting for privacy	Data synchronisation
Pebbles	✓	✓	✗	✗	✗	✗	✗	✗	✗
Rekimoto	✓	✗	✗	✗	✗	✗	✗	✗	✗
iRoom	✓	✓	✗	✗	✗	✗	✗	✗	✓
BEACH	✓	✓	✗	✓	✗	✓	✓	✗	✓
Pham	✓	✗	✗	✓	✓	✓	✗	✗	✗
SharedNotes	✓	✓	✗	✗	✗	✓	✓	✓	✓
Manifold	✗	✗	✗	✗	✗	✓	✓	✗	✓
Petrovski	✗	✗	✗	✗	✗	✓	✓	✗	✗
WebSplitter	✓	✓	✗	✓	✓	✓	✗	✓	✗

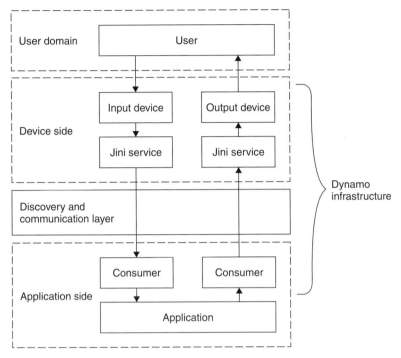

Figure 14.1. The Dynamo infrastructure.

Table 14.2. Device services provided by Dynamo.

Device Service	Description
AudioInService	Provides access to microphone, line-in, CD and mixer equipment
AudioOutService	Provides access to speaker and headphone equipment
CameraService	Provides access to webcam equipment
DisplayService	Provides access to a variety of graphical display equipment
HTTPService	Provides on-line access to world wide web documents
KeyboardService	Provides access to character input equipment
MouseService	Provides access to pointer device equipment

to use that device for interaction. In order to allow interaction to take place, the application distributes a fragment of user interface to the device service, using the linkages which have previously been established. User interaction with the application can then take place through the device using this interface fragment.

14.5.1. DESIGN OF INTERACTION METAPHOR

To help both application developers and end users to conceptualise the fragmentation, distribution and presentation of these user interface fragments on the various devices present in an environment, a suitable metaphor was required. The infrastructure thus utilises the concept of 'bubbles' to represent lightweight fragments of user interaction and their realisation across a number of input and output devices.

Bubbles are lightweight encapsulations of logically similar portions of an application's interactivity. The interactivity which is clustered together into a single bubble should be of a similar type and purpose. The elements present in a single bubble should thus be highly coupled, consistent and coherent aspects of an application's interactivity. For example, graphical visualisation elements of an application's user interface should be located in the same bubble if possible. Equally, audio output elements of an interface should also be clustered together. The aim of such groupings of interactivity is to try to promote one-to-one mappings of bubbles to interaction devices. This should thus improve the likelihood of a suitable device being found to realise a particular bubble.

The interactive elements of an application can be split into numerous separate bubbles which are then distributed to various devices for realisation (presentation to the user through the device). Splitting the interaction of an application into a number of different bubbles makes it easier to identify devices which are able to satisfy a particular bubble. This is because we are obviously more likely to find a device to fulfil the needs of a subset of application interactivity (i.e. a single bubble) rather than the entire interaction requirements of the application as a whole.

Each fragment includes a description of the bubble's presentation, internal bubble data and behavioural control logic. These elements closely match the long established model-view-controller (MVC) paradigm of user interface modularisation [Dix *et al.* 1998]. Figure 14.2 shows the distribution of three separate bubbles from an application to different realiser devices.

Once transferred to a particular device, a bubble does not have to stay on that device, but may be forwarded on to be presented by other devices as many times as the application requires. We refer to such entities as 'bubbles' to emphasise their lightweight nature and ability to 'float' between devices. In addition to this, bubbles are also 'squashy' in that their resource requirements may be reduced to meet the constraints of available interaction devices. Finally, the process of creating and distributing such encapsulations of interactivity can be envisioned as the blowing out of a stream of bubbles onto a collection of waiting devices.

Also incorporated into bubbles are privilege matrices which are used to enforce access rights for the manipulations of bubbles and their associated data. The information held within a matrix indicates the various privileges that each user has over the bubble in question. The complete set of actions present in each matrix is shown in Table 14.3.

The use of these access rights can best be illustrated with an example from the described MUI scenario. If the user has *read* access to the message then a bubble will be sent to an appropriate device to present its contents. If the message contains sensitive information then the bubble will be sent to a private device available to the user. If the message is not sensitive then the next available public device will be used.

Due to the fact that particular bubbles may be distributed to more than one device at any one time, some mechanism is required to ensure the synchronisation of multiple instances of the bubble in various locations. The infrastructure automatically synchronises these instances on the different devices. Thus, any changes which take place in the data

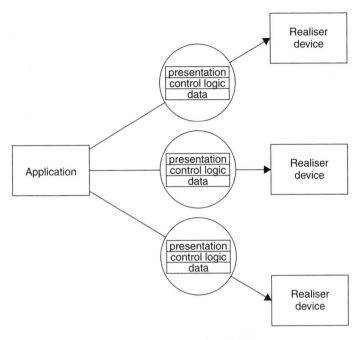

Figure 14.2. Distributing bubbles.

Table 14.3. Actions in the bubble privilege matrix.

Action	Description
Blow	Records if a user is permitted to realise the bubble on any device
Sense	Records if a user is permitted to appreciate the realisation of a bubble on a device
Alter	Records if a user is permitted to make changes to the data content of a bubble
Waft	Records if a user is permitted to move a bubble from device to device
Clone	Records if a user is permitted to make a copy of a bubble
Pop	Records if a user is permitted to remove a bubble from a device

element of one instance of a bubble are propagated to all other instances to maintain consistency.

14.5.2. BUBBLE GLOSSES

Some frameworks support the automatic generation of user interfaces from structured descriptions [Abrams *et al.* 1999]. We believe that due to the importance of user interfaces, they should be designed and built using best practices from HCI theory and techniques. In our view, the current state of the art in automatic generation is not adequate to produce interfaces that are well designed enough to be suitable for use with real world, large scale systems. For this reason we employ numerous hand-crafted alternative interfaces as a basis for providing multiple representations of a single bubble.

Each bubble is associated with one or more 'glosses' which represent alternative real-isations of the presentation facet of the bubble. The various glosses associated with a bubble will suit different types and configurations of interaction device. Each bubble has a list of possible glosses which is prioritised in order of preference. That is, the most suitable and desirable gloss is first on the list. This priority order may be altered at any time during run-time to reflect changes in circumstances within the application and its environment. The glosses of a particular bubble may vary greatly in type and form. For example, a particular bubble may have a graphical as well as an audio gloss, both of which provide alternative realisations of the bubble.

Each gloss of a bubble is associated with a number of device descriptor parameters which indicate the properties and behaviours required to realise that gloss. Using these parameters, it is possible to identify which glosses are suitable for realisation on which devices and vice versa. When a device is attempting to realise a bubble, the description parameters of all of the bubble's glosses are interrogated to find the most appropriate gloss for the devices in question. To help illustrate the use of glosses, Figures 14.3 to 14.5 show how some of the glosses used in the example scenario might look when presented on the various devices available.

A possible transcript of an audio-based gloss for the presentation of messages via a telephone call is presented below:

Voice synthesiser: Welcome to the interactive messaging service.
Voice synthesiser: You have two new messages.
Voice synthesiser: Press or say 'one' to review messages, press or say 'two' to exit.
User: One!
Voice synthesiser: Message received from Steve at 5 p.m. yesterday.

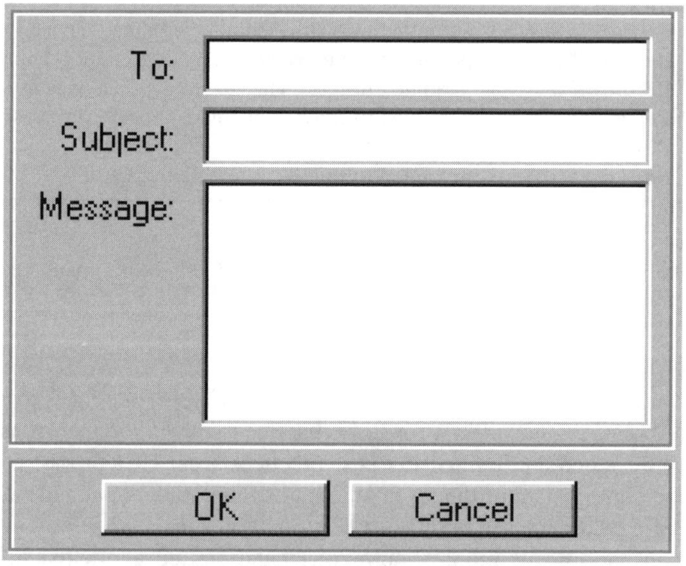

Figure 14.3. The handheld or kiosk PC creator graphical gloss.

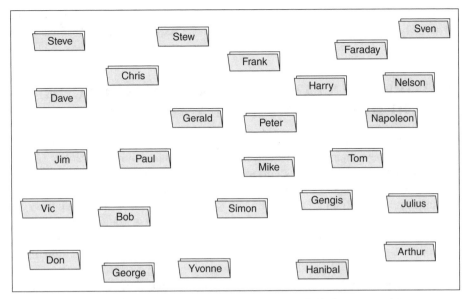

Figure 14.4. The public message board containing folded message glosses.

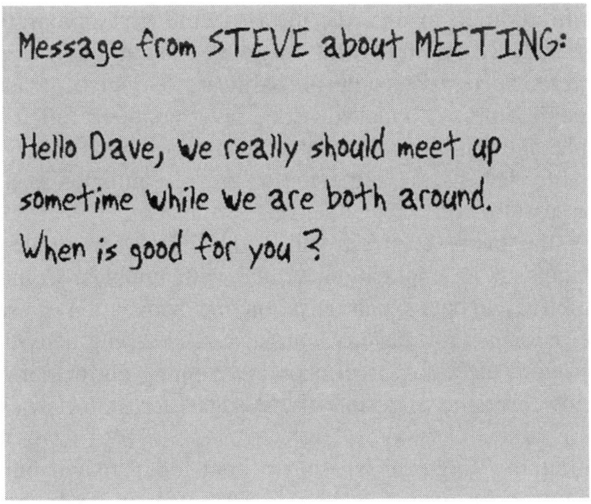

> Message from STEVE about MEETING:
>
> Hello Dave, we really should meet up sometime while we are both around. When is good for you?

Figure 14.5. The private handheld or kiosk PC graphical message gloss.

Voice synthesiser: Press or say 'one' to hear message, press or say 'two' to move on to next message.
User: One!
Voice synthesiser: Message content is as follows.…
etc…

Finally, the SMS gloss used in the creation of new messages (which cannot be illustrated visually) is basically an SMS parser which checks incoming SMS, looks for new message requests, creates new message bubbles and initiates distribution as required.

14.6. IMPLEMENTATION OF INFRASTRUCTURE AND DEVELOPMENT FRAMEWORK

The infrastructure described in this chapter has been designed to support the construction of distributed, heterogeneous multi-device, multi-user, interactive applications. A prototype version of the infrastructure is currently available and has to this date been used to construct and test a number of different multi-user interactive applications.

The infrastructure is built on top of Jini and RMI technologies from Sun Microsystems. Jini provides all of the basic facilities for the description, registration and discovery of device services. This is based around one or more centralised lookup services which store details of all currently available services. A lookup service is first identified by multicast communication; once identified, new services can be added to the registry or it can be queried to identify a suitable device for a particular purpose. The infrastructure also provides a programmer's API and set of common device services to aid in application development.

The infrastructure takes on many of the important and common tasks required to build such applications. In addition to this, the infrastructure gives programmers access to a selection of varied and diverse input and output devices through a high-level service-consumer architecture. Services execute on hardware to which devices are directly or indirectly (i.e. through a proxy) attached. Code level adapters called 'Consumers' give programmers simple and easy access to the features of remote services. In this way heterogeneous, remote devices are presented to the programmer as a consistent set of local consumer classes.

RMI is used by the bubbles infrastructure to achieve communication between the distributed components of an application written with bubbles. Communication within bubbles involves control and data synchronisation, the distribution of user interface events and the passing of messages for various internal housekeeping activities. Such communication is achieved with the aid of RMI daemons running on the various devices which handle the actual low level transmission of messages across the available interconnection mechanism.

The Java programming language is used to bind together and build additional features and facilities on top of these low level mechanisms in order to create a complete implementation of the bubbles run-time infrastructure. A developer-level API has also been provided for the Java programming language to allow applications to be built which utilise the bubbles infrastructure. This developer-level framework will be discussed in the following section.

A main aim of the infrastructure is to encompass and support the integration of a wide variety of heterogeneous devices. Depending on the resources and capabilities of a particular device, it may be integrated into the infrastructure in one of two ways. Devices with Jini and Remote Method Invocation (RMI) support can be connected directly to the

Dynamo infrastructure. Devices with no Jini/RMI support must be connected indirectly, via a special device proxy. Where relevant in this chapter, descriptions of infrastructure operation will appear for both of these types of device.

14.7. OPERATION OF THE INFRASTRUCTURE

The aim of this section is to provide a more concrete description of the infrastructure by describing the various features of dynamic operation. This should shed light on how many of the features and functions of the infrastructure are achieved in practice.

14.7.1. DYNAMIC DEVICE SERVICE REGISTRATION

The infrastructure utilises a Jini lookup service as a centralised registry for all enabled and currently available device services. In many multi-device interaction scenarios, there is a constant ebb and flow of devices as people move around and equipment is switched on and off. The use of the Jini lookup service assists the infrastructure in supporting the real-time registration, reconfiguration and removal of device services.

All Jini/RMI enabled devices integrated into the infrastructure are represented as one or more Jini services. When a device is switched on, plugged into the network or comes within range, it notifies the registry to indicate that it is able and willing to handle requests from consumers. Non Jini/RMI enabled devices must use the Java Surrogate Architecture (SA) [Sun Microsystems 2001] in order to participate in interactive applications built using the Dynamo infrastructure. When employing the surrogate architecture, a Jini/RMI enabled surrogate host server is used as a proxy for a non Jini/RMI device. The host server bridges the gap between the main Jini/RMI protocols used by the infrastructure and the proprietary communication protocols of non Jini/RMI device services. As with Jini/RMI enabled devices, the Jini registry is still used to maintain a list of all the currently enabled and available devices, however the entries in the registry refer to the proxies rather than the services themselves. When a device is switched on, plugged into the network or comes within range, it uploads its proxy onto the surrogate host. This proxy then registers with the Jini registry to indicate that it is able to handle the requests of consumers.

14.7.2. DYNAMIC DEVICE SERVICE SELECTION

When an application needs to engage in a particular form of user interaction, a device service which can provide suitable interaction is first identified by querying the Jini lookup. The infrastructure allows devices to be dynamically identified and utilised by applications on the fly. This dynamic linking makes it possible for applications to maximise interaction quality by having access to all currently available devices.

To simplify the identification and selection of devices, no application interfaces directly with the Jini registry. Instead, all applications interact with a utility class, known as the 'Arbitrator'. An application developer specifies what type of interaction they require at a particular point in an application, using query templates to specify the desired physical properties of a suitable device. At run-time, the Arbitrator will then provide a set of currently available services which are most appropriate to the needs of the application. On the basis of this information, a linkage between application and device can be made. The infrastructure then supports communication between the main application and chosen devices.

The Arbitrator abstracts over the Jini registry, but also provides additional, more advanced functionality for device selection such as:

- Selecting given combinations of device properties
- Selecting given combinations of interfaces and users
- Enforcing access privileges on user interfaces
- Handling private viewing and public announcement concerns
- Load balancing between interaction devices
- Minimum resource arbitration combined with upgrading and degrading of service

To aid in the selection of device services, each one is associated with a number of description parameters which describe their specific abilities and constraints. When the Jini registry is queried, these parameters can be used to identify services which are suitable for use by the application. The set of parameters which may be used in describing device services is extendable and can be added to by the developer as and when new parameters are required. A few of the basic device description parameters for a number of common devices are shown in Table 14.4.

14.7.3. APPLICATION SERVICE LINKAGE

Once a device service has been selected, a device consumer is then created and linked to it in order to allow an application access to the features of the service. With Jini/RMI enabled devices, communication between the service and consumer is achieved using the RMI features of Java. To achieve application-to-device communication, a consumer must first interface with the RMI code on the application side. RMI then uses the standard TCP/IP network protocols to communicate with RMI on the remote (device side) machine. The device side RMI then interfaces directly with the device service code to invoke its services. This architecture is illustrated in Figure 14.6.

Devices without support for RMI, and indeed devices with no support for Java, may still be integrated into the infrastructure. Communication with such a device's services must be performed using proprietary mechanisms since RMI-based communication is

Table 14.4. Device description parameters.

All devices	Graphic display	Pointing device	Audio output
mobile [bool]	size [int*int]	button count [int]	bit rate [int]
owner [string]	resolution [int]	accurate [bool]	encoding [string]
authorised user [string]	employment level [int]	fast [bool]	stereo [bool]
privacy type [int]			
number of users [int]			
usage range [int]			

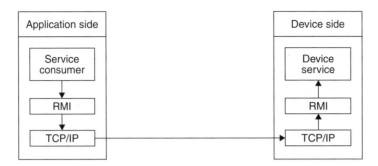

Figure 14.6. Device service with RMI support.

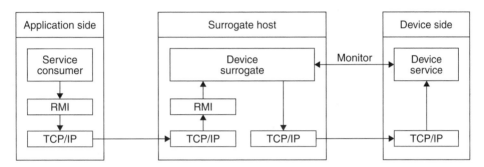

Figure 14.7. Device service without RMI support.

not possible. The architecture employed in the connection of non-RMI devices into the infrastructure is shown in Figure 14.7.

The figure shows that in order to make use of a particular service, the application side cannot contact the device side directly since no RMI support is available. Instead, a consumer interfaces with RMI on the application side as normal. However, this RMI code must connect to a surrogate 'proxy' rather than directly to the device side. This is still done using TCP/IP network protocols between the two RMI elements. All RMI requests received by the proxy must then be converted into an appropriate representation that can then be forwarded on via a raw TCP/IP connection to the service side to be handled by the relevant device services. Service monitoring by the surrogate must take place to ensure that appropriate action takes place when the device is switched off, breaks down, is disconnected, or is otherwise unavailable.

14.7.4. BUBBLE SYNCHRONISATION

A particular bubble may be distributed to more than one device at the same time, resulting in the simultaneous presentation of multiple instances of the bubble in various locations. Different instances of the same bubble presented on different devices are automatically synchronised by the infrastructure. Thus, any changes which take place in the data element

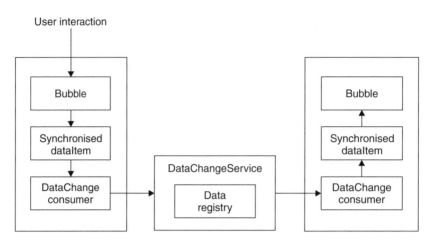

Figure 14.8. Data Synchronisation in Dynamo.

of one instance of a bubble are propagated to all other instances to maintain consistency. This data synchronisation mechanism makes use of a number of special "Synchronised-DataItem" classes (e.g. SynchronisedString, SynchronisedVector etc). These classes are adaptations of existing standard data types, but with additional data communication and integration facilities. The mechanism used to propagate data change is illustrated in Figure 14.8.

The diagram shows that when one copy of a data item is altered (via user interaction with a bubble for example), a DataChangeConsumer is first notified. This consumer passes on the notification to a centralised DataChangeService which maintains a registry of all copies of all synchronised data items. When the DataChangeService receives a notification, it fires out an update notification to all copies of the originally altered data item. This update is received by the DataChangeConsumer which is local to each copy of the data item. Once the data items have updated their internal values to reflect the propagated change, any bubbles which present the values held in the data item are also updated.

Distributed data synchronisation achieved in the manner described above has a fundamental empathy with the architecture of the infrastructure which is parallel and distributed in nature. Distributed synchronisation also provides some resilience to temporary loss and regaining of connection, common in the type of applications targeted by the infrastructure. In matters relating to the synchronisation of data values, the described infrastructure implements an open policy based mechanism. This allows developers to produce new synchronisation policies and 'plug' them into the infrastructure, thus allowing any synchronisation algorithm to be used.

14.8. INFRASTRUCTURE UTILISATION

We have thus far described the infrastructure in some detail, we now however extend this discussion by demonstrating the utilisation of the infrastructure for the example

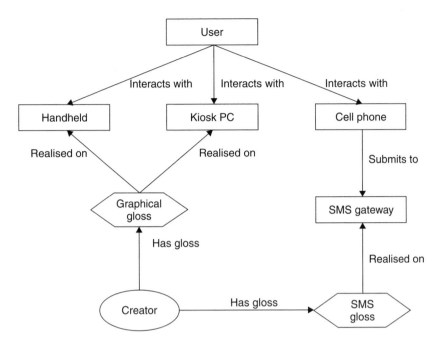

Figure 14.9. Message creation sub-architecture.

MUI scenario. The scenario has two main use cases, the creation of a new message and the viewing of an existing message. Figures 14.9 and 14.10 show the architecture of the system, split between these two use cases. This partitions the system logically and makes the architecture simpler to understand compared with a single, more complex representation. Due to the dynamic nature of device registration and selection, the architecture diagrams encompass a number of different variations for achieving the desired functionality.

Entities from the diagram are as follows:

- User – the application end user who interacts with the system.
- Handheld – a handheld interaction device (e.g. a palm pilot, an IPAQ, etc). This device has display, pointer and character entry services.
- Kiosk PC – a desktop PC provided for use by the general public. This device has display, pointer and character entry services.
- Cell phone – personal mobile phone with standard audio and SMS capabilities.
- SMS gateway – an existing subsystem that permits the transmission of SMS messages.
- Creator – a bubble which encapsulates the interaction required to create a new message.
- Graphical gloss – a presentation of the creator bubble suitable for realisation with graphical display, pointer and character entry services.
- SMS gloss – a presentation of the creator bubble suitable for realisation via SMS on the SMS gateway service.

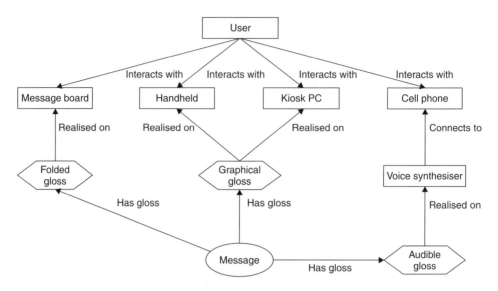

Figure 14.10. Message viewing sub-architecture.

Entities from the diagram are as follows:

- User – the application end user who interacts with the system.
- Message board – a large scale, public screen (plasma, projected etc), providing both display and pointer services.
- Handheld – a handheld interaction device (e.g. a palm pilot, an IPAQ, etc). This device has display, pointer and character entry services.
- Kiosk PC – a desktop PC provided for use by the general public. This device has display, pointer and character entry services.
- Cell phone – personal mobile phone with standard audio and SMS capabilities.
- Voice synthesiser – a gateway that transforms text into speech for rendering on a mobile phone.
- Message – a bubble which encapsulates a single message.
- Graphical gloss – a presentation of the message bubble suitable for realisation with graphical display, pointer and character entry services.
- Audible gloss – a presentation of the message bubble suitable for realisation on a mobile phone via a voice synthesiser gateway service.
- Folded gloss – a presentation of the message bubble suitable for realisation with a large scale, public display service.

14.9. APPLICATION USAGE SCENARIOS

We now demonstrate the operation of the message board system and infrastructure using a number of possible usage scenarios. For the sake of simplicity, no exceptions to the normal

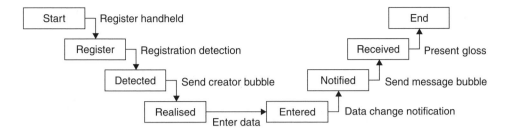

Figure 14.11. User adds message to board using a handheld.

thread of interaction with the system are presented here. This includes such occurrences as the inability to find a suitable available service, the execution of the system when a user attempts a particular action without inappropriate access privileges, the operation of the system if a user cancels the creation of a message part way through, and so on.

Stages from the process model shown in Figure 14.11 are as follows:

- Register handheld – the handheld device is switched on (or device-side components booted) causing the device's services to initialise and be registered with the Jini lookup.
- Registration detection – the application detects that a new device has been added and links it in using a new consumer.
- Send creator bubble – the application sends the message creation bubble to be realised to the handheld so that the user may create a new message.
- Enter data – the user enters data into the message creator bubble using the graphical gloss which is presented on the handheld's display.
- Data change notification – when the message has been created (and accepted), a data change notification is received by the application.
- Send message bubble – the application then sends a completed message bubble to the large scale public display (we assume the public display has previously been identified and linked into the application).
- Present gloss – the public display presents the bubble graphically, using the 'folded' gloss to maintain privacy.

Stages from the process model shown in Figure 14.12 are as follows:

- Enter SMS text – the user enters the text of an SMS message on their phone.
- Send SMS – once happy with the message, the user sends the SMS to the appropriate service centre.
- Parse message – on receipt of the message, its contents are parsed to identify distinct elements.
- Insert data – the parsed data items are inserted into a new message.
- Data change notification – when the message has been filled, a data change notification is received by the application.
- Send message bubble – the application then sends a completed message bubble to the large scale public display.

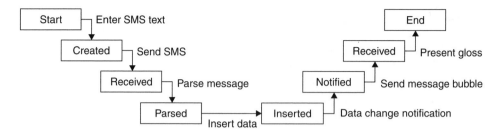

Figure 14.12. User adds message to board using a cell phone.

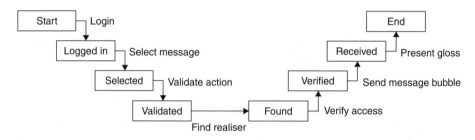

Figure 14.13. User views public message from board.

- Present gloss – the public display presents the bubble graphically, using the 'folded' gloss to maintain privacy.

Stages from the process model shown in Figure 14.13 are as follows:

- Login – the user logs into the public display with their username.
- Select message – the user then selects a folded message from the display.
- Validate action – the system checks to ensure that the action is valid (i.e. the user has access to open the folded message).
- Find realiser – since the message selected is a public one, a suitable public display is identified for the presentation of the message.
- Verify access – the system checks to ensure that the user has access to the chosen display.
- Send message bubble – the application then sends the message bubble to the public display.
- Present gloss – the public display presents the bubble using the unfolded gloss which is most suitable.

Stages from the process model shown in Figure 14.14 are as follows:

- Register handheld – the handheld device is switched on (or device-side components booted) causing the device's services to initialise and be registered with the Jini lookup.

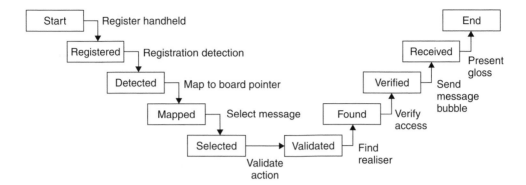

Figure 14.14. User views private message from board using handheld.

- Registration detection – the application detects that a new device has been added and links it in using a new consumer.
- Map to board pointer – the pointing service provided by the handheld is mapped to a pointer on the public display, allowing the user to interact with the public display using their own stylus.
- Select message – the user then selects a folded message from the display using the stylus-based pointing mechanism.
- Validate action – the system checks to ensure that the action is valid (i.e. the user has access to open the folded message).
- Find realiser – since the message selected is a private one, a suitable private display is identified for the presentation of the message (this may well be the display service on the handheld, if it is suitable).
- Verify access – the system checks to ensure that the user has access to the chosen display.
- Send message bubble – the application then sends the message bubble to the private display.
- Present gloss – the private display presents the bubble using the unfolded gloss which is most suitable.

Stages from the process model shown in Figure 14.15 are as follows:

- Register phone – the user dials a pre-defined number to register the phone and invoke the message checking service.
- Registration detection – the application detects that a new device has been added and links it in using a new consumer.
- Check for messages – the current set of valid messages is queried to find if there are any messages for the user.
- Find gateway – if any messages for the user are present, a voice synthesis gateway is identified.
- Send message bubble – the application then sends the message bubble to the gateway for realisation.

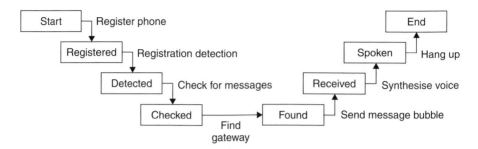

Figure 14.15. User listens to private message from board using cell phone.

- Synthesise voice – the textual contents of the message is converted into audio and piped down the phone line to the user.
- Hang up – once completed, the gateway closes the connection with the user's phone.

14.10. DISCUSSION

Due to the complexity of building distributed, multi-user, multi-device applications, we have focused much attention on making the infrastructure simple and easy to understand. When attempting to control the perceived complexity of the infrastructure, we considered the perspectives of both the application developer and the end user. The infrastructure thus presents a clean and simple API to the developer to allow ease of application construction. The end user is also presented with simple interfaces and metaphors for interacting with the infrastructure and applications. The bubble metaphor is instrumental in abstracting intricacies and complexities of the infrastructure. By providing a universal and easily understandable model of infrastructure operation, it becomes much simpler for the developers and end users to appreciate and conceptualise the behaviour of the framework. Low-level bubble transmission and data synchronisation are hidden from all concerned and performed internally by the infrastructure.

The work described in this chapter is part of an ongoing project to investigate the interaction between humans and computers through multi-user interfaces. The infrastructure which has been described in this chapter is mature enough to allow complex, real world applications to be constructed. To this date, we have built numerous different MUI applications using the infrastructure. These include a multi-user distributed road traffic jam avoidance system, a conference attendees contact board system, and the message board application from the example scenario described in this chapter. Commonalities identified during the development of these applications were factored out and fed back into the infrastructure in order to extend the support provided to the developers of such applications.

Our experiences with using the approach to date are promising and we feel that the above factors have resulted in an infrastructure which offers a wide range of essential features to the developers of future MUI applications. This is a direct consequence of the fact that the infrastructure undertakes many of the commonly experienced difficult, complex and unpleasant tasks involved in the construction of MUI applications. In addition to this, the infrastructure is an invaluable tool to aid in the investigation and examination of the type, nature and form of user interaction in MUI applications.

We are currently planning a further phase of evaluation of the dynamo infrastructure, perhaps involving the use of one or more of the previous approaches described in an earlier section of this chapter as a comparison against which to assess the new infrastructure. In addition to this, and due to the emphasis of the Dynamo project on HCI, we also intend to undertake some socio-technical studies of the applications built on top of this and other infrastructures as a further evaluation of the approach.

The development and utilisation of this new infrastructure highlighted a number of important new and complex HCI issues within the area of MUIs. At the most general level, we need to address the problem of how to extend the bubbles conceptual model to give end users enough understanding to use the system fully, while masking its underlying complexity. This model extension must involve a design process of extensive prototyping, user-testing and in-depth study. Key questions include, how do we design the interaction and dialogue between the user and infrastructure to allow bubbles to be manually wafted from device to device? And, what kind of device selection mechanism do we offer the user for them to select which devices to use for a particular interaction? Rather than requiring the user to drill down through many layers of menus, we shall be experimenting with various direct manipulation style methods which allow the user to physically choose the devices they want to use in the real world (e.g. using a bar-code scanner with unique bar-codes on each device, as used by Masui and Siio [2000]).

There is also another core set of issues concerning how to scale and separate the user interface across different form-factor devices in a manner that makes optimal use of their interactive and display capabilities. Our primary approach for this will be the development of a set of 'mapping heuristics', which will be based on analyses of extensive user tests on different device and user-interface configurations, as well as using findings from existing research. These heuristics will range from broad-grained rules, which describe the general suitability of devices to different user-interface functions and inter-device configurations, through to fine-grained rules, which describe the details of how the user-interface can spread across specific devices in particular application contexts. For example, a broad-grained heuristic for small-screen handheld devices might describe them as being suited to displaying concise text and small, iconic graphics [Brignull 2000] and private information [Greenberg *et al.* 1999], whereas a fine-grained heuristic would refer to the context of a particular application domain and would describe the details of what the UI should look like and how it should function when used with specific configurations of other devices.

14.11. CONCLUSIONS

Recently there has been considerable research investigating the potential of sharing user-interfaces across devices – from single applications such as Rekimoto's M-Draw [Rekimoto 1998] which uses a PDA as a tool palette for a multi-user interactive whiteboard application, through to generalised device integration frameworks [Pham *et al.* 2000] which utilise combinations of hardware devices from a user's environment. However, none of these existing approaches have addressed all of the important challenges posed by multi-user interfaces and multi-device ubiquitous computing environments. Such challenges include the rapid turnover of devices, the need for dynamic interface spreading and adaptive interface representations, the automated selection and integration from heterogeneous sets of devices, and support for aiding high level end user understanding.

In this chapter, we have described an application framework for the construction of multi-user, multi-device applications. Such applications place challenging requirements on any infrastructure including the need to support dynamic device flows, uncertainty, non pre-defined linking, distributed execution and unreliability of devices and interconnects. Unless a suitable infrastructure is available, the weight of these factors will fall upon the application developer.

We have described a support infrastructure and associated implementation framework for the integration of multiple interaction devices into multi-user arrangements. In doing so, we have explained the concept of device services, device registry, selection and device properties. We have also provided an overview of low-level communication between applications and interaction devices. We have presented the bubbles mechanism for the encapsulation and distribution of application interactivity and shown how such bubbles may be distributed, presented and synchronised. Finally, we demonstrated how the infrastructure could be used to construct an application from an example MUI scenario.

The developed Dynamo infrastructure provides facilities to automate the fragmentation, distribution and synchronisation of user interface fragments. To help both developers and end users conceptualise the operation of the infrastructure, an interaction metaphor has been developed in parallel with the implementation of the main features of the infrastructure. The support infrastructure also includes features which allow multiple representations of individual interface fragments. This is done to ensure that suitable interfaces are provided for presentation on different devices, in different usage contexts and for different users.

REFERENCES

Abrams, M., Phanouriou, C., Batongbacal, A.L., and Williams, S.M. (1999) UIML: an Appliance-independent XML User Interface Language. *Computer Networks*, May 1999, 31(11–16), 1695–1708.

Brignull, H. (2000) *An Evaluation of the Effect of Screen Size on Users' Execution of Web-Based Information Retrieval Tasks*. Unpublished Masters Thesis, Sussex University, Brighton, UK, September 2000.

Cooperstock, J.R., Fels, S.S., Buxton, W., and Smith, K.C. (1997) Reactive Environments: Throwing away your keyboard and mouse. *Communications of the ACM*, 40(7), 65–73.

Dix, A., Finlay, J., Abowd, G., and Beale, R. (1998) *Human Computer Interaction*, 2nd edition 398–9. Prentice Hall.

Fox, A., Johanson, B., Hanrahan, P., and Winograd, T. (2000) Integrating Information Appliances into an Interactive Workspace. *IEEE Computer Graphics & Applications*, 20(3), 54–65.

Greenberg, S., Boyle, M., and Laberge, J. (1999) PDAs and Shared Public Displays: Making Personal Information Public, and Public Information Personal. *Personal Technologies*, 3(1), 54–64.

Han, R., Perret, V., and Naghshineh, M. (2000) WebSplitter: A unified XML framework for multi-device collaborative web browsing. *Proceedings of CSCW 2000: ACM Conference on Computer-Supported Cooperative Work*, December 2000, 221–30.

Marsic, I. (2001) An architecture for heterogeneous groupware applications. *Proceedings of the International Conference on Software Engineering*, May 2001, 475–84.

Masui, T. and Siio, I. (2000) Real-World Graphical User Interfaces. *Proceedings of the International Symposium on Handheld and Ubiquitous Computing (HUC2000)*, September 2000, 72–84.

Myers, B.A., Stiel, H., and Gargiulo, R. (1998) Collaboration Using Multiple PDAs Connected to a PC. *Proceedings of CSCW'98: ACM Conference on Computer-Supported Cooperative Work*, November 1998, 285–94.

Petrovski, A., and Grundy, J.C. (2001) Web-Enabling an Integrated Health Informatics System. *Proceedings of the 7th International Conference on Object-Oriented Information Systems*, August 2001. Lecture Notes in Computer Science.

Pham, T.L., Schneider, G., and Goose, S. (2000) A Situated Computing Framework for Mobile and Ubiquitous Multimedia Access using Small Screen and Composite Devices. *Proceedings of CHI'00 Conference*, April 2000, 323–31.

Rekimoto, J. (1998) A Multiple Device Approach for Supporting Whiteboard-based Interactions. *Proceedings of CHI'98 Conference*, April 1998, 344–51.

Schuckmann, C., Kircher, L., Schummer, J., and Haake, J.M. (1996) Designing object-oriented synchronous groupware with COAST. *Proceedings of CSCW'96: ACM Conference on Computer-Supported Cooperative Work*, November 1996, 30–38.

Streitz, N.A., Geißler, J., Holmer, T., *et al.* (1999) i-LAND: An interactive landscape for creativity and innovation. *Proceedings of CHI'99 Conference*, May 1999, 120–27.

Sun Microsystems (2001) *Jini Technology Surrogate Architecture Overview*. http://developer.jini.org/exchange/projects/surrogate/overview.pdf, May 2001.

Tandler, P. (2000) Architecture of BEACH: The Software Infrastructure for Roomware Environments. *Proceedings of CSCW'00: Workshop on Shared Environments to Support Face-to-Face Collaboration*, December 2000.

Wang, W., Dorohonceanu, B., and Marsic, I. (1999) Design of the DISCIPLE synchronous collaboration framework. *Proceedings of the 3rd International Conference on Internet, Multimedia Systems and Applications*, October 1999, 316–24.

Part VI

Evaluation and Social Impacts

15

Assessing Usability across Multiple User Interfaces

Gustav Öquist[1], Mikael Goldstein[2] and Didier Chincholle[2]

[1] *Uppsala University, Department of Linguistics, Sweden*
[2] *Ericsson Research, Usability & Interaction Lab, Sweden*

15.1. INTRODUCTION

Designing the right interface for a mobile device is quite difficult, and an inadequate design can impair usability. Mobile devices have inherited a large body of interaction principles from the stationary computer desktop paradigm. Nonetheless, creating usable mobile interfaces does not necessarily imply squeezing the whole desktop into a small screen. On the contrary, if the metaphors in a desktop interface fail to apply in a miniaturized interface, they may even become counterproductive.

In this chapter, we will present a method that can be used to assess usability over different interfaces. To begin with, we will consider the different environments, or contexts, where different devices and interfaces are used. Next, we will identify some environmental variables that affect the usability of different devices and interfaces in various contexts of use. After defining these parameters, we will apply them to evaluate the usability of mobile interface solutions in a MUI. We will end with a brief discussion about our findings and some concluding remarks.

Multiple User Interfaces. Edited by A. Seffah and H. Javahery
© 2004 John Wiley & Sons, Ltd   ISBN: 0-470-85444-8

15.2. MULTIPLE USER INTERFACES: MULTIPLE CONTEXTS OF USE

The MUI concept enables unified access to information across a range of different devices. However, the most important factor here is not the devices but rather the locations and situations where the application will be used. The reason for this is that different devices are targeted for use in different locations, whereas different interfaces are targeted for use in different situations. But there is nothing that says that there must be a specific device for each location, nor that each interface will be used in the same way in each situation. The combination of location and situation creates different *contexts of use*, which are decisive in determining how usable different combinations of devices and interfaces are going to be.

Most devices are designed for specific contexts of use, which can be more or less dependent on either location or situation; a desktop computer is, for example, designed to be used primarily in a stationary location, whereas a mobile phone is designed to be used primarily in situations where calls are made. By identifying the contexts of use, it is possible to identify which interfaces are useful in the situations and locations where an application is to be used. Weiss [2002] proposes a personal computing continuum ranging from desktop to laptop to palmtop to handheld, where portability increases as device size decreases (Figure 15.1).

We will use the personal computing continuum to help identify different contexts of use for a MUI, but we will try to do it in a device-independent approach.

To begin with, we have the *stationary* context of use, which can be exemplified by the desktop computer. The main characteristic of the stationary context is that the devices do not have to be portable, although it is possible to use mobile devices in this context as well. It may seem strange to mention this context of use when discussing MUIs, but a desktop computer is very likely to be part of any MUI, at least until a mobile device can offer the same versatility as a large screen, a full-size keyboard and a full graphical user interface (GUI). There is also at least one other reason for including a desktop interface in a MUI, which is that most of us work with desktop computers on a daily basis and may therefore want to access information from there.

The first mobile context, which we will refer to as the *seated* context of use, is exemplified by the laptop user. The user is comfortably seated in front of a table and can interact with the device using both hands. The actual device does not necessarily

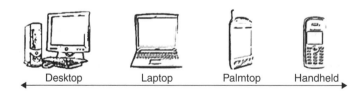

Figure 15.1. The personal computing continuum.

have to be a laptop computer; it could be any other device so long as it is mobile. The main characteristics of the seated context are that the user has both hands free if desired and that the interface can be given undivided attention. While the desktop has only the stationary context, mobile devices can be used in many different contexts.

In the second mobile context, which we will refer to as the *standing* context of use, the user holds the device with one hand and operates it with the other, such as when using a palmtop computer or a Personal Digital Assistant (PDA). Although we call this the standing context, the user does not necessarily have to stand; the device can in fact be the same as the one used in the seated context of use. The main characteristic of the standing context of use is that the user should be able to stand and operate the device with one hand, which implies that the device must be portable enough to hold in one hand.

The third mobile context, which we will refer to as the *moving* context of use, is typically associated with smaller handheld devices. The user interacts with the device with the same hand that is holding it, but does not necessarily have to move. Again, the device does not have to be different from the one used in the other mobile contexts of use. The main characteristic of the moving context of use is that the user should be able to move and have one hand free to do other things.

It should by now be evident that the more portable a device is, the more contexts it can be used in. However, although the possible contexts of use increase with portability, using the same interface on increasingly portable devices does not necessarily make the same miniaturized interface equally useable. Rather, it is simply not possible to do as much with an interface on a highly portable device as on a less portable device. This lack of direct transfer is partly because of the physical constraints imposed on the interface by the smaller device size, and partly because of the usability constraints imposed on the interface by the variations in contexts of use. It is very difficult to assess how useful a MUI is because of the diversity of contexts of use, but being aware of the impact of the different contexts on interaction is a start (Table 15.1).

Even at this early stage it is quite easy to see that an interface intended for a moving context of use will be fairly useless if the device it is implemented on is intended for a seated context. This observation may seem obvious, but in most real-life cases the level of usability of different interfaces is less clear. Not only do we need a more precise definition of what we mean by usability, we also need factors that we can use to assess the usability of different interfaces in different contexts of use.

Table 15.1. Contexts of use suitable for different devices on the personal computing continuum.

	Desktop	Laptop	Palmtop	Handheld
Stationary	Yes	Yes	Yes	Yes
Seated	No	Yes	Yes	Yes
Standing	No	No	Yes	Yes
Moving	No	No	No	Yes

15.3. MULTIPLE CONTEXTS OF USE: MULTIPLE FACTORS OF USABILITY

Usability is regarded as a composite metric composed of 'the effectiveness, efficiency, and satisfaction with which a specific user can achieve specified goals in particular environments' [ETSI 1993]. A measure of usability typically includes both objective and subjective metrics. Most usability metrics have been developed with the stationary context in mind, where the user's undivided visual attention focuses on the interface during the whole task [McFarlane 1997]. Efficiency is defined as 'the accuracy and completeness of goals achieved relative to the resources (e.g., time, human effort) used to achieve the specified goals' [ETSI, 1993]. The issues relating to portability and varying contexts of use, where attention is distracted from time to time, have thus not been at stake [Rosenberg 1998; Weiss 2002]. Although the definition of usability mentions 'particular environments', the metrics used to evaluate MUI usability need to be interleaved with additional factors relevant to multiple contexts of use. It is impossible to account for all factors that affect usability in mobile contexts, but by monitoring a few of them we may at least be able to make some well-founded predictions about usability.

To this end we have singled out four factors that have a large impact on mobile usability. We are not claiming that these are the most significant factors, since that would require an extensive empirical analysis, but we do claim that they are of importance for mobile usability. The first is the aforementioned *portability* factor, which largely governs in which contexts of use a certain device or interface may be usable. The second factor is what we call *attentiveness*, which is based on how much attention a certain interface can take for granted under various circumstances. The third factor is *manageability*, which is determined by how different interfaces can be handled by the user. The fourth, and last, factor is *learnability*, which has to do with how easily an interface can be learned. We will examine each of these factors in turn, and see how combinations of them differ for the contexts of use we have identified.

15.3.1. PORTABILITY

When on the move, we usually carry certain items, such as a wallet, keys, or a watch, and, more recently, a mobile phone. However, the number of items we are prepared to carry is limited, as every additional item, or information appliance, adds weight and demands space [Norman 1988]. To overcome this limitation several information appliances are often embedded into multi-purpose devices. For example, for many mobile users, the PDA has become a valuable companion to their mobile phone. Since additional items create inconvenience, it may be worthwhile to combine the PDA with the mobile phone (Figure 15.2).

This is analogous to the principle behind the Swiss army knife, which combines several specific tools into one general tool. However, the usefulness of any tool is dependent on a combination of design and use. Whereas it is quite uncomplicated to combine a screwdriver with a knife, combining PDA functionality with a mobile phone is less simple. Certain questions arise immediately: should the PDA be embedded into the mobile phone, or should the mobile phone be embedded into the PDA, or should the combination of both

(a)   (b)

Figure 15.2. The Sony Ericsson P800 smart phone, which can be used in (a) the standing and (b) the moving context of us.

Table 15.2. Levels of portability usable in different contexts of use.

	Plural	Dual	Single
Stationary	–	–	–
Seated	Yes	Yes	Yes
Standing	No	Yes	Yes
Moving	No	No	Yes

be a new device altogether? These questions are not trivial. The portability of the device is decisive in determining the contexts in which it may be used, which in turn affects the usability of different interfaces. We can identify three levels of portability, according to how many items comprise a portable device: *plural, dual* or *single* portability (Table 15.2). In the stationary context of use we need not worry about how many separate parts make up the device, as these components will not be transported anywhere. In the seated context of use, it is possible to have a few separate devices that can communicate with each other, so long as they are not too large or cumbersome to transport. In the standing context of use it is possible to use up to two objects; for instance, when using a PDA one often holds the device in one hand and uses a stylus (e.g. a pen) with the other. Finally, in the moving context of use, where only one hand can be used to interact with the device, the device can consist of only a single artefact.

15.3.2. ATTENTIVENESS

As portability increases, the amount of attention that an interface on the device can demand decreases. Further, the more portable a device is, the more numerous are the locations and situations of its use. The issue of limited and divided attention in combination with the effects of interruptions when using mobile devices has been pointed out by several [McFarlane 1997; Pascoe *et al.* 2000]. Users interacting with mobile devices while on the move give them less visual attention because they must also attend to a demanding environment. In addition, the conventional Human-Computer Interaction (HCI) focus on a single task does not take into account multi-tasking, which also includes the issue of interruption [McFarlane 1997]. Hjelm [2000] states that 'pen and keyboard devices won't be the terminals used for ubiquitous access to information because they require you to stop what you are doing and shift your attention to using the device.' Mobile phones, on the other hand, can be used simultaneously while one is doing other tasks, thus supporting multi-tasking and divided attention. Pascoe *et al.* [2000] also single out this multi-tasking aspect, the *using while moving* ability fieldwork users require when entering research data into a pen-based handheld computer (such as a Palm Pilot) while simultaneously observing a target object.

In the stationary and seated contexts of use, the task handled by the computer can be safely assumed to be the primary task. Desktop computers and PDAs usually require constant visual attention during the completion of a task. During *primary* task performance, the user's complete visual attention is thus focused onto the device [Pascoe *et al.* 2000]. As people have limited attention resources, they can find it difficult to visually interact with a device while simultaneously attending to a dynamic environment. Pascoe *et al.* [2000] propose the Minimal Attention User Interface (MAUI) for fieldworkers observing the behaviour of target objects in a mobile context, in which device interface interaction is understood as a *secondary* task and visual observation of the target object (e.g., a herd of giraffes grazing acacia leaves) is the primary task, demanding full visual attention. These authors have actually pinpointed the modus vivendi of the mobile user group in general, not only the fieldworker in particular. The mobile user is using his device while moving. This implies that his visual attention is focused on a primary task while he is moving around, which requires the better part of his undivided visual attention, meaning that the device can demand only *minimal* attention. They suggest that it is important to be able to shift the human-computer interaction to other sensory channels (e.g., touch and hearing), since the visual sense is primarily occupied. Providing a means to allow for modality switching – that is, switching attention from vision to the less occupied auditory and tactile senses – may ease interaction [Norman 1998]. We can identify different levels of attentiveness according to the mode of attention: *primary, secondary*, or *minimal* attentiveness (Table 15.3).

Table 15.3. Levels of attentiveness in different contexts of use.

	Primary	Secondary	Minimal
Stationary	Yes	Yes	Yes
Seated	Yes	Yes	Yes
Standing	No	Yes	Yes
Moving	No	No	Yes

15.3.3. MANAGEABILITY

The desktop computer is in one sense a multipurpose device; it can be used for several different purposes, although it is limited to the stationary context of use. On the other hand, mobile phone interaction is typically governed by the very portable one-handed moving context of use, tailored to fit voice communication. For the one-handed interaction paradigm, direct GUI interaction is exchanged for indirect, and character input is accomplished by using a many-to-one approach to enter characters on a keypad. In the seated context of use, the kinship to the desktop computer interaction methods is fairly straightforward. Direct interaction is still practiced, but the mouse is exchanged for a stylus and a touch sensitive screen and the right mouse key is exchanged for a long tap. However, most elements are miniaturized. A tiny soft-keyboard format, akin to the hard full-size, provides familiarity but hampers input speed since it only supports 'hunt and peck' interaction. Consequently, the capacity for one-handed usage is important since mobile users often prefer to use the fingers of the hand holding the device to tap the screen instead of using a stylus. In fact, one-handed operation allows them to operate the device with one hand whilst continuing to perform primary tasks with the other (Figure 15.3).

In such circumstances, the hand that holds the device also has to perform the interaction tasks: a user can easily hold the device in one hand whilst leaving the thumb free for manipulating the screen or buttons. When data input is necessary the second hand becomes crucial for either writing input text or selecting items. One has to stop and focus on the task during high-precision pen-based interaction. Thus, using the second hand means stopping the primary task [Chincholle *et al.* 2001]. However, one-handed manipulation also presents some drawbacks. In fact, when using one-handed navigation to interact, the precision is quite poor, especially when using the thumb. It can also be difficult to reach the whole display since most thumbs can only reach 2/3 of the display surface on a PDA. But one-handed interaction has almost all the advantages of direct manipulation and is fast and simple to use. The use of a two-handed interaction device (such as a PDA and a stylus) while on the move is often cumbersome [Kristoffersen and Ljungberg 1999], since it is quite difficult to accurately point and select a small target object on a small screen under such circumstances. The device is likely to be slightly unstable, making pen-based input more difficult to accomplish. Thus, the user has to stop and focus on the task when making high precision stylus-based manipulation. Furthermore, when the

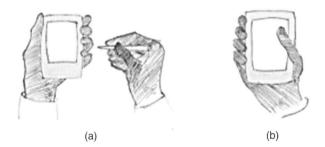

(a)   (b)

Figure 15.3. Examples of (a) two-handed and (b) one-handed usage.

user manipulates an object on the display with the stylus, the stylus and the hand holding the stylus might cover parts of the screen. This makes two-handed interaction devices more difficult to use, especially in unstable environments such as on moving trains, cars, etc. We will identify different levels of manageability according to the stability that is achieved: thus we have two-handed *stable*, two-handed *unstable* or one-handed *unbalanced* manageability (Table 15.4).

15.3.4. LEARNABILITY

Mobile users have little patience for learning how to operate new services. They don't focus on their device in the same way as when they are sitting in front of their desktop computer. The concept of learnability (as well as flexibility) is sometimes considered a factor contained within usability [ETSI 1993]. This refers to the fact that a distinction is made between an untrained (novice) and trained (expert) user. When switching to the target interface, one usually assumes that the interaction paradigm will be the same as the one learned on previous interfaces. However, this is not necessarily true. In fact, the MUI interaction paradigm is one of its most distinct features, forcing users to change their approach after moving to it from previous interfaces. This may cause severe problems since well-established metaphors cannot be transferred successfully to the new interface [Ericsson *et al.* 2001].

In fact, for some types of input paradigms, learning time may constitute several hours, e.g., the Half-Qwerty keypad [Matias *et al.* 1994] and the Chording glove [Rosenberg 1998]. Cooper [1995] distinguishes between three types of interface design paradigms: *technological, metaphorical*, and *idiomatic*. The technology paradigm builds on understanding, whereas the metaphoric paradigm is based on intuition, and the idiomatic paradigm is based on providing a swift learning environment to accomplish tasks. The amount of time required to learn how to use an interface depends in large part on the type of interface. According to Cooper [1995], all idioms have to be learned but good idioms only need to be learned once. Metaphorical design relies on previously well known interaction techniques that can be transferred to the new interface, and this design actually requires the least learning.

The user usually has acquired some previous knowledge of various kinds of interfaces before learning how to interact with the new target interface to conduct a task. For example, most users have a proficiency in interacting with a stationary PC. Some users also have acquired a proficiency in the art of touch-typing using all ten fingers. The same

Table 15.4. Levels of manageability in different contexts of use.

	Stable	Unstable	Unbalanced
Stationary	Yes	Yes	Yes
Seated	Yes	Yes	Yes
Standing	No	Yes	Yes
Moving	No	No	Yes

goes for interaction with a mobile phone and a PDA. The issue here is the *additional* learning that has to take place, in terms of both time and effort, in order for the user to gain proficiency in accomplishing the task on the target MUI. We can identify different levels of learnability according to the interface paradigm that may be used: thus we have *technological, metaphoric* or *idiomatic* learnability (Table 15.5).

15.3.5. INDEXICAL FACTORS OF USABILITY FOR DIFFERENT CONTEXTS OF USE

It should by now be quite evident, at least for the readers who like to look for correlations, that the factors we have presented are not mutually exclusive, and that each of them affects the other in some way or another. The factor that is most dominant is, not too surprisingly, portability. Yet analysis of portability alone does not present us with the whole usability picture; nor do any of the other factors in isolation. The combination of portability, attentiveness, manageability, and learnability may, however, give us a clear perspective. We will now examine the different mobile factors and create a profile for each context of use, thus creating a template with which we can assess interfaces. If an interface matches a template, then it can be assumed to be usable in that context of use; if it does not, the factors that do not match can tell us which usability problems are likely to arise in that context of use.

In the stationary context of use, the device does not have to be portable. However, apart from portability, the stationary and seated contexts of use are similar. In both contexts an interface will have the primary mode of attention; moreover the user has stable use of both hands when interacting with the device. In the stationary context of use, learnability may be technological, metaphorical and idiomatic. Since the metaphoric paradigm is most commonly used on devices with larger screens, we used that paradigm for both contexts of use in the profile (Table 15.6), but there is no reason that any of the other learnability paradigms cannot be used as well.

Table 15.5. Levels of learnability in different contexts of use.

	Technological	Metaphoric	Idiomatic
Stationary	Yes	Yes	Yes
Seated	Yes	Yes	Yes
Standing	No	Yes	Yes
Moving	No	No	Yes

Table 15.6. Factors of usability indexical for different contexts of use.

	Portability	Attentiveness	Manageability	Learnability
Stationary	Plural	Primary	Stable	Technological
Seated	Plural	Primary	Stable	Metaphoric
Standing	Dual	Secondary	Unstable	Metaphoric
Moving	Single	Minimal	Unbalanced	Idiomatic

Several factors distinguish the standing from the seated context of use. Portability has increased, allowing a separation into two objects (e.g., a PDA and a stylus); moreover attentiveness has been reduced as the interfaces used in this context may form a secondary focus of attention. Manageability is also more unstable, since the users holds the device with one hand and operates it with the other. The technological paradigm requires the most additional learning effort in this context [Cooper 1995]; an idiomatic interface may be used, as can a metaphorical interface, although its integrity can no longer be assured due to the degree of interface minimization.

When we go from the standing to the moving contexts of use, all mobile factors change. Portability demands that the device be a single artefact; attentiveness is minimal due to the fact that the user should now be able to move, and manageability is reduced to one hand and is therefore unbalanced. Furthermore, the paradigm of learnability is limited to the idiomatic. It may seem that the more portable a device is, the less versatile it is. However, the more portable a device is, the more contexts it can be used in – which is a kind of versatility in itself.

15.4. ASSESSING USABILITY OF MOBILE INTERFACES

We will now apply our methods for the assessment of usability to a few different interfaces. First we will describe them, and then we will analyse their portability, attentiveness, manageability and learnability. When we have assessed these factors, we will compare them to the profiles for different contexts of use in order to determine the most appropriate context and to anticipate which usability problems may occur in other contexts. We have chosen to assess interfaces that are used both for *input* and *output* of information, as these functions are the most fundamental in any type of interface. Input and output functionality can be safely assumed to exist in any MUI application.

We have chosen to include three different interaction paradigms in the assessment, all of which are analysed for input and output, giving us six different interfaces to assess. We will refer to the interface paradigms as the *traditional*, the *dynamic*, and the *ubiquitous*. In the traditional paradigm the interfaces imitate the interaction used on desktop computers, whereas interfaces of the dynamic paradigm redefine interaction to better suit the mobile devices. The ubiquitous paradigm is exemplified by interfaces that are inspired by Weiser's [1991] vision of the computer for the 21st century, in which computing is integrated invisibly into the fabric of our everyday lives (Table 15.7).

Table 15.7. Exemplified solutions for input and output on mobile devices.

	Traditional	Dynamic	Ubiquitous
Input	Senseboard	Tegic T9	FJG Keypad
Output	SmartView	Adaptive RSVP	TactGuide

For input, we exemplify the traditional paradigm with the Senseboard text entry interface [Goldstein *et al.* 1999; Alsiö and Goldstein 2000], which is based on sensors picking up finger depressions when the user is touch typing on a virtual full-size QWERTY keyboard. The dynamic paradigm is exemplified by the Tegic T9 text entry interface, which reduces the multi-tapping input process on mobile phone numeric keypads to single taps, based on predictions of the words that might be typed. Ubiquitous input is exemplified by the Finger-Joint-Gesture palm-keypad [Goldstein and Chincholle 1999; Goldstein *et al.* 2001], which is a generic interface based on sensors picking up the hand's gestural input.

For output, the traditional paradigm is represented by the SmartView Internet browser [Milic-Frayling and Sommerer 2002], which displays miniaturized versions of web pages as they would have been displayed on a full size screen. Dynamic output is exemplified by Adaptive RSVP [Öquist and Goldstein 2002], in which texts are successively displayed as small chunks on the screen at a pace based on predictions of how fast words and sentences may be processed by the reader. As an example of a ubiquitous output interface, we have chosen TactGuide [Sokoler *et al.* 2002], which guides the mobile user to a physical location by using directional cues applied through a tactile interface.

15.4.1. MOBILE INPUT INTERFACES

At a glance, entering input on mobile devices does not seem like a daunting problem. However, the limited input capabilities on mobile devices restrict the usability of most of the entry techniques developed for larger devices. Recent years have seen much research on mobile input solutions, mainly for text entry. One of the main reasons for the interest in this area is the unexpected success of short text messaging (SMS) on mobile phones. Not only has a good text entry technique become a competitive argument when selling mobile phones, the use of SMS is also commonly seen as the door opener to more elaborate mobile computing in the future. In addition, PDAs have been developed offering traditional desktop applications such as email clients and word processors, which require more efficient text input than earlier generations of mobile devices [MacKenzie and Soukoreff 2002]. Of the three input solutions we will examine, the first two address text entry solutions, whereas the third is a more generic solution for handling input to a mobile device. Let us begin with the traditional input.

15.4.1.1. Senseboard: A Traditional Input Interface

Senseboard (www.senseboard.com) facilitates traditional keyboard typing without the physical limitations of a full size QWERTY keyboard. Senseboard consists of two pads positioned in the user's palms. Each pad captures the motion of the fingers and the hand, thus enabling keyboard typing without a keyboard. The design is based on the fact that users who have acquired the skill of touch-typing can accomplish this without the keyboard being present [Goldstein *et al.* 1998, 1999; Alsiö and Goldstein 2000]. The first proof of concept prototype was named the Non-Keyboard QWERTY touch-typing input paradigm (Figure 15.4). The fuzzy input of each touch-typing finger press is fed into a language model, the entered character string is parsed into word units by pressing the thumb (space bar), and these word units are parsed using a sliding tri-gram analysis

Figure 15.4. The Non-Keyboard QWERTY proof of concept prototype.

for a predefined language on the lexical level. At the sentence level, by using syntactic analysis in combination with frequency ranking analysis, it is possible to predict the intended input accurately.

In a Wizard of Oz study, users who had previously acquired proficiency in full-size QWERTY touch-typing could input a recited (at 44 wpm) orally presented long (Swedish) text with a character error rate of 12.3% [Goldstein *et al.* 1999]. This figure can be compared to the error rates for entering a similar text using either a full-size QWERTY keyboard (6.4%, benchmark condition), the Nokia 9000 Communicator (33.5%), or the miniaturized soft QWERTY of a Palm Pilot (62%, stylus input). The high character error rates were due mainly to omitted characters. MacKenzie *et al.* [1999] obtained an input speed of approximately 20 wpm for a short (English) sentence entered on a miniaturized soft QWERTY keyboard using a stylus.

The Senseboard interface is typically intended for a seated context of use. The portability factor for the Senseboard interface is plural, since the typing sensors are separate artefacts. The attentiveness factor is primary since the interface demands the user's full attention. The manageability factor of the interface is stable; the user needs a table or a similar flat surface to type. The learnability factor is metaphorical or idiomatic if the user knows proper QWERTY keyboard typing, but otherwise it is technological. The Senseboard interface thus matches the seated context of use, for which it is intended.

15.4.1.2. Tegic T9: A Dynamic Input Interface

'Text on nine keys' (T9) is a technology developed by Tegic Communications (www.tegic. com). T9 originated as a spinoff from an augmentative communication tool which enabled people with disabilities to type text by using eye movements. The problem with the tool was that the eyes can target only eight specific areas clearly, creating a level of uncertainty, while at the same time the system needs to be very quick. The developers addressed the problem by grouping letters together onto fewer target areas, and creating an extremely compressed linguistic database that recognizes commonly used words.

Figure 15.5. A typical device that benefits from Tegic T9, the Sony Ericsson T300 mobile phone. Characters are entered by using pushing once on the keys 2–9, while key 1 is used for spacing between words.

They quickly realized that the technology could be useful for a much broader audience (Figure 15.5).

In T9 characters are entered by pressing numeric keys, *once* for each letter, exactly as when entering characters using a full-size QWERTY keyboard. This is known as dictionary-based disambiguation. T9 automatically compares the entered key presses to a linguistic database to determine the correct word when the Space key (Key 1) is depressed. For example, entering the (English) word "HELLO" is accomplished using only five key presses with T9 (when entering this word by traditional multi-tap you have to perform 13 key presses). Pressing the keys 4 (GHI), 3 (DEF), 5 (JKL), 5 (JKL) and 6 (MNO) once, followed by Space (Key 1), generates the most probable word. If more than one word matches a key sequence, alternatives appear in order of decreasing probability by pressing the Option key (below the Yes key on a Sony Ericsson phone, see Figure 15.5).

By computing the KSPC (keystrokes per character) measure [MacKenzie 2002] for a given text entry technique, it is possible to assess the efficiency relative to the QWERTY keyboard which has a KSPC of 1.0. KSPC measures the average number of keystrokes required to generate one character of text for a given text entry technique in a given language. KSPC for dictionary-based disambiguation is 1.0072, which is very close to 1.0. The multi-tap mode on a mobile keypad, on the other hand, renders a KSPC of 2.0342.

The Tegic T9 interface is typically intended for moving contexts of use. The portability factor for the T9 interface is singular, and it is intended for devices with a small input area. The attentiveness factor is minimal, since it is possible to stop typing and resume. The manageability factor of the interface is unbalanced; it is intended to be used with one hand. The learnability factor is idiomatic; the user just presses the buttons with the labelled character. T9 thus matches the moving context of use, for which it is intended.

15.4.1.3. Finger-Joint-Gesture Keypad: A Ubiquitous Input Interface

The Finger-Joint-Gesture (FJG) palm-keypad is a generic concept [Goldstein and Chincholle 1999; Goldstein *et al.* 2001]. It anticipates that the mobile phone monolith, incorporating keypad, screen, microphone, loudspeaker, and battery, will be fragmented into tailor-made chunks that will fit each task smoothly. With the evolving technology of Bluetooth in 1999, one prerequisite for future fragmentation became a fact. The FJG palm-keypad glove neither employs a physical keypad, nor requires a flat surface to generate the characters. Instead, the inside of the phalanges of the index, long, ring and little fingers are used as "keys" and the thumb as an operator. The loudspeaker is placed on the tip of the index finger of the prototype glove and the microphone is placed in the palm (Figure 15.6).

The spatial layout of the FJG palm-keypad is designed to be similar to the traditional design of the keypad employed on mobile phones (compare with Figure 15.5). The idea is based on Rosenberg's [1998] Chording Glove concept, where Shift (hard) buttons were placed on the side of the index finger and the thumb was used as an operator. The use of the inside of the finger phalanges as keys creates a natural (4 × 3) 12-key telephone palm-keypad. The FJG palm-keypad glove thus employs previously acquired source metaphors and a spatial layout that the user is well acquainted with. The FJG palm-keypad glove provides a way to enter the digits of the telephone keypad, and includes gesture function keys that can be combined in order to generate new modes and functions.

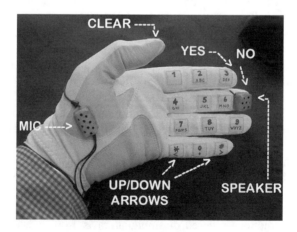

Figure 15.6. The Finger-Joint-Gesture palm-keypad prototype.

The applications are thus not restricted to the telephone keypad and accompanying telephone functions: e.g., by switching mode, which is done by using one of the function keys, the cursor keys 'Left–Right' and 'Up–Down' can be implemented as well. In a laboratory experiment [Goldstein *et al.* 2001] using 18 subjects, the traditional hard phone keypad was compared to the Finger-Joint-Gesture mock-up prototype using the well-known 9-word text sentence *the quick brown fox jumped over the lazy dogs* using multi-tap. Error-corrected text input speed was of the same magnitude (approximately 6 words/min) for both types of keypad, whereas digit input (of the same sentence) was significantly higher for the hard phone keypad (9.7 vs 8.2 wpm).

The Finger-Joint-Gesture palm-keypad interface is intended for a moving context of use. The portability factor for the Finger-Joint-Gesture interface is singular; the device is in itself the whole interface. The attentiveness factor is minimal, as the interface is always present. The manageability factor of the interface is unbalanced; it is intended to be used with one hand. The learnability factor is idiomatic; the user simply presses the buttons with the appropriate character. The Finger-Joint-Gesture palm-keypad thus matches the moving context of use, for which it is intended.

15.4.2. MOBILE OUTPUT INTERFACES

Few in the wireless business today argue that the mobile Internet should be something that exists aside from the regular Internet. Users of mobile devices have been reluctant to use protocols, such as the Wireless Application Protocol (WAP), which offer degraded content compared to what they have learned to expect from regular Internet access. However, the usefulness of accessing a web page with a traditional desktop layout via a mobile device is seriously limited by the obvious mismatch between small screens and large page layouts, even on mobile devices with relatively large screens such as PDAs [Jones *et al.* 1999]. In order to solve the dilemma, several solutions have been proposed. Some strive to keep the original web page format, a strategy exemplified in the traditional output interfaces, whereas others, such as dynamic output solutions, seek to restructure the original web format. As mobile interfaces offer more than Internet access, the third solution presented deals with problems of navigation through physical space.

15.4.2.1. SmartView: A Traditional Output Interface

Among those who want to preserve the traditional format two trends can be clearly distinguished: those who *condense* the original content to reduce the space it occupies on the screen, and those who *compress* the original content to fit the screen. The condensing approach is usually based upon applying text summarization techniques (see Mani and Maybury [1999] for an overview) to single or multiple web pages, thus yielding suitable summaries to present to the user [Buyukkokten *et al.* 2001]. Although this approach reduces the required screen space, the condensing process may cause discrepancies in how the content is interpreted depending on which interface is used for access. There is also a certain value in providing the users with content as it was originally produced and intended to be perceived, which allows the users to judge for themselves what is valuable or not.

This latter concern informs the philosophy behind the compression approach, as exemplified by the SmartView browser [Milic-Frayling and Sommerer 2002], which delivers all of the original content to the user. We have chosen to include the SmartView browser as an example of traditional output since it reproduces the original web page and also combines most of the strategies that have been applied to compression in previous research. The SmartView browser initially allows the user to view the web page as an overview thumbnail partitioned into sections based on the graphical layout, which is assumed to indicate the logical decomposition of the original document; each of the individual sections can then be selected by the user to be viewed independently from the rest of the document in a detailed view (Figure 15.7).

SmartView's first step in the creation of the web page overview is to render a snapshot of the page as it would have looked on a large screen; this snapshot is then scaled down to a thumbnail image that fits the screen width of the mobile device. The next step is to establish the logical sections of the page, which is done by analysis of the geometric properties of the original document structure. Depending on the size and arrangement of tables and cells, an algorithm determines whether a table or cell is to be bookmarked as a logical section or whether processing is to be continued recursively. The result of the analysis is a vector of nodes referring to the logical sections of the original web page. The thumbnail image is then displayed with superimposed regions indicating the segments of the page determined during analysis. If a section is requested for viewing, a document corresponding to the page fragment is created by extracting the section represented by the node, along with all of its contents. This extracted segment is then wrapped by code representing the path from the root of the document model down to the node. In this manner a minimal yet structurally consistent and detailed view is created. If a link is selected in the detailed view, the linked page is processed by SmartView, resulting in a thumbnail overview of the corresponding page with indicated page decomposition [Milic-Frayling and Sommerer 2002].

The SmartView interface is typically intended for a standing context of use. The portability factor of the SmartView interface is typically that of the PDA, which is dual. The

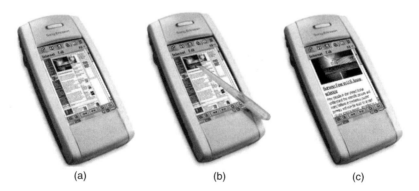

(a)   (b)   (c)

Figure 15.7. The SmartView browser presents a web page on a mobile device with (a) the compressed page overview (b) a selected section and (c) a detailed view of the selection.

attentiveness factor is secondary since the device does not demand primary attention all the time. The manageability factor of the interface is unstable; the user typically holds the device with one hand and operates the interface with a stylus in the other. The learnability factor is metaphorical. The SmartView thus matches the standing context of use.

15.4.2.2. Adaptive RSVP: A Dynamic Output Interface

One approach to overcome the constraints caused by limitations in screen space, which are usually pertinent to portable devices, is to swap a spatial dimension for a temporal one, and thus present the information in smaller units at a time [Bruijn and Spence 2000]. Rapid Serial Visual Presentation (RSVP) and Leading are the two major techniques that have been proposed for dynamic text presentation. Leading, or the Times Square Format, scrolls the text on one line horizontally across the screen, whereas RSVP presents the text as chunks of words or characters in rapid succession at a single visual location (Figure 15.8). Both formats offer a way of reading texts on a very limited screen space [Mills and Weldon 1987]. Comparisons between the formats have so far been inconclusive [Kang and Muter 1989], but at normal reading speeds RSVP appears to be more efficient. From a physiological perspective, RSVP also seems more natural to use, as the text moves successively rather than continuously [Öquist and Goldstein 2002]. More importantly for mobile usability, however, RSVP reduces the need for physical interaction, in the form of paging or scrolling, when reading from small screens. RSVP also permits other content, such as images or sounds, to be added to the text presentation with exact timing in respect to what the user reads [Öquist and Goldstein 2002; Goldstein *et al.* 2002].

Sicheritz [2000] compared ordinary reading from a book to reading on a PDA via RSVP, with results showing that neither reading speed nor comprehension differed. However, the NASA-TLX (NASA Task Load Index) [Hart and Staveland 1988] revealed significantly higher task load ratings for the RSVP conditions. One explanation of the

Figure 15.8. A simple RSVP browser on a handheld device.

high cognitive load may have been that each text chunk was exposed for the same fixed duration of time, which clearly opposes Just and Carpenter's findings that 'there is a large variation in the duration of individual fixations as well as the total gaze duration on individual words' when reading text from paper [Just and Carpenter 1980]. Adaptive RSVP attempts to mimic the reader's cognitive text processing pace more adequately by adjusting each text chunk exposure time according to certain characteristics of the text appearing in the RSVP text presentation window [Öquist and Goldstein 2002]. By assuming Just and Carpenter's [1980] *eye-mind hypothesis,* i.e., that the eye remains fixed on a text chunk as long as it is being processed, the needed exposure time of a text chunk can be assumed to be proportional to the predicted gaze duration. The variations in exposure times are calculated on the basis of the linguistic properties of the text being displayed, such as word lengths, sentence lengths, word frequencies, word distributions, etc. [Öquist and Goldstein 2002]. In one sense dynamic text presentation with adaptation can be thought of as a predictive text output interface, just as Tegic T9 is a predictive text input interface.

In a balanced repeated-measurement experiment employing 16 subjects, two variants of adaptive RSVP were benchmarked against regular RSVP with fixed exposure times and traditional text presentation. All conditions were performed on a Compaq iPAQ 3630 PDA, and both short (approximately 250 words) and long texts (approximately 4000 words) were included in the experiment. For traditional text presentation, MS Explorer was used for short texts whereas MS Reader was used for long texts. For short texts, all RSVP formats increased reading speed by 33% compared to the MS Explorer format, with no significant differences in comprehension or task load. For long texts, no significant differences were found in reading speed or comprehension, but the null hypothesis regarding no difference in task load was rejected. For regular RSVP, five out of six task load factors became significantly higher than with the MS Reader. However, when adaptive RSVP was used, only one task load factor for each of the adaptation algorithms was significantly higher [Öquist and Goldstein 2002].

The adaptive RSVP interface is typically intended for a moving context of use. The portability factor is typically that of the handheld device, which is singular. The attentiveness factor is secondary; although the text presentation advances automatically, it may be easily halted. The manageability factor of the interface is unbalanced; the interface is intended for handheld devices that can be operated with one hand. The learnability factor is idiomatic; the text presentation may be started, stopped, reversed, or fast forwarded. Adaptive RSVP would thus match the moving context of use if it were not for the attentiveness factor. This factor does not render the interface useless in the moving context, but it is quite likely that usability issues are going to arise as a result of the attentiveness mismatch.

15.4.2.3. TactGuide: A Ubiquitous Output Interface

When receiving output from a mobile device today, the most commonly used human sense, or modality, is probably the visual. But, perhaps more than other devices, mobile devices are easily adapted for using other modalities as well. Mobile phones already combine sight, hearing and touch in the way they communicate, and their users are therefore quite accustomed to devices that blink, beep and vibrate. Since mobile devices are handheld, they may also make use of the hand's sensitivity to touch, and thus use the tactile sense as a channel of communication. Currently, the use of tactile sentences in mobile interaction is more or less limited to call notification and enhancements in games. Therefore, we have chosen to include TactGuide [Sokoler *et al.* 2002] as an example of how tactile output can be used more intelligibly. TactGuide is a mobile navigation tool interface that literary 'points you the way' to a selected destination by subtle tactile directional cues. The paths are not very long, but are difficult to follow: for example, the route to a certain bookstore in a large mall; or the route back to your car in the mall's parking lot. The original interface was based on a handheld device with four holes positioned around a raised dot, with a metal peg directly underneath each hole. The direction to the destination was indicated by one of the four pegs being raised through its corresponding hole (Figure 15.9). The user's thumb receives two tactile inputs when it is placed on the display: one from the center dot, and one from the peg indicating the direction to the destination: forward, backward, left, or right [Sokoler *et al.* 2002].

The TactGuide interface was primarily designed to complement the use of other senses, most prominently the visual input of the real world as it is traversed in the process of finding the route from one point to another. TactGuide was never intended to be the sole indicator of the direction one should take, but rather as a personal sign indicating the right direction at each junction. The interface relies upon three sources of information: the location of the destination, the location of the user, and the orientation of the user. Such information can be obtained through the Global Positioning System (GPS) or Local Positioning Systems (LPS) relying on network triangulation or Bluetooth beacons. It would be straightforward to implement the TactGuide interface on a cellular phone by substituting the centre peg with key 5 on the numeric keypad, and adding tactile cues to numeric keys: 2 for forward, 8 for backward, 4 for left, and 6 for right.

The TactGuide interface is typically intended for a moving context of use. The portability factor for the TactGuide interface is singular; the device is in itself the whole interface. The attentiveness factor is minimal: the interface is used when the user wishes. The manageability factor of the interface is unbalanced; it is intended to be used with one hand.

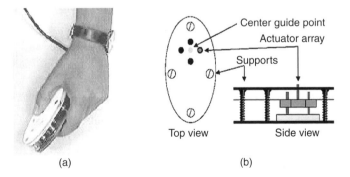

(a)   (b)

Figure 15.9. (a) The TactGuide prototype and (b) a schematic view indicating the tactile cues [Sokoler *et al.* 2002].

The learnability factor is idiomatic; the user simply follows the subtle tactile cues. The TactGuide thus matches the moving context of use.

15.5. DISCUSSION

We will now briefly review and summarize the results from the usability assessment and discuss the assessment method we have used. Our primary motivation for assessing the usability of different interfaces in this chapter was not to find out how usable they were, but rather to exemplify how usability could be assessed over multiple interfaces and to illustrate how different interfaces have been adapted to the mobile environment. The assessment method in itself seems to have worked, as we were able to clearly distinguish the factors of usability pertinent to the different interfaces. The different factors of usability for each interface were then matched against the templates for their intended contexts of use. Five out of six interfaces matched their contexts: Senseboard matched the seated context of use; SmartView matched the standing context of use; Tegic T9, the FJG keypad, and the TactGuide matched the moving context of use (Table 15.8).

One output interface mismatched on one factor: the attentiveness required of the Adaptive RSVP interface did not match with the intended moving context. Yet, as mentioned earlier, a mismatch between a single factor of use and an indexical factor of context does not necessarily imply that the interface is rendered useless, but rather that caution should be exercised, as it can be seen as a sign of potential usability problems. As a matter of fact, in the case of the Adaptive RSVP interface, attentiveness actually does cause user complaints from time to time. Using a camera to monitor the user's attention and then controlling the text presentation accordingly has been suggested as a solution to the attentiveness mismatch [Öquist *et al.* 2002].

Table 15.8. Summary of the usability assessment. The usability factors marked with an asterisk (*) do not match their intended context of use.

	Interface	Senseboard	Tegic T9	FJG Keypad
	Portability	Plural	Single	Single
	Attentiveness	Primary	Minimal	Minimal
Input	Manageability	Stable	Unbalanced	Unbalanced
	Learnability	Metaphoric	Idiomatic	Idiomatic
	Context of use	Seated	Moving	Moving
Interaction paradigm		Traditional	Dynamic	Ubiquitous
	Portability	Dual	Single	Single
	Attentiveness	Secondary	Secondary*	Minimal
Output	Manageability	Unstable	Unbalanced	Unbalanced
	Learnability	Metaphoric	Idiomatic	Idiomatic
	Context of use	Standing	Moving	Moving
	Interface	SmartView	Adaptive RSVP	TactGuide

The summary of the usability assessment (Table 15.8) gives a detailed overview of how different interfaces may be used, and also indicates if there are any mismatches between the intended context of use and the indexical context of use. In addition, it shows which interfaces will be useful for inclusion in a MUI application intended for a certain context of use. For example, if you need to include a mobile input interface in a MUI application that will be used in a standing context, you then simply locate the interface which can be used in a standing context – in this case, Tegic T9.

The taxonomy for the MUI usability assessment introduced in this chapter does not really offer any answers in itself. The value of any taxonomy can only be assessed by the value it renders to the interpreter of the phenomena it attempts to clarify. The framework is intended primarily as a support to structure reasoning about MUI usability, which in turn can be useful when identifying and understanding issues related to MUI usability. The assessment of usability offered by validating the contexts of use, environmental variables, and usability factors proposed in the taxonomy can never be substituted for proper empirical usability evaluations. Nonetheless, the taxonomy can be used to assess the appropriateness of a particular mechanism for its intended use, and to predict problems and difficulties that will arise if it is used inappropriately. The foremost merit of the taxonomy we have proposed, however, is that it has added a structure and a vocabulary to support our reasoning.

This chapter has focused on mobile devices, but there are, of course, many other devices that could be used in a MUI application as well. These do not necessarily have to be mobile: for example, an interactive TV or an information kiosk could be used in MUI applications. The usability assessment method introduced in this chapter can be extended to assess usability over other such MUIs as well. The most straightforward way of doing so is to add additional contexts of use, distinguish the usability factors that are pertinent to them, and create new indexical templates for each contexts of use to match against. There are, of course, many mobile interaction techniques that we haven't covered in this chapter: for example, auditory interfaces, tilt-based interfaces, speech-based interfaces, and so on. Yet we hope that the different types of interfaces we have examined have offered some insight into the diversity of interaction methods that can be used in a MUI. We also hope that we have offered some guidance on how different devices and interfaces may be combined in different contexts of use, with the ultimate aim of achieving highly usable MUIs.

15.6. CONCLUSIONS

Only by utilizing combinations of interfaces and devices that function well within the user's environment is it possible to achieve a high usability in a MUI application. In this chapter, we have outlined four typical contexts of use of MUIs: the stationary, the seated, the standing, and the moving context. Next, we characterized the different contexts of use by enumerating the four environmental factors that we believe affect usability most: portability, attentiveness, manageability, and learnability. For each of the contexts of use we could identify combinations of environmental factors that were indexical for different contexts of use. By assessing portability, attentiveness, manageability, and learnability, it became possible to compare the different environmental factors to the indexical factors for specific contexts of use.

This result means that it is possible to assess in which context of use a certain MUI application can achieve usability and if a MUI application can achieve usability in its intended context of use. By defining and assessing usability over six different mobile interfaces, from three different interaction paradigms, we have found that the usability assessment provides some guidance. Furthermore, the method identified a potential usability problem in one interface and indicated the environmental variable that caused it.

REFERENCES

Alsiö, G., and Goldstein, M. (2000) Productivity Prediction by Extrapolation: Using workload memory as a predictor of target performance. *Behaviour & Information Technology*, 19(2), 87–96.

Bruijn, O., and Spence, R. (2000) Rapid Serial Visual Presentation: A space-time trade-off in information presentation. In *Proceedings of Advanced Visual Interfaces, AVI'2000* (eds V.D. Gesù, S. Levialdi and L. Tarantino), 196–201. New York: ACM Press.

Buyukkokten, O., Garcia-Molina, H., and Paepcke, A. (2001) Seeing the Whole in Parts: Text summarization for web browsing on handheld devices. *Proceedings of WWW'2001*, Hong Kong, China, May 2001.

Chincholle, D., Eriksson, M., and Viefhues, B. (2001) Towards Direct Manipulation Interfaces and Eyes-free Controls for Optimizing Interaction with a Mobile Multimedia Player. *Proceedings of HFT'2001*, Bergen, Norway, November 2001, 269–73.

Cooper, A. (1995) *About Face: The Essentials of User Interface Design*. New York, NY: IDG Books.

Ericsson, T., Chincholle, D. and Goldstein, M. (2001) Both the Device and the Service Influence WAP Usability. *Proceedings of IHM-HCI2001, Usability in Practice, volume II* (eds J. Vanderdonckt, A. Blandford and A. Derycke), Toulouse, Cépaduès-Editions, 79–85.

ETSI Technical Committee Human Factors (1993) Guide for Usability Evaluations of Telecommunications Systems and Services. *European Telecommunications Standards Institute Technical Report*, ETR/095/DTR/HF-03001. Sophia Antopolis, ETSI.

Goldstein, M., Book, R. Alsiö, G., and Tessa, S. (1998) Ubiquitous Input for Wearable Computing: QWERTY keyboard without a board. In *Proceedings of First Workshop on Human Computer Interaction with Mobile Devices* (ed. C. Johnson), Glasgow, Scotland. Available at: http://www.dcs.gla.ac.uk/~johnson/workshops_old/mobile.html (February 2003).

Goldstein, M., Book, R., Alsiö, G., and Tessa, S. (1999) Non-keyboard QWERTY Touch Typing: A portable input interface for the mobile user. *Proceedings of the ACM Conference on Human Factors in Computing Systems, CHI'1999*, 32–9. New York: ACM Press.

Goldstein, M., and Chincholle, D. (1999) The Finger-Joint-Gesture Wearable Keypad. *Proceedings of Interact'1999*, Workshop on Mobile Devices, Edinburgh, Scotland, August 1999.

Goldstein, M., Chincholle, D., and Backström, M. (2001) Assessing Two New Wearable Input Paradigms: The Finger-Joint-Gesture palm-keypad glove and the invisible phone clock. *Personal Technologies special issue on mobile devices*, 45–55. London, Springer.

Goldstein, M., Öquist, G., and Björk, S. (2002) Evaluating Sonified Rapid Serial Visual Presentation: An immersive reading experience on a mobile device. *Proceedings of User Interfaces for All, UI4ALL'02*, Paris, France, October 2002.

Hart, S.G., and Staveland, L.E. (1988) Development of NASA-TLX (Task Load Index): Results of empirical and theoretical research. *Human Mental Workload* (eds P.A. Hancock and N. Meshkati), Amsterdam, North-Holland, 139–83.

Hjelm, J. (2000) *Designing Wireless Information Services*. New York, NY, John Wiley & Sons.

Jones, M., Marsden, G., Mohd-Nasir, N., Boone, K., and Buchanan, G. (1999) Improving Web Interaction on Small Displays. *Computer Networks*, 31, 1129–37.

Just, M. and Carpenter, P.A. (1980). A Theory of Reading: From eye fixations to comprehension. *Psychological Review*, 87(4), 329–54.

Kang, T.J., and Muter, P. (1989) Reading Dynamically Displayed Text. *Behaviour & Information Technology*, 8(1), 33–42.

Kristoffersen, S., and Ljungberg, F. (1999) "Making Places" to Make IT Work: Empirical explorations of HCI for mobile CSCW. *Proceedings of Group'99*. ACM Press.

MacKenzie, I.S. (2002) KSPC (Keystrokes per Character) as a Characteristic of Text Entry Techniques. *Proceedings of Mobile HCI 2002, Human Computer Interaction with Mobile Devices* (ed. F. Paterno), 195–210. Berlin: Springer.

MacKenzie, I.S., and Soukoreff, R.W. (2002) Text Entry for Mobile Computing: Models and methods, theory and practice. *Human-Computer Interaction*, 17, 147–98.

MacKenzie, I.S., Zhang, S.X., and Soukoreff, R.W. (1999) Text Entry using Soft Keyboards. *Behaviour & Information Technology*, 18, 235–44.

Mani, I., and Maybury, M.T. (1999) *Advances in Automatic Text Summarization*. Boston, MA: MIT Press.

Matias, E., MacKenzie, I.S., and Buxton, W. (1994) Half-QWERTY: Typing with one hand using your two-handed skills. *Companion of the CHI '94 Conference on Human Factors in Computing Systems*, 51–2. New York: ACM Press.

McFarlane, D.C. (1997) Interruption of People in Human-Computer Interaction: A general unifying definition of human interruption and taxonomy. *Naval Research Laboratory Report*, NRL/FR/5510-97-9870, Washington, DC, NRL.

Milic-Frayling, N., and Sommerer, R. (2002) SmartView: Flexible viewing of web page contents. *Proceedings of WWW'2002*, Honolulu, USA, May 2002.

Mills, C.B., and Weldon, L.J. (1987) Reading text from computer screens. *ACM Computing Surveys*, 19(4), 329–58.

Norman, D.A. (1988) *The Psychology of Everyday Things*. New York, NY: Doubleday.

Norman, D.A. (1998) *The Invisible Computer. Why good products can fail, the personal computer is so complex and information appliances are the solution*. Cambridge, MA: MIT Press.

Öquist, G., and Goldstein, M. (2002) Towards an Improved Readability on Mobile Devices: Evaluating Adaptive Rapid Serial Visual Presentation. In *Proceedings of Mobile HCI 2002, Human Computer Interaction with Mobile Devices* (ed. F. Paterno), 225–40. Berlin: Springer.

Öquist, G., Goldstein, M., and Björk, S. (2002) Utilizing Gaze Detection to Stimulate the Affordances of Paper in the Rapid Serial Visual Presentation Format. In *Proceedings of Mobile HCI 2002, Human Computer Interaction with Mobile Devices* (ed. F. Paterno), 378–81. Berlin, Springer.

Pascoe, J., Ryan, N., and Morse, R. (2000) Using while Moving: HCI issues in fieldwork environments. *ACM Transactions on Computer Human Interaction*, 7, 417–37.

Rosenberg, R. (1998) *Computing without Mice and Keyboards: Text and graphic input devices for mobile computing*. Doctoral dissertation, Department of Computer Science, University College, London.

Sicheritz, K. (2000) *Applying the Rapid Serial Presentation Technique to Personal Digital Assistants*. Master's Thesis, Department of Linguistics, Uppsala University, Sweden.

Sokoler, T., Nelson, L., and Pedersen, E.R. (2002) Low-Resolution Supplementary Tactile Cues for Navigational Assistance. In *Proceedings of Mobile HCI 2002, Human Computer Interaction with Mobile Devices* (ed. F. Paterno), 369–72. Berlin: Springer.

Weiser, M. (1991) The Computer for the 21st Century. *Scientific American*, 265(3), 94–104.

Weiss, S. (2002) Handheld Usability. New York: John Wiley & Sons.

Iterative Design and Evaluation of Multiple Interfaces for a Complex Commercial Word Processor

Joanna McGrenere

Department of Computer Science[1], University of British Columbia, Canada

16.1. INTRODUCTION

Desktop applications such as the word processor and the spreadsheet have become increasingly complex in terms of the number of options available to the user. With every new release, these applications expand in terms of the functionality that they offer. This phenomenon, sometimes called creeping featurism [Hsi and Potts 2000; Norman 1998] or bloatware [Kaufman and Weed 1998], is pervasive: having a long feature list is now seen as essential for products to compete in the marketplace. The issue of bloat is more than simply one of 'maxing out' system resources such as disk space, memory, and processor speed. These constraints are real, but bloat is also an issue of human constraints. Heavily-featured applications are visually more complex than lesser-featured applications.

Multiple User Interfaces. Edited by A. Seffah and H. Javahery
© 2004 John Wiley & Sons, Ltd   ISBN: 0-470-85444-8

Menus have multiplied in size and number, and toolbars have been introduced to reduce complexity, but they too have grown in a similar fashion.

Our research was motivated by a concern for the users of today's complex productivity applications. We assumed that if we, as expert users, were struggling, then average users and novice users must be really struggling. There was very little actual research, however, that indicated whether or not this was the case.

We noticed at the time we began our work in 1998 that the terms bloat and bloatware were appearing with some regularity in the computer literature and in the popular press. Although these terms were never clearly defined, they certainly implied that users were having a negative experience of functionality-filled software. But again there was very little research evidence to show that all users were experiencing complex software as bloated. If all users do not experience complexity this way, as bloat, then we wondered what were the factors that impacted the user's experience? Is it, for example, expertise or the number of functions that are used?

Our main research objectives were three-fold: (1) to gain a systematic understanding of users' experiences with complex software; (2) to move toward a new interface model that is derived from this understanding; and (3) to evaluate the new interface model in light of the problems that users experience.

In this chapter we describe research that was conducted to address the above three objectives and the methodology used to eventually arrive at a multiple interfaces design solution for a complex commercial word processor. We conducted three studies, one was a pilot study and the other two were full user studies. An overview of our three studies is shown in Figure 16.1.

In Study One we conducted a broad-based assessment of user needs. We worked with 53 users of MSWord 97. Based on our findings from Study One, we created our first multiple-interfaces prototype for MSWord 2000 that contained one personalizable interface. This was informally evaluated in our Pilot Study with four users. Personalization was achieved

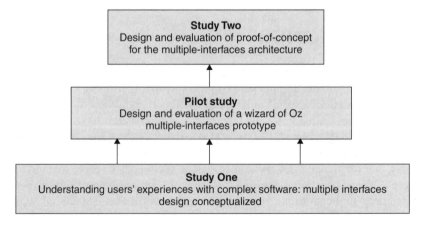

Figure 16.1. Research overview showing the sequence of studies that were conducted and how the results of earlier studies framed later studies.

through Wizard of Oz methodology. The results from the Pilot Study were promising and encouraged us to iterate on the design of the prototype, remove the wizard, and conduct a formal evaluation with 20 users. That was our Study Two.

One of the things that all three studies have in common is the MSWord application. For practical reasons it made sense to focus on one application, however, the interface design that was prototyped and the results of the evaluations that we conducted are intended to generalize to other heavily-featured productivity applications that are used by a diversity of users.

Study One and Study Two have already been reported in some detail separately in the literature [McGrenere and Moore 2000; McGrenere *et al.* 2002]. The goal of this chapter is not to duplicate those publications, but rather to document these two studies together, to include the pilot study, and to specifically highlight the full process of arriving at our multiple interfaces design. By documenting these three studies together we will necessarily be omitting much of the detail and be focusing on the methodology and selected results. In particular our research serves as a good case study of user-centred design methodology. That methodology espouses early and continual focus on users and iterative design and evaluation. It is a cornerstone of the field of human-computer interaction.

We would like to point out at the outset of this chapter that we use the term 'multiple user interfaces' somewhat differently than how it has been defined in this book. We use the term to describe two or more interfaces that have different amounts of functionality for the same application on the same device. By contrast, multiple user interfaces is used more broadly in this book to refer to different interfaces or views for different devices used over a network for the same application or data repository, for example, an email application that has different interfaces for each of the desktop, mobile phone, and PDA client devices. The term 'multiple user interfaces' seems appropriate for either or both of these notions, since they address different dimensions of the problem of adapting the interface to the specific needs of the user and the context in which the user works.

16.2. DESIGN SOLUTIONS TO COMPLEX SOFTWARE

Despite the lack of research into the user's experience of complex software, there have been a number of alternative interface designs to the 'all-in-one' style interface in which the menus and toolbars are static and every user, regardless of tasks and experience, has the same interface. These design solutions have appeared in both the research literature and in commercial products and they tend to fit into one of two categories: (1) ones that take a level-structured approach [Shneiderman 1997], and (2) ones that rely on some form of artificial intelligence.

A level-structured design includes two or more interfaces, each containing a *predetermined* set of functions. The user has the option to select an interface, but not to select which functions appear in that interface. Preliminary research suggests, however, that when an interface is missing even one needed function, the user is forced to the next level of the interface, which results in frustration [McGrenere and Moore 2000]. There are a small number of commercial applications that provide a level-structured interface (e.g., Hypercard and Framemaker). Some applications, such as Eudora, provide a level-structured approach *across* versions by offering both Pro and Light versions. Such product versioning, however, seems to be motivated more by business considerations than by an attempt to meet user needs.

The Training Wheels interface to an early word processor is a classic example of a level-structured approach that appears in the research literature. By blocking off all the functionality that was not needed for simple tasks, it was shown that novice users were able to accomplish tasks significantly faster and with significantly fewer errors than novice users using the full version [Carroll and Carrithers 1984]. Despite the promise of this early work, the transition between the blocked and unblocked states was never investigated.

The broad goal of intelligent user interfaces is to assist the user by offloading some of the complexity [Miller *et al.* 1991]. Adaptive interfaces are one form of intelligent interface; they rely on computational intelligence to automatically adjust in a way that is expected to better suit the needs of each individual user. In practice, however, an interface that changes automatically often results in the user perceiving a loss of control.

There is a quasi third category, namely adaptable or customizable interfaces. These interfaces allow users themselves to personalize the interface in a way that is suitable to them. The main problem with customizable interfaces is that the mechanisms for customizing are often powerful and complex in their own right and therefore require time for both learning and doing the customization. Thus, only the most sophisticated users are able to use them. (Mackay found the latter to be true in the case of UNIX customization [Mackay 1991].) Customization has not typically been designed for the purpose of reducing complexity, but rather for making sophisticated changes to the interface. It is for that reason that we have described adaptability/customization as only a quasi design solution to complex software.

An adaptive interface can be contrasted with an adaptable interface in terms of how much control the user has over the interface adaptation [Fischer 1993]. There has in fact been a debate in the user interface community about which of these two approaches is best. Some argue that we should be focusing our efforts on the design of interfaces that give users a sense of power, mastery and control, whereas others believe that if we find just the right adaptive algorithm, users won't have to spend any time adapting their own interfaces [Shneiderman and Maes 1997]. This debate has been mostly theoretical to date in that there has been very little comparison of the two alternative designs in the research literature.

MSWord 2000 makes a significant departure in its user interface from MSWord 97 by offering menus that adapt to an individual user's usage [Microsoft 2000]. When a menu is initially opened a 'short' menu containing only a subset of the menu contents is displayed by default. To access the 'long' menu one must hover in the menu with the mouse for a few seconds or click on the arrow icon at the bottom of the short menu. When an item is selected from the long menu, it will appear in the short menu the next time the menu is invoked. After some period of non-use, menu items will disappear from the short menu but will always be available in the long menu. Users cannot view or change the underlying user model maintained by the system; their only control is to turn the adaptive menus on/off and to reset the data collected in the user model. We will return to the adaptive interface of MSWord 2000 in our Study Two, described in Section 16.5.

Two examples in the research literature that incorporate intelligence are Greenberg's work on Workbench, which makes frequently-used commands easily accessible for reuse [Greenberg 1993] and the *recommender* system that alerts users to functionality currently being used by co-workers doing similar tasks [Linton *et al.* 2000].

No user testing has been reported in the literature for any of the interfaces given above except for Training Wheels.

16.3. STUDY ONE

Study One fulfilled our first research objective, namely, to gain a more systematic understanding of users' experiences with complex software. It also provided specific direction for our second objective, which was to move to a new interface model. This study was the result of a collaborative effort with Dr. Gale Moore, a sociologist at The University of Toronto.[2]

16.3.1. METHODOLOGY

The sample consisted of 53 participants selected by the researchers from the general population. All participants were users of MSWord 97. While this was not a simple random sample, participants were selected with attention to achieving as representative a sample of the general adult population as possible. That is, we paid particular attention to achieving representation in terms of age, gender, education, occupation and organizational status.

Participants completed a lengthy questionnaire prior to meeting with the researcher. It included a series of questions on work practices, experience with writing and publishing, the use of computers generally, and the use of word processors specifically. Throughout the questionnaire open-ended responses were encouraged and space provided. During the one-on-one on-site interviews an identification instrument was used to collect data on the familiarity and use of functions. Given our focus on the user we defined functions from the perspective of the user rather than using a traditional Computer Science definition. Functions were defined as *visually specified affordances* and therefore toolbar buttons and final menu items made up the great majority of the 265 functions we considered. For each function, participants were asked:

1. Do you know what the function does? And if so,
2. Do you use it?

Responses to question one were scored on a two-point scale: *familiar* and *unfamiliar*. Responses to question two were scored on a three-point scale: *used regularly, used irregularly*, and *not used*. Participants were told that familiarity with a function indicated a general knowledge of the function's action but that specific detailed knowledge was not required. A regularly-used function was defined as one that was used weekly or monthly and an irregularly-used function was one that was used less frequently.

We concluded with an open-ended in-depth interview. This was used to both ground and extend the quantitative work. Here specific issues that had been raised during the functionality identification were probed and participants were encouraged to talk broadly about their experiences with word processing in general, and MSWord, in particular.

Participating in this study required approximately one to two hours of each participant's time.

16.3.2. SELECTED RESULTS

Figure 16.2 shows some of the quantitative data that was collected on function use and familiarity. We can see that there are a number of functions that weren't used or used by only few. For example, in Figure 16.2a we see that 42 functions were not used by any of our participants and 118 functions were used by 25% or fewer of our participants. Putting these two counts together tells us that more than half of the functions were used by 25% or fewer of our participants. And there were very few functions that were used regularly – only 12 functions were used regularly by 75% or more of the users (Figure 16.2b). What's interesting here is that the familiarity data is much more evenly distributed (Figure 16.2c), which suggests that there might be more going on than simply users being overwhelmed by a whole bunch of unknown and unused functions. The capture of this familiarity data is one of the novel aspects of our study.

Through reliability analysis of questionnaire responses we were able to construct a Feature Profile Scale.[3] This scale identifies individual differences with respect to the perception of heavily featured software.

The *feature-keen* are at one end of the scale. These users:

- want complete software (not light versions),
- want the most up-to-date software, and
- believe that all interface elements have some inherent value (whether or not they are actually used).

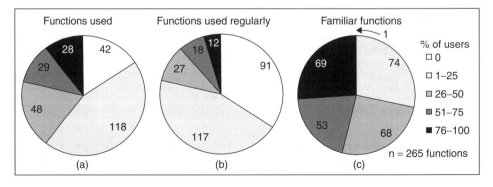

Figure 16.2. Number of functions that were (a) used (regularly or irregularly), (b) used regularly, and (c) familiar to our participants (n = 53). (Reproduced by permission of Canadian Information Processing Society).

At the other end of the scale are the *feature-shy*. These users:

- don't necessarily have to have complete software,
- tend to be suspicious of upgrades, and
- only want the interface elements that they use.

The *feature-neutral* are, just as the name suggests, less opinionated with respect to their perception of heavily featured software.

The graph in Figure 16.3 shows that these individual differences are independent of computer expertise in that there is no pattern to the data; the different user profiles (feature-shy, feature-neutral, and feature-keen) are distributed across the different levels of computer expertise. Although not shown here, the individual differences were also found to be independent of the number of familiar and used functions.

To state this another way, our findings suggest that it is not the case that expert participants who use a relatively large number of functions are always the users who want to have feature-filled software. Nor is it the case that novice users who typically use fewer functions are the ones who always want to have a simple interface with few functions. Had we not conducted this study we would likely have assumed a naïve design solution – one that gives experts a feature-filled version of MSWord and that gives novices a feature-reduced version of MSWord. We learned through this research that such a design is not the right solution. It will not satisfy all users, or even a majority of users.

Detailed analysis of the interview transcripts was carried out in order to contextualize the quantitative data. We do not report that analysis here, but rather provide two quotations which breathe some life into the previous graphs.

First we hear what a senior technical expert had to say about MSWord. Note that this participant was familiar with 86% of the functions and actually used 38% of them, which was relatively high compared to our other participants. He reported having used MSWord for six years and was a daily user of MSWord.

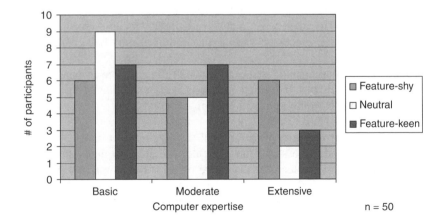

Figure 16.3. Distribution of computer expertise across the Feature Profile Scale. (Reproduced by permission of Canadian Information Processing Society).

*I want something much simpler. . . . I'd like to be able to customize it to the point that
I can eliminate a significant number of things. And I find that very difficult to do. Like
I'd like to throw away the 99% of the things I don't use that appear in these toolbars.
And I find that you just can't – there's a minimum set of toolbars that you're just
stuck with. And I think that's a bad thing. I really believe that you can't simplify
Word enough to do it.*

This can be contrasted with what another participant who was a junior consultant had
to say. She reported familiarity with 43% of the functions and the use of 30%. She used
MSWord daily and had also used it for six years.

*I like the idea of knowing that in case I needed to do something, that that stuff is
there. And again, I think it goes back to the personality thing I was talking about
where, you know, there's [sic] people that are options people. . . .. I love to know that
options are there, even if I never use them. I really like knowing that it does all
that stuff.*

These quotations shed some light on the diversity of opinion. Some users simply like
to know that options are available and seem empowered by having additional features
to learn, whereas other users are frustrated by having excess options in the menus and
toolbars that are not being used.

The general sentiment expressed in the interviews with respect to the number of functions available can be summarized into the following three observations:

Observation 1: Many participants expressed frustration with having so many unused
functions. The dominant reasons for frustration were the desire for something simpler
and to reclaim screen real estate. To counter this, some participants seemed perfectly
content to have a vast selection of functions.

Observation 2: Although some participants would be content with a 'light' version of
MSWord, the dominant feeling was not to have unused functions removed from the
application entirely. The main reasons against a light version were the apprehension
of a total loss of unused functions, and the perception of only being able to work at a
certain limited level.

Observation 3: Some participants used exploration of the interface as a means of learning
the software. They felt that if unused functions were eliminated entirely, this would
limit their ability to learn through exploration.

So what does this all mean for bloat? Recall that the term bloat had been used very
loosely in both the popular press and the computer literature to imply that most people
were overwhelmed by all the features that were present. But this is not what we found
in our study. Based on both the quantitative and qualitative data we collected, we were
able to redefine the term bloat with respect to functions used and wanted. In particular,
we discovered both an objective and subjective component to bloat. *Objective bloat* we
define to be the set of functions not used by any users. These functions really should

be eliminated and ideally prevented from occurring altogether. More interesting is *subjective bloat* which we define as the set of unwanted functions that varies from user to user. What's important to note is that for any user, subjective bloat is not simply the complement of the set of used functions. Some users want functions even if they do not use them.

Some may question the usefulness of this redefinition. We believe the danger of using the term bloat too broadly is that it suggests the naïve design solution to complex software which we have already dismissed as one that simply will not work. Our goal was to provide a more nuanced definition to this term as a first step to arriving at a robust design solution to the problem of heavily-featured productivity software.

The results from our Study One suggested that the philosophy of design needed to move away from 'enabling the customization of a one-size-fits-all interface' to supporting the creation of a truly personalizable interface. The personalization solution would need to be lightweight and low in overhead for the user, yet not limit or restrict their activities. We postulated multiple interfaces as one way to accommodate both the complexity of user experience and their potentially changing needs. Individual interfaces within this set would be designed to mask complexity and ideally to support learning. We recognized that continual access to the underlying formatted document or text would need to be preserved.

Multiple interfaces design, conceptualized from Study One, raised a number of important research questions:

(1) Will users grasp the concept of multiple interfaces?
Certainly from our perspective it seemed to be an intuitive design, but this had to be evaluated in some fashion.
(2) Is there value to a personalized interface?
Some of the early research in intelligent user interfaces made the implicit assumption that having a personalized interface would be valuable – researchers assumed the value existed and worked on finding just the right algorithm to adapt the interface to the individual user's needs. The results of this early work were not terribly successful, but how should this be interpreted? Was it having a personalized interface that was not useful or was the method/algorithm for achieving the personalization the problem. We felt it was important to evaluate this question in its own right, which is why we used Wizard of Oz methodology to accomplish the personalization within our Pilot Study.
(3) If there is value in having a personalized interface, even for only a subset of users, how can the construction of the interface be facilitated?

Our Pilot Study and Study Two address these three research questions.

16.4. PILOT STUDY

Our pilot study focussed on our first two research questions above, namely whether or not users would be able to grasp the concept of multiple interfaces and whether in fact there was value to having a personalized interface.

Our first prototype included three interfaces between which the user could easily toggle. It was implemented entirely in Visual Basic for Applications (VBA) in MSWord 2000. The three interfaces were as follows:

Default Interface: This contained the full functionality offered in an 'out-of-the-box' version of MSWord 2000.

Minimal Interface: This contained a small subset of the functionality available in the Default Interface, namely, the 10% of the functions from the default interface that were reported as most frequently used in Study One.

Personal Interface: This contained just those functions that the user wanted.

The general goal was to accommodate those users who wanted a simplified interface but with easy access to all functions just one click away. Figure 16.4 shows a screen capture of the prototype. It is important to note that the minimal interface and the default interface remained static; it was only the personal interface that changed for each user. There was no way for users to personalize their own personal interface in this first prototype. Rather it was the researcher who made the personalizations. When the prototype launched, the minimal interface was the interface that was visible.

16.4.1. IMPLEMENTATION

Our goal was to evaluate our prototype in a field setting with participants who were already users of MSWord 2000. For that reason, our prototype was implemented so that it did not

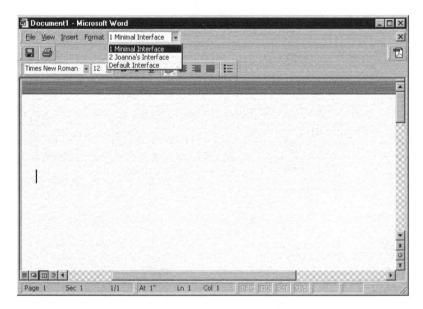

Figure 16.4. Multiple interfaces prototype for the Pilot Study. Here the minimal interface is showing. A toggle on the menu bar allows users to easily switch between the three interfaces.

interfere with any customization that participants may have already made to their MSWord interface. It was also designed to be easily installed on top of an existing installation of MSWord. This was accomplished by placing the required VBA code in a specialized document template file that was loaded into MSWord on startup. If necessary, a user could have removed the prototype by simply deleting this template file and re-launching MSWord. The information about function availability in the personal interface was stored in a flat file enabling the prototype to be effectively stateless; this facilitated the quick reconstruction of a personal interface should a problem with the software have occurred.

There were approximately 700 lines of VBA code required for this first version of the prototype. Despite what that might imply, creating the prototype was not straightforward. A number of approaches were tried before we found one that worked. The second version of the prototype (described in further detail later in this chapter), was significantly more complex and required approximately 5000 lines of code.

16.4.2. OBJECTIVES AND METHODOLOGY

Our objectives for this study were basic and straightforward and our methodology was designed to match the objectives. In particular, we wanted to explore user response to the prototype interface system, to collect real command usage data over an extended period of time, to test the stability of the prototype and the software logger, and to learn what was going to be easy/difficult, from a methodological point of view, about evaluating a prototype such as ours in a field setting.

There were four participants, two of whom were unbiased in that they were unaware of the research objectives. These participants were both female, middle-aged, administrative assistants, who were regular MSWord users and were generally proficient with computers. The remaining two participants were on the research team. An obvious apparent conflict is that the author of this chapter performed both the role of the researcher and a user in this pilot study. In any formal study, acting in such a dual role would be problematic. In our pilot study, however, the objectives were very basic and the usage data was based on real tasks done over an extended period of time which would have taken considerable effort to manufacture. Having two extra participants even though they were aware of the design rationale behind the multiple-interfaces prototype was seen to add value to the informal evaluation.

The methodology for the study involved having a short initial meeting with each of the participants during which the researcher installed the prototype and the software logger. The prototype was briefly demonstrated to the participant and the participant was asked which menu items and toolbar items she would like in her personal interface. Participants were encouraged to initially select only items that they expected to use regularly. The researcher then met with each participant every week or two to see if she would like any adjustments to her personal interface, and if she were to have the option to have the prototype removed and go back to the regular MSWord interface, would she choose to have it removed. The modification of the personal interface by the researcher was the Wizard of Oz component of this study. These one-on-one sessions were usually very brief, on the order of five minutes. Participants each used the prototype for approximately two months during the summer of 2000.

16.4.3. SELECTED RESULTS

Detailed usage data was collected through software logging and we were therefore able to quantify usage behaviours such as how much time was spent in MSWord, how much time was spent in each of the three interfaces, how often the participant switched between interfaces, which functions were used and when, and how the personal interfaces grew over time. We summarize the key findings derived from both the informal conversations during the regular research-participant sessions and the quantitative data collected from the software logs:

- All participants grasped the concept of multiple interfaces very easily. Beyond the initial installation session there was very little modification to any of the personal interfaces, indicating that users used a fairly stable set of functions.
- Participants wanted functions based on expected future use, not based on recency of use.[4] For example, midway through the study both of the unbiased participants made heavy use of a function that was not included in their personal interfaces. This high-frequency function use was documented in the software logs and therefore apparent to the researcher. When these participants were asked independently if they would like any modifications to their personal interfaces, they both declined. When the researcher specifically mentioned the highly-used function, both participants indicated that it was functionality that they used infrequently during the year and that it was best to just use it from the full interface.
- For technical reasons participants were required to start and stop the software logger. This overhead was in fact the biggest complaint that they had about their involvement in the study. The real damage of having a user-driven software logger was that the two unbiased participants did not differentiate the prototype from the software logger in that they thought that you couldn't have one without the other. Thus, they were really evaluating both together as one system. It certainly pointed to a weakness in the study methodology that needed to be rectified in the second study.
- There was one system crash – luckily it was on one of the unbiased participant's machine towards the very end of the study. We later found that it was related to a bizarre glitch in the VBA programming environment.
- For three out of the four participants, the minimal interface did not add any real value. Two of the participants asked to have their personal interface visible on launch rather than the minimal interface part way into the study – after this point they essentially ignored the minimal interface. For a third user, the minimal interface was almost identical to her personal interface and she ended up somewhat confused as to why she had both of these interfaces.
- At the end of the study, participants were given the option to continue using the prototype. Three out of the four participants chose to keep the prototype interface. They actually did continue to use the prototype. One participant was ambivalent about the prototype throughout the study and chose to have it removed once the study concluded. The two unbiased participants completed the Feature Profiling questions from our Study One. Interestingly enough, the ambivalent participant was found to be feature-keen and

the participant who chose to continue to use the prototype was feature-shy. This finding provided early support for our personality profiling and indicated a match between our multiple interfaces prototype and personality type.

The results of the Pilot Study encouraged us to iterate on the design of the prototype and to do a formal evaluation. This was our Study Two.

16.5. STUDY TWO

Our high-level goals for this study were twofold. Our first goal was to understand how users experienced the novel aspects of the multiple interfaces prototype. This goal followed directly from our Pilot Study. Questions of interest included:

- Will users have a positive experience with multiple interfaces?
- How will users use the interfaces? For example, will they spend most of their time in their personal interfaces or in the full interface?
- How many functions will they add to their personal interfaces?

Capturing the users' experience needed to be accomplished in a significantly more systematic fashion than was done in our Pilot Study.

Our second goal was to compare our user-adaptable design with the adaptive design in MSWord 2000. We were specifically interested to know which of the two interface designs users would prefer and why, and how the two designs would compare with respect to users' ability to control, navigate, and learn the software.

The design of the prototype was modified slightly for Study Two. We eliminated the minimal interface because it didn't appear to provide much value for our Pilot Study participants. On startup, our new prototype launched right into the user's personal interface. The personal interface initially contained only six functions. We also changed the name of the default interface to the full interface to reflect more accurately the content of this interface. Screen captures for the modified prototype are shown in Figure 16.5.

The biggest modification to the prototype was the addition of an easy-to-use mechanism whereby users could personalize their own interfaces. The mechanism is shown in Figure 16.6.

What makes our design unique is the *combination* of three design elements, rather than any single design element:

(1) Two interfaces, one that is personalized (the personal interface) and one that is the full set of functions (the full interface), and a switching mechanism between interfaces that requires only a single button click.
(2) The personal interface is *adaptable* by the user with an easy-to-understand adaptation mechanism.
(3) The personal interface begins small and, therefore, unless the user adds many functions, it will remain a minimal interface relative to the full interface.

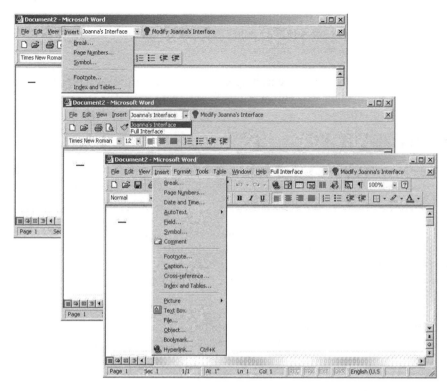

Figure 16.5. User opens the Insert menu in the personal interface, toggles to the full interface, and re-opens Insert menu. For this user the Insert menu has many more items in the full interface than in the personal interface. (Reproduced by permission of ACM Inc).

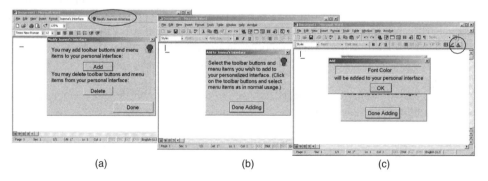

         (a)   (b)   (c)

Figure 16.6. Process for adding a function to the personal interface – in this example the Font Colour function is added. This is accomplished by clicking on the 'Modify Joanna's Interface' button, which pops up a dialogue box (a). After selecting Add (or Delete), a second dialogue box appears (b). All buttons or menu items selected while this dialog box is present are added (or deleted) after a confirmation (c). Clicking on Done Adding returns to normal mode. (Reproduced by permission of ACM Inc).

16.5.1. METHODOLOGY

The individual differences that we first identified in Study One appeared to play a role in our Pilot Study, and so we included these individual differences as an independent variable in Study Two. We had 10 feature-keen and 10 feature-shy participants.

In order to participate, users had to meet a number of criteria: they had to be regular MS Word 2000 users, they had to have it installed on their machine, they had to use it on one machine only, they had to have been using it for at least one month prior to the study, and they had to live within a half hour drive from our research lab. Participants were primarily solicited through a call for participation posted to numerous electronic newsgroups serving people at the University of Toronto, and Toronto residents in general. Interested participants had to fill out an online questionnaire that screened for individual differences (feature-keen and feature-shy) and the criteria mentioned above.

Figure 16.7 shows the timeline of our field study. For four weeks participants used our prototype, which we called MSWord Personal. They then returned to MS Word 2000 for two weeks. During this time the researcher conducted three on-site one-on-one meetings with each participant. At the first meeting the prototype and the software logger were installed. Given the problems we experienced with software logging technology in the Pilot Study, we used a different software logger for this study which did not require operation by the participant. (This software logger had not been available to us during our Pilot Study.) At the second meeting, four weeks into the study, the prototype was uninstalled. The user was not aware that this was going to take place. At the third meeting, six weeks into the study, the logger was also uninstalled, the participant's machine was restored to its original state prior to the study, and a semi-structured interview was conducted. Throughout the study a series of online questionnaires was also completed, Q1 through Q8. These questionnaires collected data for other dependent variables that included user satisfaction, and the perceived ability to navigate, control and learn the software.

The logistical constraints in conducting a field study precluded the counterbalancing of word processor conditions. The formal design of our study was a 2 (personality types, between subjects) × 3 (levels, levels 1,3 = MSWord 2000, level 2 = MSWord Personal, within subjects) design where level 2 was nested with 5 repetitions. This design is best characterized as a quasi-experimental design [Campbell and Stanley 1972].

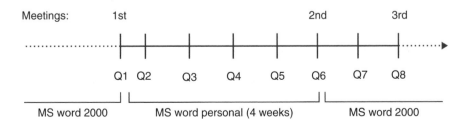

Figure 16.7. Timeline of the Study Two protocol. (Reproduced by permission of ACM Inc).

16.5.2. SELECTED RESULTS

We first concentrate on our goal to capture the users' experiences of the multiple-interfaces design. Selected results are provided. These results are derived from the logging data and the semi-structured interviews. For technical reasons we are missing some of the logging data from one of our participants, so for analyses that rely on the logging data, we have N = 19, rather than N = 20.

Overall positive experience: The majority of participants had a positive experience of MSWord Personal. They liked having their own interface but were strongly in favour of easy access to the full set of functions.

Amount of time spent in the personal interfaces: 75% of the participants spent 50% or more of their time in their personal interface which strongly suggests that it provided added value to the participants.

Functions added to the personal interfaces: For any given participant, if a function was used on 25% or more of the days that word processing occurred, there was a 90% or greater chance that the participant added the function to his/her personal interface. The likelihood of adding a function increased as the frequency of use increased. In other words, the most frequently used functions were those that were added to participants' personal interfaces. This is certainly what we expected to occur and is an indicator that participants were able to personalize according to their individual usage.

Approach to personalization: Analysis was done to uncover the approach users took to personalizing and using the two interfaces. In particular, we looked at whether participants tended to add functions up-front towards the beginning of their time using MS Word Personal, or in a more continuous manner as they required the functions (*up-front* versus *as-needed*). We also looked at whether participants added all the functions they would ever expect to use, or just the most frequently-used functions (*all* versus *frequently-used*). In the end, we weren't able to identify an approach that dominated all the other approaches. We found that six participants used the *up-front* strategy and 13 participants used the *as-needed* strategy. Relative to which functions were added, 12 participants added *all* functions they expected to use and seven participants added only the *frequently-used* functions. Seven participants gave up on their desired approach to personalization. They did not give up entirely on using their personal interfaces, but rather they altered their strategy midway through the study. None of the participants who took the approach of adding functions up front gave up, suggesting that this was a more effective strategy than the as-needed strategy. We strongly suspect that if the personalizing mechanism had been less clunky,[5] the number of participants who gave up would have been even lower.

Customization triggers: We tried to determine what triggered users to modify their personal interfaces. We found that 77% of the total number of functions added over the four weeks were added within the first two days – so there appeared to be an initial-bulk addition. The second most dominant trigger was the immediate need for a function.

Differences between the feature-keen and the feature-shy: Counter to our expectations, there were no substantial differences found between how the feature-keen and the feature-shy interacted with MSWord Personal and what they had to say about it.

We next summarize the results of the comparison between MSWord Personal and the adaptive interface of MSWord 2000. The data are derived from responses to the online questionnaires. Here we did find some statistically significant differences between the feature-keen and the feature-shy participants. We highlight only a few of these differences.

Figure 16.8 shows the results for the satisfaction, navigating, control, and learning dependent variables. The x-axis represents the progression of time through the online questionnaires (Q1 to Q7). The y-axis shows response ratings on a Likert scale. Taking the variable satisfaction as an example, the statement that appeared in the questionnaires was 'the software is satisfying to use'. A response of '1' meant 'strongly disagree' and a '5' meant 'strongly agree'.

We focus on the comparison of the Q1 and Q6 data points. This comparison captures the users' reported levels of each dependent measure after one month or more of MS Word 2000 (Q1) compared to one month of use of MS Word Personal (Q6). Additional comparisons are summarized in Table 16.1, which shows the results from a Q6 versus Q7 comparison and a comparison of Q2, Q3, Q4, Q5, Q6.

In addition to reporting statistical significance we report effect size, eta-squared (η^2), which is a measure of the magnitude of the effect of a difference that is independent of sample size. Landauer notes that effect size is often more appropriate than statistical significance in applied research in human-computer interaction [Landauer 1997]. The metric for interpreting eta-squared is: .01 is a small effect, .06 is medium, and .14 is large.

The analysis found that there was a significant cross-over interaction for satisfaction ($F(1, 18) = 4.12$, $p < .06$, $\eta^2 = .19$) prompting us to test the simple effects for each group of participants independently. The comparison was not significant for the feature-keen participants, however, the increase in satisfaction was borderline significant for the feature-shy ($F(1, 9) = 3.645$, $p < .10$, $\eta^2 = .29$). This suggests that the feature-keen did not experience any significant change in satisfaction between MSWord 2000 and MSWord Personal, however, the feature-shy did experience an increase in satisfaction.

A very similar result was found for control. There was a significant cross-over interaction for control ($F(1, 18) = 4.38$, $p < .06$, $\eta^2 = .20$). Testing the simple effects found the

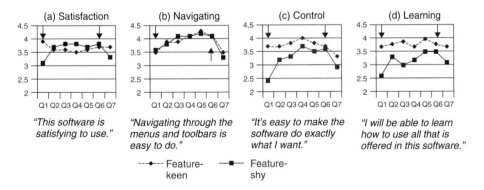

Figure 16.8. Satisfaction, navigating, control, and learning. Graphs and original statements are given (N = 20). (Reproduced by permission of ACM Inc).

Table 16.1. Comparison of independent variables over time.

Q1 vs Q6	Independent Variables		
	Version (V)	Personality (P)	V X P
Satisfy	1.27	1.12	4.12 **
Navigate	5.76 ***	.03	.05
Control	4.38 **	6.21 ***	4.38 **
Learn	4.13 **	4.07 **	2.64
Q6 vs Q7			
Satisfy	.85	.18	.85
Navigate	8.02 ***	.07	.16
Control	5.89 ***	.70	.44
Learn	3.08 *	1.33	1.11
Q2 – Q6			
Satisfy	.27	.28	.27
Navigate	2.38 *	.00	.41
Control	2.02	2.32	.64
Learn	1.56	1.90	1.10

*p < .10 **p < .06 ***p < .05

comparison to be non-significant for the feature-keen participants, however, the feature-shy perceived a significant increase in control with MSWord Personal ($F(1, 9) = 11.17$, $p < .01$, $\eta^2 = .55$).

In terms of navigation, there was a very strong main effect, whereby both groups of users sensed a greater ability to navigate MSWord with the Personal version rather than with the 2000 version ($F(1, 18) = 5.76$, $p < .05$, $\eta^2 = .24$).

With respect to learnability, there was a main effect of personality type ($F(1, 18) = 4.07$, $p < .06$, $\eta^2 = .18$) whereby, regardless of version, the feature-keen felt better able to learn the functionality offered than did the feature-shy participants.

These results are quite powerful. In all cases there was either a main effect showing improvement for both groups of users or there was improvement for the Feature Shy without a negative effect on the Feature Keen. In other words, changing the design of the interface can positively impact the experience of one group of users without negatively impacting another group.

In the final debriefing interview participants were asked if they could explain how the "expandable" (adaptive) menus worked. Seven of the 20 participants had to be informed that the short menus were in fact adapting to their personal usage. Participants were then asked to rank according to preference MSWord Personal, MSWord 2000 with adaptive menus, and MSWord 2000 without adaptive menus (the standard 'all-in-one' style interface). Figure 16.9 shows that 13 participants ranked MSWord Personal ahead of either form of MSWord 2000. Aggregating across all of the feature-shy and feature-keen

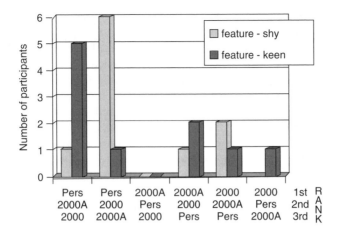

Figure 16.9. Ranking three different interfaces for MSWord: Word Personal (Pers), Word 2000 without adaptive menus (2000), and Word 2000 with adaptive menus (2000A) (N = 20). (Reproduced by permission of ACM Inc).

participants reveals an interesting difference: only two of the feature-shy ranked adaptive before all-in-one as compared to seven of the feature-keen. This can perhaps be explained by the fact that six of the seven participants who were unaware of the adapting short menus were feature-shy participants. This is an indicator that lack of knowledge that adaptation is taking place contributes to overall dissatisfaction with an adaptive application.

Prior to our work comparisons between adaptive and adaptable interfaces had been mostly theoretical. This study allowed us to compare one instance of each of these design alternatives in the context of a real software application with real users carrying out real tasks in their own environments. Results favoured the adaptable design but the adaptive interface definitely had support. With respect to the adaptable design, users were capable of personalizing according to their function usage and those who favoured a simplified interface were willing to take the time to personalize.

16.6. SUMMARY AND CONCLUSIONS

In this chapter we have documented the iterative design, implementation, and evaluation of multiple interfaces for a commercial word processor. This research began out of a concern for how users were coping with the complexity of everyday productivity applications. We had our own beliefs about where the problems might lie, but rather than generating designs based on those intuitions we began with what the users themselves had to say. Study One was an exploratory study designed to uncover users' experiences with their word processor, MSWord. Our one-on-one sessions with each of the 53 participants were both structured and open ended. We systematically reviewed functions and captured both expertise and work practice through a questionnaire. We also spoke to each participant. In that free-form exchange, participants told us their thoughts about their word processor and we probed aspects of the participants' experiences. Out of this exploratory study emerged the concept of multiple interfaces for a word processor and a profiling scale that captured individual differences with respect to heavily-featured software. The multiple-interfaces design was not well specified at this stage, but included two or more interfaces with varying amounts of functionality, between which the user could easily switch, and access to the underlying document was to be preserved.

In the Pilot Study we created our first multiple-interfaces prototype. The choice to continue working with MSWord was a natural one. We explored implementation issues for both programming the prototype itself and for the software logging technology. Rather than implement a fully functioning prototype, we created one that would support three interfaces, but with which the user him/herself would not be able to modify the interfaces. In this way, evaluation of the prototype required a Wizard of Oz study design, although the wizard in our case was not hidden from the participants. Our goal was to test out the concept and the technology as soon as possible and gather early feedback before proceeding too far with the design. It was a 'fingertip' length study in that we as researchers needed to evaluate the design in the context of our own use. For this reason, two of the researchers were included as pilot participants, in addition to two unbiased users of MSWord. From the Pilot Study we learned that the minimal interface provided little benefit and was in fact problematic for one of the participants. We also learned that having multiple interfaces seemed to provide more value for feature-shy users than it did for feature-keen users.

The results from the Pilot Study were promising and encouraged us to redesign the prototype and perform a formal evaluation with 20 participants, our Study Two. The minimal interface was removed and an easy-to-use customization facility was added to the prototype. Software logging technology was found that ran transparently to the user and this replaced the problematic logger that was used in the Pilot Study. Our goal at this point was to compare our user-adaptable multiple-interfaces prototype to the adaptive interface of MSWord 2000. In order to maximize ecological validity it was important to evaluate and compare these interfaces in a field setting where data was being captured in the context of the participants' real work. Doing a fully-counterbalanced experiment in the field was not feasible so we had to settle for a quasi-experimental design, which still allowed us to make some statistical comparisons. We used both quantitative and qualitative techniques in order to capture as full a picture of the participants' experiences as possible. Using this comprehensive data-capture proved extremely valuable in our analysis. One example of this is our discovery that the feature-keen and the feature-shy did not differ greatly in terms of how they approached the task of personalizing or what they had to say about having a personal interface, yet the feature-shy experienced a statistically significant increase in their level of satisfaction while using the multiple-interfaces prototype.

At each stage of the research we formulated our research questions and attempted to match the methodology to the questions. At certain stages in-depth qualitative methodology was required and at other stages quantitative methodology was appropriate. At some stages it was appropriate for the researchers to informally evaluate their own use of the technology, while at other stages this would have been clearly inappropriate.

In general the results from this body of research are promising in that 65% of the participants from our final study chose our design over the all-in-one style interface and the adaptive interface. We have little doubt that this number would have been higher with some small design modifications to the personalization mechanism. These results certainly encourage further exploration in the design space of multiple interfaces. One potential way of streamlining personalization would be to add a mechanism that provides usage information and allows the user to directly add features that have been used frequently or recently. This would move the design in the direction of user-assisted personalization (a mixed-initiative interface [Horvitz 1999]), where the user has ultimate control but also benefits from user-modelling technology.

ACKNOWLEDGEMENTS

Funding for this research was provided by IBM Canada through a graduate fellowship for Joanna McGrenere, by the Natural Sciences and Engineering Research Council of Canada, and by Communications and Information Technology Ontario. Mary Czerwinski assisted with both the design of Study Two and the statistical analysis. Microsoft Corporation provided the logging technology used in both the Pilot Study and in Study Two and the questionnaire used to assess expertise in Study Two.

REFERENCES

Campbell, D.T., and Stanley, J.C. (1972) *Experimental and Quasi-Experimental Designs for Research*. Chicago, IL: Rand McNally & Company.

Carroll, J., and Carrithers, C. (1984) Blocking Learner Error States in a Training-Wheels System. *Human Factors*, 26(4), 377–389.

Fischer, G. (1993) Shared knowledge in Cooperative Problem-Solving Systems: Integrating Adaptive and Adaptable Components. In M. Schneider-Hufschmidt, T. Kuhme and U. Malinowski (Eds), *Adaptive user interfaces: Principles and practice*, 49–68. North Holland: Elsevier Science.

Greenberg, S. (1993) *The Computer User as Toolsmith: The use, reuse, and organization of computer-based tools*. Cambridge: Cambridge University Press.

Horvitz, E. (1999) *Principles of Mixed-Initiative User Interfaces*. Proceedings of ACM CHI 99, 159–66.

Hsi, I., and Potts, C. (2000) Studying the Evolution and Enhancement of Software Features. *Proceedings of International Conference on Software Maintenance*, 143–51.

Kaufman, L., and Weed, B. (1998) Too Much of a Good Thing? Identifying and resolving bloat in the user interface. A CHI 98 workshop. *SIGCHI Bulletin*, 30(4), 46–7.

Landauer, T. (1997) Chapter 9: Behavioral research methods in human-computer interaction. In M.G. Helander, T.K. Landauer, and P.V. Prabhu (Eds), *Handbook of Human-Computer Interaction (2nd ed.)*, 203–27. Amsterdam: Elsevier Science.

Linton, F., Joy, D., Schaefer, P., and Charron, A. (2000) OWL: A recommender system for organization-wide learning. *Educational Technology & Society*, 3(1).

Mackay, W.E. (1991) Triggers and Barriers to Customizing Software. *Proceedings of CHI'91*, 153–160.

McGrenere, J. (2002) *The Design and Evaluation of Multiple Interfaces: A solution for complex software*. Doctoral dissertation, University of Toronto, Canada.

McGrenere, J., and Moore, G. (2000) Are We All in the Same "Bloat"? *Proceedings of Graphics Interface 2000*, 187–96.

McGrenere, J., Baecker, R.M., and Booth, K.S. (2002) An Evaluation of a Multiple Interface Design Solution for Bloated Software. *Proceedings of ACM CHI 2002, ACM CHI Letters* 4(1), 163–70.

Microsoft (2000) *Product Enhancements Guide*. http://www.microsoft.com/Office/evaluation/ ofcpeg.htm.

Miller, J.R., Sullivan, J.W., and Tyler, S.W. (1991) Introduction, in J.S. Sullivan and S.W. Tyler (Eds) *Intelligent User Interfaces*, 1–10. NY: ACM Press.

Norman, D. (1998) *The Invisible Computer*, 80. Cambridge, MA: MIT Press.

Shneiderman, B. (1997) *Designing the User Interface: Strategies for effective human-computer interaction (3rd ed.)*. Reading, MA: Addison-Wesley.

Shneiderman, B., and Maes, P. (1997) Direct Manipulation vs Interface Agents: Excerpts from debates at IUI 97 and CHI 97. *Interactions*, 4(6), 42–61.

FOOTNOTES

1. The research was conducted as the doctoral research of Joanna McGrenere at the University of Toronto [McGrenere 2002]. Dr. Ronald Baecker (University of Toronto) and Dr. Kellogg Booth (University of British Columbia) co-supervised this research. Dr. Gale Moore (University of Toronto) served as a supervisor for Study One and was actively involved throughout the research.

2. This study, here referred to as Study One, fits into a broader research project called *Learning Complex Software*, of which Moore is the project leader.

3. This scale was motivated by unpublished profiling work done by Microsoft that was described at a CHI '98 conference workshop [Kaufman and Weed 1998].

4. Research has shown recency of command use to be the best predictor of future use [Greenberg 1993] although it is not a perfect predictor.

5. Some users complained that the confirmation dialog box (shown in the third screen-shot of Figure 16.6) made selecting functions cumbersome. In the next design iteration this dialog box should be removed.

Inter-Usability of Multi-Device Systems – A Conceptual Framework

Charles Denis and Laurent Karsenty

IntuiLab SA

17.1. INTRODUCTION

Due to the diversity of multi-device services, users can access certain services across a range of contexts in time and space. Some fields of application, such as personal information management (e-mails, diary, address book, etc.), travel planning, real-time information management (weather, stock exchange, news, etc.) and e-banking are particularly suited to this kind of use. For these domains, using more than one device offers an added value to the services and tends to become a real requirement [Watters *et al.* 2003].

By studying how this variety of devices can be used together, we extend our understanding of human-computer interaction and usability. Earlier studies on the usability of multi-device services were mainly related to the characteristics of mobile devices and their impact on the user's performance [Jones *et al.* 1999; Dillon *et al.* 1999; Han and Kwahk 1994]. Such observations can help to improve the usability of mobile devices compared

Multiple User Interfaces. Edited by A. Seffah and H. Javahery
© 2004 John Wiley & Sons, Ltd ISBN: 0-470-85444-8

to traditional desktop applications, but this is not sufficient to guarantee the usability of multi-device systems. A usability study on a multi-device system requires us to focus on the transitions between devices: switching from one device to another should be as seamless as possible. Thus, a service cannot be considered on a given device separately from other devices.

The usability of a multi-device system must be analysed for each platform, taking into account each possible form of transition between the available devices. Based on the original concept of inter-operability, which is defined as the possible uses of software programs on several computer platforms, we introduce the concept of *inter-usability* to designate the ease with which users can reuse their knowledge and skills for a given functionality when switching to other devices. The notion of inter-usability can be compared to the notion of horizontal usability (see Chapters 1 and 2).

The objective of this chapter is to present a framework for achieving inter-usability between devices. Our approach is based on two components: (i) a theoretical analysis of the cognitive processes underlying device transitions, and (ii) an exploratory empirical study of the problems in using functionalities across multiple devices. The exploratory study consists of ten interviews with users of multi-device services such as electronic mail services, diaries and address books. An analysis of the users' comments illustrates some of the problems in the use of multi-device services.

17.2. INTER-USABILITY: A CONCEPTUAL FRAMEWORK

17.2.1. PRINCIPAL PROCESSES INVOLVED IN TRANSITIONS BETWEEN DEVICES

A multi-device environment is by definition heterogeneous. The technical characteristics of each device offer varying capacities in terms of data, functions, input and output methods, interaction styles and procedures. When users encounter a service on a new device for the first time, or when they want to carry out an operation on a given device, they must apply problem-solving methods. These problem-solving methods help the user understand and use the service when the user interface differs from the one they are already familiar with. In this kind of situation, users usually search for familiar analogous situations [Holyoak and Koh 1987]. When they find a suitable analogy, they transfer the knowledge to the new context. The transferred knowledge can be used to generate inferences or to apply a familiar procedure to the new context.

However, the analogies usually need to be adapted before they can be used [Holyoak *et al.* 1994]. This adaptation is possible only if users can easily access knowledge of the details of their present situation. In particular, this knowledge allows them to check whether the pre-requisites for an action are fulfilled in the current context.

In the context of a multi-device service, this means that users will be able to adapt their knowledge of the service if they understand the technical characteristics of each device and the effects of these technical characteristics on the user interface. We can assume that some users have this knowledge, and can therefore adapt easily to each device. But the majority of users do not have the required knowledge and will probably encounter certain difficulties.

Our empirical study provides some examples of this knowledge effect. For example, in receiving e-mails on a mobile telephone, long e-mails are not loaded all at once. Users receive only the first lines of the message and can then decide whether or not to download the rest. Users react in different ways to this feature, as it differs from the way e-mails are received on a PC. Thus, a user who understands that this feature is due to the lower transmission speed of the GPRS might think that it is an advantage. On the other hand, a user who does not have this technical knowledge does not understand why some of their messages are abbreviated. This user considers it a drawback of the service.

The user's expertise level on each device is an important dimension in inter-usability. We must also consider the user's expertise level on the service. Based on research in cognitive psychology [Holyoak and Koh 1987; Novick 1992], we expect that in searching for their usual service on a new device, new users will be more sensitive to *surface features*, which refer to visual presentation and terminology. In contrast, expert users should make more use of their knowledge of the underlying *structure* of the service, to identify what they are able to do and how. For example, let us consider a function that is available on a PC with a larger screen but is not immediately visible on a mobile device. A new user will likely be confused by the function's lack of visibility on the mobile device, whereas an expert user, making use of their knowledge of the service, will more easily imagine possible locations for the function.

Seamless transitions between devices can help the user retrieve their knowledge of a service and adapt it to a new device. But to make these transitions seamless, users must have a suitable representation of the state of the data. This representation is based on the user's *memory of the last operations performed*. This representation is particularly important if, during a device switch, a user wants to continue a task begun on another device or re-use the results. For the transitions to be as seamless as possible, users must believe that the multi-device system shares their own memory of the data state. They then know that they do not need to repeat a series of operations to recover a given state; they can take advantage of the shared context.

A multi-device interaction scenario illustrates the advantage of this kind of transition. Through a multi-device on-line reservation service, a user has purchased a ticket from Paris to London from Air France, with a departure on the 25th of March 2003 at 8 am and a return flight the 26th of March at 5.30 pm. This reservation was made from their PC. Once in London, the user realises that their meeting will not be finished in time for the 5.30 pm flight. They connect to their reservation service from their mobile phone. The service welcomes them, mentioning that they have a return flight from London planned for 5.30 pm. Then the service asks what their request is. The user replies: "I would like a later flight back to Paris". Due to the sharing of context between the user and the system, the user does not need to re-enter the departure airport, the airline company preference and the return date. The user can thus feel that their service is 'following' them wherever they go.

The significance of shared memory in multi-device systems indicates that the temporal dimension of system use is important. In particular, we must distinguish between short and long breaks in inter-device transitions. In short breaks, we can assume that users will remember the expectations linked to their last operations. However after a long break, they may have forgotten the exact state of their previous activities. They might therefore not understand some of the system's behaviours. This situation concerns systems with long response periods in particular [Dix 1994], such as can be the case in stock exchange services. In some cases, the consequences of a user's operation appear only after a few hours or days. In this case, the system's feedback can only be understood if the user can remember the last operation they carried out, or the context that brought them to carry out the operation.

When there is a change of device, there is usually also a change in the surrounding environment. Furthermore, we know that situations involving mobility lead to a greater variety in the contexts of use [Rodden *et al.* 1998]. Contrary to the use of a fixed work-station, where users have a greater familiarity and certainty about the environment and resources, a mobile activity often involves unfamiliarity with the situation and a feeling of lack of skill regarding the environment [Perry *et al.* 2001]. Consider the case where a user has last performed a task in a mobile context. If they want to resume that task in a different context, they have fewer clues to remember the mobile context than if they had continued the activity on the same device and in the same context. A multi-device system should thus help them to retrieve not only the state of their activity, but also the original context of this state.

In summary, from an analysis of the different cognitive processes in inter-device transitions, we can conclude that from the user's point of view, service continuity involves two dimensions:

• *Knowledge continuity*, based on the retrieval and adaptation of knowledge constructed from the use of one or more devices;
• *Task continuity*, based on the memory of the last operations performed with the service, independently from the device used, and the belief that this memory is shared with the system.

Each of these forms of continuity is based on a series of requirements. These will be examined in the following pages.

17.2.2. REQUIREMENTS FOR KNOWLEDGE CONTINUITY

To maintain knowledge continuity, ideally all devices should present the service in the same way and allow access to the same data and the same functions. This goal is not realistic given the technical constraints of mobile devices. It may not even be desirable, at least for certain services for which users might only want to access a subset of functions in certain recurrent situations.

In the following pages, we will examine the usability difficulties that can be caused by inter-device inconsistencies. These difficulties are addressed on three levels: (i) the terminology and visual appearance of the user interface, (ii) the data and available functions and (iii) the procedures.

17.2.2.1. Visual Appearance and Terminology

Differences in surface features of the user interface between devices can cause usability problems. Such inconsistencies can interfere with the analogical transfer process by preventing users from transferring their initial understanding of the service to the new device. This type of difficulty generally leads to the production of mistakes or at least a sub-optimal use of the service.

Graphical differences between devices can operate at two different levels:

- Differences in *spatial organisation* of information mean that an object is not in the same place in two different versions of the service. In this case, users will have to make an effort to locate the object. At best, this will increase their workload; at worst, if they can't locate the object quickly, they could conclude that the related function is unavailable on the new device.
- Differences in the *shape* of an interface object can cause users to fail to associate the object with its function.

To give an example from our study, a user sending e-mails from his mobile telephone complained that he could not add his correspondent's electronic address directly from his address book, as he was used to doing on his PC. In fact, this function was also available on his telephone, however the button did not have the same appearance as the one on the PC. This lack of visual consistency led the user to conclude that the function he had been looking for was unavailable on the telephone.

We must emphasise that not all graphical differences are a source of difficulty for users. For example, a change of size, orientation or colour is generally not a problem. The problem occurs when users do not have enough visual clues to judge whether two objects are similar.

With voice interfaces, users find themselves lost at first in determining what knowledge of their service they can reuse or adapt. The guidance offered by the system at the beginning of the service must take this difficulty into account and try to provide them with the necessary clues so that they can find their bearings. These clues being solely verbal, the terminology should be chosen carefully.

Terminological differences can be a source of continuity problems with any kind of user interface. When an object (button, hypertext link, menu item, etc.) is labelled inconsistently between devices, the user must follow a reasoning process to establish whether the object has the same function as its instance in another version of the service. Here we find a well-known problem in graphical user interfaces, which could be amplified in multi-device services, since the presentation context of a term often changes between devices. Users thus have even fewer clues to help them interpret the meaning of an inconsistent term.

17.2.2.2. Partition of Data and Functions

A multi-device service is generally not available in its entirety from each device, due in particular to the technical constraints of mobile devices, which prevent access to large quantities of data and certain complex functions. A mobile device can thus offer only a partial access to the service. This restriction of the service can apply to data as well as functions.

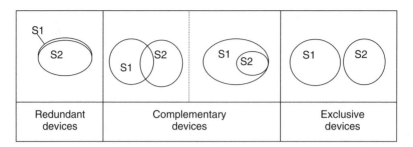

Figure 17.1. Degrees of device redundancy.

The devices can then be (see Figure 17.1):

- *Redundant*: All the devices give access to the same data and functions. A new user can fail to understand this redundancy due to the different appearance or structure of the service.
- *Exclusive*: Each device gives access to different data and functions. This configuration is quite rare. We can find an example in [Robertson *et al.* 1996]: the multi-device system is composed of an interactive TV and a PDA and the PDA is used as a remote control for cable television services. Here, there could be a problem if the user believes that they can access the same data or use the same functions on different devices.
- *Complementary*: The devices have a zone of shared data and functions, but at least one of the devices gives access to data or functions that are unavailable on the other device(s). This is the most common configuration. It combines the problems of redundant and exclusive devices.

The partition of data and functions can be responsible for discontinuity during the transition between two devices. For example, we encountered an electronic mail service user who knew that when he had to interrupt writing a message, his PC's mail application offered to store his data. This user could also manage his electronic messages with his mobile telephone. On the mobile device, the automatic function for storing messages was unavailable. Not knowing about this functional inconsistency, he noticed too late that the message would not be able to be stored, which he deeply regretted.

17.2.2.3. Procedures

The fact that a function is available on two devices is not enough to guarantee good continuity in a transition between these devices. In addition, the function should be accomplished in the same way. When a goal requires different actions between devices, the user must be able to suppress the analogical transfer. If the system does not help users understand the new process, there is a risk the users will give up before finishing the task.

For example, many personal time-management applications allow the user to associate a date with meetings, tasks, notes and links to the address list. Some of this information is not available in the same way on a mobile device. For example, the tasks could be available in the list of tasks but not under a specific date. Some users can then think that when using their mobile device, it is not possible to synchronise tasks entered with the time management software.

Procedural consistency between different devices is not always the best solution. For example, when using a telephone in a mobile context, voice interaction is preferable to a graphical user interface, even if users are most familiar with a screen-based approach for the same task. Moreover, for many services, natural language would be more appropriate than a constrained form of interaction based on isolated words and hierarchical menus, even though these graphical techniques would be close to the interaction style that users are generally familiar with. However, if natural language is used only with the mobile telephone, few users will know how to exploit it efficiently, because they will try to apply the procedures they are familiar with on a PC. Thus the system should provide support to help users learn a different procedure on a new device.

17.2.3. REQUIREMENTS FOR TASK CONTINUITY

In a change of device, it may be necessary for the user to continue or retrieve a previous task. To ensure task continuity, the user must recover the state of the data and the context of the activity.

17.2.3.1. Recovery of the State of Data

When an interrupted activity is resumed on a different device, the main difficulty for users is to remember the state of their data at the time of interruption. A seamless transition between devices requires the data recovery to be immediate and consistent with the users' expectations. One of the challenges for multi-device services is to translate the state of the data in a way that users would expect.

An example from our study illustrates this need. On returning from business trips, a user was used to reloading on his PC the e-mails he had already consulted on his mobile telephone, leaving a copy on the server. The user's motivation was to be able to deal with the more important mail back at the office, taking more time to do so. The messages that had been reloaded appeared as unread, in the same way as the 'real' new messages. In these conditions, the user admitted having difficulty in quickly distinguishing the messages he had read in the office before leaving on the business trip, the ones read on the mobile telephone during the business trip, and those which were really new since returning from the trip. This divergence between the restored state of the data and the expected state is thus responsible for an interruption in continuity of the task of managing a message service.

This example shows that two factors have to be considered in the state of the data. We must consider the absolute state of the data, independent of device, but it is equally necessary to take into account the state of the data *on each device*.

17.2.3.2. Recovery of the Activity Context

Failure to remember the operational context of the last action on the previous device can also be a problem in transitions between devices.

When users interrupt their activity, their last action is located at a particular level of a hierarchy of goals and sub-goals that constitutes their operational context. If the interruption is sufficiently long, users can lose the memory of this local and/or global planning. This problem is not specific *in principle* to the situations in which a multi-device service is used, but a change of device can amplify the problem since the information used in the activity often changes appearance between devices.

This is the case, for example, if after leaving a document in a particular state on a PC word processor, users cannot find the same work setting when re-opening the document on their Pocket PC (for example the cursor is now at the start of the document or the toolbar is no longer active). Because users do not recover the work setting on the new device, they can have difficulty recovering their operational context. They risk for example not remembering that they were about to modify the sentence where the cursor was situated, whereas if they had found the document open at the same page and the cursor on the same line, this configuration would have made the goal recovery easier.

We can assume that context retrieval varies with the duration of interruption between two sessions. If the interruption is short enough, the recovery of the state of the data should be sufficient for most users to remember where their task left off. However, after a long interruption, this could be insufficient. It will be important to further study contextualisation needs associated with a change of device, in particular the length of time after which loss of context can be expected.

17.3. DESIGN PRINCIPLES FOR INTER-USABILITY

The problems we have referred to show that when changing devices, it is important for users to be able to easily transfer and adapt their knowledge of the service and the representation they have of their task. This need meets certain obstacles in the case of multi-device services, due to the different design constraints between devices. Although it is highly desirable for the user interfaces to be very similar between devices, in practice this goal is not realistic.

We should thus promote consistency of design while permitting certain inconsistencies due to operational constraints, utility and efficiency criteria. But in this case, the multi-device service design should provide help functionality so that users can know as soon as possible what they should do and how, whatever the device used; if necessary, the functionality should also help users understand the limitations imposed by the devices. The pursuit of this objective will lead to *transparent* interfaces as described by [Maass 1983]: 'A transparent system makes it easy for users to build up an internal model of the functions the system can perform for them.' Transparency is by definition a dynamic notion, since user expertise on the system evolves over time. The help given to the user must thus be *adapted* to their experience with the service.

Consistency, transparency and adaptability are the three main principles on which inter-usability should be based. Given that inter-usability includes two dimensions, knowledge and task continuity, we can apply each of these three principles to each dimension (see Figure 17.2).

	Dimension of inter-usability	
Inter-device design principle	Knowledge continuity	Task continuity
Inter-device consistency		
Transparency		
Dialogue adaptability		

Figure 17.2. Multi-device systems analysis grid.

17.3.1. INTER-DEVICE CONSISTENCY

To support knowledge continuity and task continuity, a multi-device service should maximise inter-device consistency as long as this does not contradict technical or operational constraints. This consistency can be addressed on four levels:

- *Perceptual consistency*: When possible, the appearance and structure of the information should be similar on different devices. With graphical interfaces, this similarity applies to objects and the spatial organisation of information. With a voice interface, similarity applies to the order in which the information is presented – this order should be consistent with the order in the visual interface(s).
- *Lexical consistency*: The objects of the user interface should have the same label across devices.
- *Syntactical consistency*: To attain a given goal, the same operations should be performed across devices. It is possible to modify a task for a different device so as to make it more efficiently adapted to that device, but in this case, the system should be able to accept different sets of operations for the same task.
- *Semantic consistency*: Services should be similar across devices. In other words, the partition of data and function should be redundant between devices. In the same way, the effect of the operations should be as similar as possible across devices. To ensure task continuity, the state of the data in the last operations performed by the user should be reflected on all devices. Moreover, so as to recover the context of their tasks, users should be able to recover the state of their activity and/or the history of their last operations.

As we have said, it is unlikely that maximum consistency will be possible in multi-device systems. A certain amount of inconsistencies cannot be avoided. Inconsistencies are not always a problem (users can recover easily from some of them, without any particular help). We have referred to the case of interface objects that vary in physical parameter (colour, orientation, size) without causing any particular difficulties. In addition, we have mentioned that users may expect some inconsistencies. For instance, users don't expect to perform complex manipulations of graphics on their mobile phone. However, it must be stressed that the knowledge available today does not allow us to predict which kinds of inconsistencies lead to usability problems. Nevertheless, there are certain inconsistencies from which the user cannot recover, in particular those for which users need to access specific knowledge to understand and adapt to the inconsistency. In this case, transparency is required.

17.3.2. TRANSPARENCY

Transparency can be defined as a property of the man-machine dialogue allowing users to construct an accurate representation of the system so as to interact efficiently with it. This property can be based on guidance methods appropriate to the user's expertise level and system functionalities [Karsenty 2002; Maass 1983; Norman 1988]. In practical terms, transparency must allow users to immediately know what they can do (the data and accessible functions), how they can do it (the procedures) and why the system reacts as it does (its states, how it works and its limits).

In the case of multi-device services, transparency should ideally reuse knowledge from the devices and procedures the user is already familiar with. For example, to help a new user express a request in natural language over the telephone, the system should behave differently depending on whether the user has used natural language on other devices. If the user has experience with natural language on other devices, the only necessary help would consist in informing the user of the details of the spoken requests, for example, the ability to interrupt the system's messages at any time. If the user is inexperienced with natural language services, the system should also help the user understand that they can link several parameters in one expression (by saying, for example, "I would like a Paris to London ticket for tomorrow morning").

Moreover, the transparency principle should apply differently depending on whether the multi-device service is redundant, exclusive or complementary:

- With *redundant* devices, due to the differences in presentation, user might fail to understand that they can do the same things on different devices. The help functionality must therefore assist users to understand that the same data and same functions are available on all of the devices.
- With *exclusive* devices, the opposite problem can occur: users might believe that they can do identical things when this is not actually the case. The help functionality must thus assist them to correctly create a representation of the specifics of each device.
- With *complementary* devices, both types of problems are possible.

In addition, when system constraints prevent the user from retrieving the state of the data in their previous operations, or the context of their previous activity, transparency involves informing the user of their previous operations or context.

Some other issues in the application of transparency strategies in multi-device services include the following:

- In transmitting information about the system's properties, transparency generates an additional cognitive cost compared to what is strictly necessary to accomplish a task. This additional cost can sometimes be difficult to accept, in particular in mobile situations where users are in a hurry or distracted. It is therefore important to adapt transparency strategies to the context of use. In some cases, before explaining the properties of the mobile device, it may be preferable to wait until users have returned to a more relaxed and quiet desktop environment.

- Adapting transparency to familiar devices and procedures requires the construction of a centralised *user model*, able to be updated and consulted from each device. The software architecture of multi-device services needs to take this requirement into account.

17.3.3. ADAPTABILITY

System transparency is a dynamic notion since it depends on the user's representation of the system, which itself evolves in time and varies with the context of use. The following are some of the parameters that can be used in adapting a multi-device system to the user's profile:

- *The device(s) already used.* Adapting a multi-device system requires knowledge of which devices the user is already familiar with, since their knowledge of the service has been acquired by using these devices.
- *The amount and frequency of use of each device.* These parameters can be used to track the learning progress of the user so as to permit adaptation of the system, thus supporting task continuity. More precise indicators will probably be needed to adapt help functionality to the user, by tracking the amount and frequency of use of each functionality in a service for each type of device. If a function can be executed with several different procedures, the system should also be aware of the procedures that the user has already applied and the devices with which they are associated.
- *The last operations performed with the service.* These operations constitute a context that users might need to access in order to resume an interrupted activity.
- The time elapsed between the previous access and the current access to the service, for each device.

With these different parameters, a multi-device service will be able to adapt its behaviour so as to enable knowledge continuity and task continuity in most cases:

- For *knowledge continuity*, adaptation mainly involves varying the guidance level and amount of explanation of how the system works and the service limits for a given device. New users usually expect the system to take the initiative of offering detailed guidance. The guidance must help them learn about the different functions available and the types of possible objects in each interaction context. New users can also need explanations of system behaviours and methods for avoiding errors. In contrast, expert users require less help, as they have usually acquired a relatively precise representation of the service. For example, they do not require an explanation of the way the system works.
- For *task continuity*, system adaptation can involve the contextualisation of data, particularly by reminding the user of the actions that originated the current state of the data. For example, in the case of a stock exchange service, if a user re-accesses a service after making purchase orders and decides to consult their share portfolio, it would be useful to display the share values in the portfolio, periodically reminding the user of

their actions in the portfolio. Although not yet proven, it is possible that reminders of previous actions would be more useful if they were accompanied by information about the associated device and date of each action.

The adaptations discussed here are different from those in context-aware computing [Chen and Kotz 2000]. In the latter case, the service adapts itself to the user's local environment and to the characteristics of the device. The objective is more often to provide information relevant to the current situation. It is thus different from the objective of ensuring knowledge and task continuity, which requires adaptation to the users' cognitive characteristics.

17.4. CONCLUSION

Multi-device systems change our understanding of usability and the use of computers. They lead us to take a particular interest in the transitions between devices. During these transitions, users must be able to quickly transfer or adapt their knowledge of the service, and if necessary, continue their activity. To support this transfer, the design must not only ensure the usability of each individual device; it must also be inter-usable.

This goal requires the design of a multi-device service to be considered as a whole, probably starting with an abstract description of the interface objects [Vanderdonckt *et al.* 2001]. However, to date, there are still many unanswered questions about inter-usability. Future studies will need to better analyse how users adapt when faced with different forms of inter-device inconsistencies. Studies are also needed to address transparency strategies adapted to mobile contexts, and to specify the criteria for adapting the degree of transparency to the user's needs. Finally, empirical studies will be needed to better define the user's needs in situations where activities are resumed after a change of device.

ACKNOWLEDGEMENTS

This research is supported by France Telecom R&D, contract n°02 1B A24. Thanks to Valérie Botherel and Franck Panaget who provided us with continuous feedback on this research. Thanks also to Ahmed Seffah for his contribution in preparing this chapter and Daniel Engelberg for his help in editing it.

REFERENCES

Chen, G., and Kotz, D. (2000) A Survey of Context-Aware Mobile Computing Research. *Technical Report TR2000-381*, Department of Computer Science, Dartmouth College, November 2000.
Dillon, A., Richardson, J., and McKnight, C. (1999) The Effect of Display Size and Text Splitting on Reading Lengthy Text from the Screen. *Behaviour and Information Technology*, 9 (3), 215–27.
Dix, A. (1994) *Que Sera Sera: The problem of the future perfect in open and cooperative systems*. Proceedings of HCI'94, People and Computers IX (eds G. Cockton, S.W. Draper and G.R.S. Weir), August 1994, Glasgow, Scotland, 397–408. Cambridge University Press.

Han, S.H., and Kwahk, J. (1994) *Design of a Menu for Small Displays Presenting a Single Item at a Time*. *Proceedings of the Human Factors and Ergonomics Society 38th Annual Meeting*, October 24–8, Nashville, Tennessee 360–64. HFES Society, Santa Monica, USA.

Holyoak, K.J., and Koh, K. (1987) Surface and Structural Similarity in Analogical Transfer. *Memory and Cognition*, 15 (4), 332–40.

Holyoak, K.J., Novick, L.R. and Melz, E. (1994) Component Processes in Analogical Transfer: Mapping, pattern completion, and adaptation, in *Advances in Connectionist and Neural Computation Theory, Vol. 2: Analogical Connections* (eds K.J. Holyoak and J.A. Barnden) 113–80. Norwood, NJ: Ablex.

Jones, M., Marsden, G., Mohd-Nasir, N. *et al.* (1999) *Improving Web Interaction on Small Displays*. Proceedings of eighth International World Wide Web Conference (WWW8), May 11–14, Toronto, Canada, 1129–37.

Karsenty, L. (2002) Shifting the Design Philosophy of Spoken Natural Language Dialogue: From invisible to transparent systems. *International Journal of Speech Technology*, 5, 147–57.

Maass, S. (1983) Why Systems Transparency? in *The Psychology of Computer Use*, (eds T.R. Green, S.J. Payne and G.C. Ven Der Veer), 19–28, Academic Press, London.

Norman, D.A. (1988) *The Psychology of Everyday Things*. Basic Books, New York.

Novick, L.R. (1992) The Role of Expertise in Solving Arithmetic and Algebra Word Problems by Analogy, in *The Nature and Origins of Mathematical Skills* (ed J.I.D. Campbell), 155–88, Elsevier, Amsterdam, The Netherlands.

Perry, M., O'Hara, K., Sellen, A. *et al.* (2001) Dealing with Mobility: Understanding access anytime, anywhere. *ACM Transactions on Computer-Human Interaction (TOCHI)*, 8(4), 323–47, ACM Press, New York.

Robertson, S., Wharton, C., Ashworth, C., and Franzke, M. (1996) *Dual device user interface design: PDAs and Interactive Television*. Proceedings of the Conference on Human Factors in Computing Systems (CHI'96), Vancouver, British Columbia, Canada, 79–86. ACM Press, New York.

Rodden, T., Cheverst, K., Davies, N., and Dix, A. (1998) *Exploiting Context in HCI Design for Mobile Systems*. Proceedings of the First Workshop on Human Computer Interaction with Mobile Devices, May 21–22, Glasgow, Scotland, 12–17.

Vanderdonckt, J., Limbourg, Q., and Florins, M. (2001) *Synchronized Model-Based Design of Multiple User Interfaces*. Proceedings of the Workshop on Multiple User Interfaces over the Internet: Engineering and Applications Trends, IHM-HCI: French/British Conference on Human Computer Interaction, September 10–14, 2001, Lille, France.

Watters, C., Duffy, J., and Duffy, K. (2003) Using Large Tables on Small Display Devices. *International Journal of Human-Computer Studies*, 58, 21–37.

Subject Index

Multiple User Interfaces. Edited by A. Seffah and H. Javahery
© 2004 John Wiley & Sons, Ltd ISBN: 0-470-85444-8